Science and Technology for Development
(Pergamon Policy Studies-38)

Pergamon Titles of Related Interest

Diwan/Livingston *Alternative Development Strategies and Appropriate Technology*

Thomas *Integration of Science and Technology with Development*

Rothko Chapel Colloquium *Toward A New Strategy for Development*

Carman *Obstacles to Mineral Development: A Pragmatic View*

Goodman/Love *Management of Development Projects*

Menon *Bridges Across the South*

Science and Technology for Development

The Role of U.S. Universities

Robert P. Morgan

with
Ellen E. Irons
Eduardo A. Perez
Theodore N. Soule
Ava K. Fried

Published in cooperation with the
Center for Development Technology
Washington University, St. Louis

Pergamon Press
NEW YORK • OXFORD • TORONTO • SYDNEY • FRANKFURT • PARIS

Pergamon Press Offices:

U.S.A.	Pergamon Press Inc., Maxwell House, Fairview Park, Elmsford, New York 10523, U.S.A.
U.K.	Pergamon Press Ltd., Headington Hill Hall, Oxford OX3 0BW, England
CANADA	Pergamon of Canada, Ltd., 150 Consumers Road, Willowdale, Ontario M2J, 1P9, Canada
AUSTRALIA	Pergamon Press (Aust) Pty. Ltd., P O Box 544, Potts Point, NSW 2011, Australia
FRANCE	Pergamon Press SARL, 24 rue des Ecoles, 75240 Paris, Cedex 05, France
FEDERAL REPUBLIC OF GERMANY	Pergamon Press GmbH, 6242 Kronberg/Taunus, Pferdstrasse 1, Federal Republic of Germany

Copyright © 1979 Pergamon Press Inc.

This material is based upon work supported by the National Science Foundation under Grant No. INT 78-08292. Any opinions, findings, and conclusions or recommendations expressed in this report are those of the authors and do not necessarily reflect the views of the National Science Foundation.

All Rights reserved. No part of this publication may be reproduced, stored in a retrieval system or transmitted in any form or by any means: electronic, electrostatic, magnetic tape, mechanical, photocopying, recording or otherwise, without permission in writing from the publishers.

Printed in the United States of America

To the people of the Third World and the

people of the United States.

To Nancy, Tom and Jon.

Contents

List of Figures and Tables	ix
Acknowledgments	xiii
Introduction	xv
List of Acronyms	xxiii

Chapter

1	The Legislative Mandate	1
	The Policy Thread in U.S. Foreign Assistance Legislation	2
	Creation of Agencies with International Objectives	4
	Legislation Pertaining Specifically to Education or Universities	9
	Legislation Pertaining to Visas, Passports, Immigration, and Exchange of Persons	17
	Other Aspects of the Mandate	19
	Analysis	20
2	Engineering	24
	Types of Involvement	24
	Mechanisms	32
	Engineering Cases	33
	Engineering Technology Activity	52
	ASEE International Activities Survey	54
	Issues	55
	Current Thinking	62
	Concluding Remarks	65

Chapter		
3	Agriculture	68
	Overview of U.S. University Involvement	69
	Recent Involvement in Agricultural Development	76
	Examples of Programs	88
	Evaluation of U.S. University Involvements	101
	Current Issues and Questions	118
	Concluding Remarks	136
4	Science	137
	Types of Involvement	138
	Mechanisms	163
	Current Thinking on Science Involvements	164
	Analysis	172
	Issues	172
5	Future Roles for U.S. Universities	178
	Types of Involvement	178
	Mechanisms for Future U.S. University Involvement	190
	Criteria for Successful U.S. University Involvement	206
	Scenarios	213
6	Policy Issues and Options	218
	Funding	218
	Objectives	223
	Why Universities?	232
	Bureaucratic vs. Professional Approach	237
	Evaluation	239
	Legislation Options	241
7	Summary, Conclusions and Recommendations	248
Appendixes		267
Notes		311
Bibliography		349
Index		383
About the Authors		399

List of Figures and Tables

Figure		Page
I.1	Involvement of U.S. universities in science and technology for development	xii
1.1	Some legislation pertinent to the role of U.S. universities in science and technology for development.	3
1.2	Objectives and functions of proposed foundation for International Technological Cooperation (FITC).	6
1.3	Core program activities under Title XII that meet criteria for university participation.	13
2.1	Recommendations of NAS/NAE report on role of U.S. engineering schools in development assistance.	26
2.2	Other international activities related to Georgia Tech's 211(d) grant.	39
3.1	Generalized structure of international agricultural research for developing nations.	79

Figure

		Page
3.2	Schematic of flow of funds to U.S. universities for involvements in international agricultural activities.	87
3.3	Major contracts for INTSOY program.	97
3.4	Selected international projects involving the College of Agriculture and Natural Resources of Michigan State University (June 1977-June 1978)	100
4.1	International centers of research excellence.	139
4.2	Science involvements arranged by the mechanism used.	165
4.3	Factors relating to science involvements.	173
5.1	Role and needs of LDC universities.	180
A.1	Workshop participants.	268
A.2	Agenda of UN Conference on Science and Technology for Development (UNCSTED)	270

Tables

1.1	Estimated Title XII Levels for FY 1976-FY 1979 as Submitted by the President to the Congress	14
2.1	Summary of Present and Future Opportunities for U.S. Engineering Schools to Render Development Assistance.	25
2.2	Foreign Students in Engineering in the U.S.: Home Region and Academic Level, 1973-1974	29
2.3	Primary Sources of Funds for Foreign Students Studying in the U.S., 1976-1977	31

Figures and Tables   xi

Table		Page
2.4	211(d) Grant Impact on Georgia Tech International Development Program	40
2.5	Kabul Afghan-American Program Summary of Participant Programs.	44
2.6	Placement of IIT/K Graduates.	47
2.7	ASEE International Activities Survey Program Elements	57
3.1	U.S. University Programs in International Agriculture	72
4.1	Applications and Awards to U.S. Participants in the Senior Fulbright-Hays Program, 1948-1975.	160
4.2.	U.S. Scholars Recommended for Senior Fulbright-Hays Awards to Study in LDCs in the Natural and Applied Sciences.	161
4.3.	Factors Distinguishing Successful from Unsuccessful Links.	174
A.1	Legislative Options to Facilitate Enhanced International Science and Technology Cooperation Involving Universities	289
A.2	U.S. University Personnel Involved in BOSTID Projects, 1970-76.	299
A.3	U.S. Spending on R&D for Development.	304
A.4	National Science Foundation Budget Submission to the Congress for International Activities, FY 1979	305
A.5	Statistical Summary AID-Financed University Contracts and Grants Active during the Period 04/01/77-09/30/77.	307

Acknowledgments

The work reported in this book was carried out during the spring and summer of 1978 under National Science Foundation (NSF) sponsorship as part of U.S. preparations for the 1979 United Nations Conference on Science and Technology for Development (UNCSTED). The book closely follows the final report prepared for NSF on September 1, 1978 and is reasonably current through that date. In preparing the book for wider publication, we have made minor modifications to the NSF report in order to reflect recent developments. Perhaps the most important of these has been the steps taken by the U.S. government toward the establishment of the Institute for Scientific and Technological Cooperation (ISTC), formerly called The Foundation for Technological Cooperation (FITC). By the time this book appears, the fate of the ISTC in the legislative and budgetary process should be decided.

We wish to express our thanks to the many people who aided us in this work. Our NSF program officer, Gordon L. Hiebert, and our NSF grant specialist, Martin V. Geary, provided fine support and counsel, as did Simon Bourgin of the State Department and Edward MacCordy of Washington University. Our office support staff, Lynn Lawson, Donna Williams, and Emily Pearce, have our deepest respect and admiration for producing the NSF final report and the two interim reports that preceded it. We are grateful also to the staff of Pergamon Press for the interest, speed, and care with which they produced this book. Thanks also go to Charles G. Kick for his assistance in preparing the Index.

An integral part of this study was a small workshop held on July 13 and 14, 1978, at which a draft study paper was discussed and reviewed by a peer group from universities, government agencies, and non-governmental organizations. The participants in our workshop provided constructive, helpful comments and information. They are

M. R. Barry, M. Blase, W. D. Buddemeier, W. Eilers, T. Fox, M. Gomez, B. Hazeltine, F. A. Long, B. Lucas, H. Miller, K. N. Rao, A. Segal, V. Walbot, R. M. Walker, W. Wight, A. Wilburn, and M. Witunski. A detailed summary of the workshop purposes, organization, and discussions is included as Appendix A, which also contains the full names and affiliations of the participants in figure A.1.

M. J. Moravcsik provided detailed comments on the initial progress report and the Draft Study Paper. Useful comments on our initial progress report were received from C. Barker, C. T. Hill, S. Bourgin, and B. Lucas. K. N. Rao, M. Blase, and W. D. Buddemeier spent considerable time and effort providing information and answering questions. Early individuals too numerous to mention responded to requests for information and assistance.

The responsibility for the accuracy and wisdom of the contents of this book rests with the authors.

Introduction

STUDY OBJECTIVES

The purpose of this investigation is to examine the past, present, and future role of U.S. universities in helping to build an indigenous science and technology (S&T) base in developing countries. It is hypothesized that U.S. universities constitute an underutilized resource which could be a significant factor in such an effort. This study is relevant to the 1979 United Nations Conference on Science and Technology for (International) Development (UNCSTED) agenda item 2 dealing with institutional arrangements and new forms of international cooperation in the application of science and technology. In particular, agenda item 2(d) calls for the "strengthening of international cooperation among all countries and the design of concrete new forms of international cooperation in the fields of science and technology for development." In the opinion of V. J. Ram, principal technical advisor to UNCSTED Secretary-General da Costa, this agenda item will constitute a principal focus for progress and action at the conference. (1)

This study was carried out in response to a National Science Foundation program solicitation that was developed in close cooperation with the office of the U.S. coordinator of preparations for UNCSTED, Ambassador Jean Wilkowski. Projects supported under this solicitation were to contribute to U.S. preparations for UNCSTED in two ways: 1) by production of study papers for use by the U.S. delegation to UNCSTED that address issues and options related to conference agenda items, and 2) by promoting and improving consensus or clarification of important differences on these issues. (2)

Through examination of key documents and other information,

we have sought to 1) analyze past U.S. university involvement, 2) analyze the legislative mandate for that involvement, and 3) explore current thinking of knowledgeable individuals with regard to the role of the U.S. university in S&T for development. We have focused on the study of three fields: engineering, agriculture, and science. Based on these analyses, we have classified various forms of U.S. university involvement, identified conditions for success or failure, and discussed limitations to U.S. university involvement. We have also sought to identify, wherever possible, efficient and effective roles of U.S. universities in strengthening S&T infrastructures in less developed countries (LDCs), and to summarize successful experiences.

A major outcome of this study is an analysis of future roles of universities in helping to build an indigenous LDC S&T base, which includes a discussion of types of involvement, mechanisms (both bilateral and multilateral) for facilitating such involvement, and possible legislative and other changes to implement these mechanisms. Key policy issues and options that influence the future role of U.S. universities are discussed and the impact of U.S. university involvement on both the U.S. and LDCs is examined.

BACKGROUND AND LIMITATIONS OF STUDY

Strengthening the Indigenous Science and Technology (S&T) Base

According to a resolution adopted by the Second Preparatory Committee for UNCSTED at the conclusion of a meeting in Geneva on February 3, 1978: "The General Assembly decided to convene the UNCSTED with the objective, in particular, of strengthening the technological capacity of developing countries to enable them to apply science and technology to their own development." Thus, a central concern of this conference involves the building of an indigenous science and technology base in developing countries. We have kept this concern foremost in our analysis. Furthermore, education and training are prominently mentioned in the list of issues for consideration of draft discussion papers at every level: national, regional, and international. Of particular concern are systems for education and training in the LDCs as well as, at the international level, "appropriateness of programs for education and training of personnel from developing countries in developed countries, migration of talent and skills for developing countries, need for real concern about research and development needs of developing countries. . . . " (3)

Introduction

Thus, we find a clear focus for our study from the developing country point of view, namely the desire to build an indigenous science and technology capacity or base. However, our policy stand must also consider the extent to which U.S. involvement in this process is compatible with our foreign policy objectives and our national interests. To what uses will a greatly strengthened indigenous S&T base in LDCs be put? Will such uses be beneficial or detrimental to U.S. interests?

U.S. technical assistance and cooperation have been based on several rationales through the years which have received varying emphases. Both existing and recently proposed legislation stress the meeting of basic human needs, assisting the poorest of the poor, and growth with equity. It does not automatically follow that the strengthening of an indigenous S&T base is compatible with these objectives. Within a developing country the ability to utilize an indigenous S&T base to meet various objectives will depend on the internal political, social, and economic system, the degree of understanding of the nature of S&T, and the degree of creation of ideological and material infrastructures that transmit and support the S&T base. We have been aware of these considerations throughout the study.

In addition, there is growing concern, particularly on the part of U.S. labor, that developing countries are becoming increasingly competitive with the U.S. for markets and products, thereby resulting in employment dislocations and losses in the U.S. However, increasing the indigenous S&T base, in addition to helping LDCs to meet basic needs, could also serve to increase markets for U.S. products overseas. This is a very complex issue which we have not been able to analyze in any detail within the context of our study. However, it is part of the overall environment in which our study must be placed.

In this investigation, we focus on the development of an indigenous S&T base in LDCs because this topic looms as a central element of developing country concerns at UNCSTED. The U.S. university role is also of central concern because of the long history of U.S. university international involvement and because of the important role the university plays in our own science and technology infrastructure.

Definitions

In using the word <u>universities</u>, we include: universities with graduate programs, four-year colleges and universities, two-year community colleges, and two- and four-year engineering technology programs. However, we have given less attention to the role of community and engineering technology programs.

For <u>science</u> and <u>technology</u> we take as a starting point the following definitions presented by Moravcsik:

<u>Science</u> is an activity resulting in knowledge and understanding about the world around us.

<u>Technology</u> is an activity resulting in procedures for building and creating things, in prototypes and models of products, in gadgets and inventions.

We limit science to <u>natural</u> science and include both <u>basic</u> and <u>applied</u> science. The above definition of technology is hardware oriented, in contrast to definitions that also include "software" or "social technology" as part of technology, as is increasingly common in considering "appropriate technology" (see chapter 7). (5) Although we generally follow Moravcsik's definitions, we are aware of the broader context in which the word "technology" is often used, to include such factors as management skills, financial and credit systems, etc. In considering certain university programs such as the University of Wisconsin's Land Tenure Center, it is necessary to consider technology as having this broader context.

In this report, the phase "science and technology" (S&T) and the phrase "research and development" (R&D) appear, sometimes together and sometimes separately. UNESCO defines S&T activities as "any systematic activities closely concerned with generating, disseminating and applying technological knowledge" and which include: 1) research and experimental development (R&D) activities; 2) scientific and technological service (STS) activities, and 3) innovation and diffusion of innovation activities. (6) The <u>UN World Plan of Action for the Application of Science and Technology to Development</u> defines a country's S&T activities as: a) fundamental research, b) applied research, c) experimental development, and d) related scientific and technological activities. Thus, although research and development is an important element of science and technology, it is but one element, particularly when considering university involvement with its important education and training function. (7)

<u>Development</u> has many different meanings. Its use in this study goes beyond the purely economic. In the official U.S. statement to the Second Preparatory Committee Plenary Section on UNCSTED, Jean Wilkowski set the context for the U.S. position by describing her concept of development:

. . . to mean the development of human resources. . . <u>of people</u>. It is not governments but people who are hungry, and people who need work. Thus, we need technology that provides tools that are relevant to people: that help improve their diets, that reduce health hazards, that provide educational opportunities and shelter, so as to help break poverty's vicious circle of malnutrition, unemployment and excessive birth rates. (8)

In addition, the secretary-general of the conference, Joao da Costa, explains development and its interrelationship to science and technology in an even broader sense. He suggests that development should be considered:

> ... not only for overall economic growth strictu sensu, not even for the satisfaction of basic human needs, but also in a larger context implying an holistic approach. (Utilization of valuable existing cultural patterns, preservation of human values, participation of people in shaping the basis of their own existence, capacity of autonomous goal-setting and decision making, quality of life, human rights, peace, etc.) (9)

Another insight into the nature of UNCSTED was provided by V. J. Ram who pointed out that this was not a conference on science and technology; nor was it a conference on development. Rather, it was concerned with science and technology for development, with the emphasis on development, technology, and science in that order. (10)

We have chosen to analyze past U. S. university international involvement in three fields, namely engineering, agriculture, and science. Engineering is a profession concerned with the uses of science and technology. In U. S. engineering schools, most research is what Moravcsik would classify as applied scientific research as contrasted to technology research (see chapter 4). Thus, there may be some overlap between the engineering and science categories. In addition, science serves as an important underpinning for both engineering and agriculture. Furthermore, some interdisciplinary S&T work is not neatly classifiable. Nevertheless, this division into three categories served to focus the tasks of individual members of our study team and enabled us to examine three important areas, which cover a major portion of U. S. university involvement, in some detail. Definition of the scope included under engineering, agriculture, and science is given in chapters 2, 3, and 4, respectively.

DATA GATHERING AND ANALYSIS

Data Gathering

U. S. university involvement in international activity has been extensive. There is a great deal of printed material and material that has not been documented but which resides with knowledgeable practitioners. However, not all of the university involvement has been focused on science and technology for development. Furthermore, the involvements may be characterized generally as not having had much independent evaluation.

In performing our analysis, we have utilized a variety of sources. First, there are reports that seek to aggregate and evaluate overall experiences (see Bibliography). These reports are useful in providing an overview of our three main study areas. They suffer to some extent from lack of rigorous evaluation and from lack of LDC perspective.

Eleven journals provided key articles on agriculture, science and technology, engineering, "brain drain," institution building, and international education. They are A. I. D. R&D Abstracts, Bulletin of the Atomic Scientists, Current Literature on Science of Science, Development Digest, Engineering Education, Impact of Science on Society, Interciencia, International Development Review, International Educational and Cultural Exchange, Minerva, and Technos.

A search strategy was developed to retrieve on-line bibliographic information from the Congressional Information Service/Index (CIS/Index) and the Educational Resource Information Center (ERIC) via the Washington University Computer Search Service. The CIS/Index yielded a total of 21 citations and the ERIC search resulted in a total of 189 citations.

Letters and telephone calls to about 150 organizations and individuals yielded valuable information. In May 1978, the American Society for Engineering Education (ASEE) surveyed its member organizations to determine the nature and extent of ongoing international involvement. We analyzed this data for ASEE and included some of the results in chapter 2.

Analytical Framework

We have chosen to organize some of the analysis of this study by <u>types</u> of involvement of U. S. universities in S&T for development and <u>mechanisms</u> for bringing about these involvements.

Introduction

Figure I.1 summarizes involvements of U.S. universities under four main categories: 1) Institution-building, which is further divided into two main subcategories: a) type of LDC institution and b) type of U.S. involvement; 2) Cooperative research and development, including cooperative R&D programs between individuals or institutions in the U.S. and LDCs; 3) U.S. resource base development, concentrating on building capability within the U.S. that is relevant to development efforts, and 4) Education and training including foreign nationals studying in both degree and non-degree programs in the U.S. as well as U.S. students and faculty who go to LDCs.

We define the term mechanisms very broadly to encompass a variety of means of bringing about and implementing an involvement. Mechanisms can include organizational forms (e.g. a consortium of universities in the U.S. linked to an LDC institution to aid in institution building). They can also include programs and legislation which support various kinds of involvements. Mechanisms can be either bilateral or multilateral, and are considered in more detail in chapters 2 through 5.

Organization of the Report

Chapter 1 considers the past legislative mandate for U.S. university involvement in S&T for development. Chapters 2, 3, and 4 are devoted to examining U.S. university activity in three fields: engineering, agriculture, and science. Within the analysis of each field, we describe specific past involvements, current thinking, and analyze field-specific issues. Chapter 5 examines future roles for U.S. universities, including types of involvement, mechanisms for involvement, new forms of cooperation among U.S. institutions, and sketches three scenarios. Criteria for success or failure, conditions for success, and limitations to U.S. university involvement are also considered. Chapter 6 contains a discussion of overarching policy issues and options. Eight legislative changes are outlined for expanding U.S. university involvement in international S&T cooperation. The principal conclusions of the study are contained in the summary and conclusions chapter at the end of this report. A summary of the workshop discussions is included in Appendix A.

I. Building Institutions in Developing Countries

   A. Types of DC Institutions
      1. Universities, Technical Schools
      2. Formal and Nonformal Educational Institutions
      3. Public Agencies
      4. Private Enterprises
      5. Cooperatives
      6. Research Institutes
      7. Regional Centers

   B. Types of U.S. Involvement
      1. Universities and Colleges
      2. Community Colleges and Technology Programs
      3. Consortia of B1, B2
      4. Subgroups within B1, B2 (departments, institutes, centers)
      5. Individuals (consultants, faculty, students)

II. Cooperative Research and Development

   A. Sister-Institution Arrangement
   B. Consortia
   C. Individuals
   D. Research Focused on Development Problems
   E. Funding for One or Both Sides of Links

III. U.S. Resource Base Development

   A. 211(d) Legislation
   B. Title XII Legislation
   C. Develop Capability for R&D on Development Problems
   D. Develop Courses
   E. Develop Curricula
   F. Special Training Programs
   G. Summer Programs
   H. Cooperation between U.S. Universities and Other U.S. Organizations

IV. Education or Training

   A. Non-U.S. Students to U.S.
      1. Degree Programs (undergraduate and graduate)
      2. Nondegree Programs (usually mid-level career people)

   B. U.S. Students, Faculty, and LDCs
      1. Exchange Programs
      2. Peace Corps Volunteers
      3. Graduate Students (doing thesis/dissertation research abroad)
      4. Consultancies

Fig. I.1. Involvement of U.S. universities in science and technology for development

List of Acronyms

AAAS	American Association for the Advancement of Science
AC	Advanced Country
ACIOP	Advisory Committee on International Organizations and Programs
ACTI	Advisory Committee on Technological Innovation
AFGRAD	African Graduate Fellowship Program
AID	Agency for International Development
ALAD	Arid Lands Agricultural Development Program
ASEE	American Society for Engineering Education
AT	Appropriate Technology
ATI	Appropriate Technology International
AUB	American University of Beirut
AUSUDIAP	Association of U.S. University Directors of International Agricultural Programs
AVRDC	Asian Vegetable Research and Development Center
BIFAD	Board for International Food and Agricultural Development
BISE	Board on International Scientific Exchange
BOSTID	Board on Science and Technology for International Development
BSCS	Biological Sciences Curriculum Study
CARE	Cooperative for American Relief Everywhere
CDT	Center for Development Technology
CFTRI	Central Food Technological Research Institute
CGIAR	Consultative Group for International Agricultural Research
CIC	Committee on Institutional Cooperation
CIDA	Canadian International Development Agency
CIMMYT	International Maize and Wheat Improvement Center
CIS	Congressional Information Service

CNPq	Brazilian National Research Council
CSUCA	Superior Council of Central America
CUSURDI	Council of United States Universities for Rural Development in India
DC	Developing Country
DHEW	Department of Health, Education and Welfare
DLF	Development Loan Fund
DOC	Department of Commerce
DOE	Department of Energy
DOI	Department of the Interior
DOL	Department of Labor
DOT	Department of Transportation
DRI	Denver Research Institute
DTICA	Direccion Tecnica Interamericana Cooperative de Agricultura de Chile
ECA	Economic Cooperation Administration
ECPD	Engineers Council for Professional Development
EDC	Education Development Center
EPA	Environmental Protection Agency
ERIC	Educational Resource Information Center
EWA	Education and World Affairs
FAA	Foreign Assistance Act
FAO	Food and Agriculture Organization
FITC	Foundation for International Technological Cooperation
FOA	Foreign Operations Administration
FORGE	Fund for Overseas Research Grants in Education
FMME	Fund for Multi-Management Education
FUNBEC	Fundacio Brasileira para o Densenvolvimento do Ensino de Ciencias
GAO	General Accounting Office
IAESTE	International Association for the Exchange of Students for Technical Experience
IAPAR	Fundacao Instituto Agronomico do Parana
IBECC	Brazilian Institute for Education, Science and Culture
ICA	International Communication Agency
ICA	International Cooperation Administration
ICARDA	International Center for Agricultural Research in Dry Areas
ICED	International Council for Educational Development
ICIPE	International Center for Insect Physiology and Ecology
ICSU	International Council of Scientific Unions
IDCA	International Development Cooperation Administration
IDF	International Development Foundation

Acronyms

IDI	International Development Institute
IDRC	International Development Research Centre
IFS	International Foundation for Science
IGY	International Geophysical Year
IIAA	Institute of Inter-American Affairs
IIASA	International Institute for Applied Systems Analysis
IIE	Institute of International Education
IITA	International Institute of Tropical Agriculture
IIT/K	Indian Institute of Technology/Kanpur
INCAP	Institution of Nutrition of Central America and Panama
INELEC	National Institute of Electricity and Electronics
INP	International Nutrition Policy and Planning Program
INTSOY	International Soybean Program
IROP	Institute of Research in Overseas Programs
IRRI	International Rice Research Institute
ISTC	Institute for Scientific and Technological Cooperation
JCAD	Joint Committee on Agricultural Development
JRC	Joint Research Committee
KAAP	Kabul Afghan-American Program
KIAP	Kanpur Indo-American Program
KIST	Korean Institute of Science and Technology
KU	Kabul University
LARS	Laboratory for the Applications of Remote Sensing
LASPAU	Latin American Scholarship Program of American Universities
LDC	Less Developed Country
LTC	Land Tenure Center
MIT	Massachusetts Institute of Technology
MNC	Multinational Corporation
MSA	Mutual Security Agency
MUCIA	Midwest Universities Consortium for International Activities
NAE	National Academy of Engineering
NAS	National Academy of Sciences
NASA	National Aeronautics and Space Administration
NASULGC	National Association of State Universities and Land-Grant Colleges
NCAT	National Center for Appropriate Technology
NCP	Nutrition Center of the Philippines
NDEA	National Defense Education Act
NIEO	New International Economic Order
NIETC	National Institute for Education and Technical Cooperation

NIH	National Institutes of Health
NINDB	National Institute of Neurological Diseases and Blindness
NOAA	National Oceanic and Atmospheric Administration
NRC	National Research Council
NSF	National Science Foundation
OAS	Organization of American States
OMB	Office of Management and Budget
OPEC	Organization of Petroleum Exporting Countries
OST	Office of Science and Technology
PCV	Peace Corps Volunteer
PRDYCT	Regional Scientific and Technological Program
PSSC	Physical Science Secondary Curriculum
PUC	Catholic University of Peru
PVO	Private Voluntary Organization
R&D	Research and Development
RANN	Research Applied to National Needs
RIIC	Rural Industries Innovation Centre
RITA	Rural Industrialization Technical Assistance
S&T	Science and Technology
SECID	South East Consortium for International Development
SEED	Scientists and Engineers in Economic Development
SFCRP	Special Foreign Currency Research Program
SITE	Satellite Instructional Television Experiment
STICA	Servicio Technico Interamericana de Cooperation Agricola
STS	Scientific and Technological Service
TAB	Technical Assistance Bureau
TAP	Technology Adaptation Program
TCA	Technical Cooperation Administration
TSRTP	Tropical and Subtropical Research Training Program
UCLA	University of California at Los Angeles
UNCSTED	United Nations Conference on Science and Technology for Development
UNCTAD	United Nations Conference on Trade and Development
UNDP	United Nations Development Program
UNESCO	United Nations Educational, Scientific and Cultural Organization
UNI	National Engineering University of Peru
UNICEF	United Nations International Children's Educational Fund
UNU	United Nations University
UPADI	Pan American Federation of Engineering Associations
USAID	United States Agency for International Development
USDA	United States Department of Agriculture
USET	United States Engineering Team
VITA	Volunteers in Technical Assistance

1 The Legislative Mandate

Legislation pertinent to U.S. university involvement in science and technology for development serves a variety of functions. First, there is legislation that sets policy and directions for the types of involvements and their objectives. Central to this function has been the Foreign Assistance Act of 1961 and its subsequent amendments. Elements of this legislation also serve to establish maximum funding levels available for certain government programs that universities might be involved in. Second, legislation creates agencies related to foreign assistance and cooperation activities that might interact with universities. The prime example of this is the Agency for International Development which provides grants and contracts to U.S. universities for various international activities. This category also includes legislation enabling other federal agencies with primarily domestic missions, such as the National Science Foundation and the Department of Energy, to become involved in international activity. Third, there is legislation specifically pertaining to education or to university involvement. For example, within the Foreign Assistance Act (FAA), Section 211 of Title II has been relevant to universities, particularly since it was amended in 1966 to provide funds to establish "resource bases" at U.S. universities for international development work. (1) Also in this category falls the International Education Act of 1966 (2) and the National Defense Education Act of 1958. (3) Finally, there is legislation that affects the international exchange of personnel - faculty and students. The most recent activity in this area involved the creation in April 1978 of the International Communication Agency. (4) This category includes regulations governing the issuance of visas and passports, and affecting immigration.

We use these four functions as an organizational framework for the analysis of the overall impact of the past legislative mandate on

U.S. university involvement. A summary of the most pertinent legislation is presented in figure 1.1. In this chapter, we deal with a variety of legislation including that most specifically pertaining to education or to university involvement. Specific changes in legislation which might be made to affect the future role of U.S. universities are presented in chapter 6.

THE POLICY THREAD IN U.S. FOREIGN ASSISTANCE LEGISLATION

The initial policy thrust of the Foreign Assistance Act of 1961 emphasized long-range assistance to promote economic and social development. (5) An interpretation of this directive gave priority to assistance in the form of large-scale projects, chiefly in the areas of industrial factories and equipment, transportation facilities, and irrigation projects. (6)

As the sixties progressed, there was increasing emphasis on technical assistance, (7) which gave rise to FAA Amendment 211(d) which directly affected universities. (8) In 1966, agricultural research also was stressed as a priority, and cooperative undertakings between U.S. universities and research institutions, and LDC institutions were encouraged. (9) In 1969, the Technical Assistance Bureau was established in AID to concentrate on technical aid. (10)

One other important policy change was the "New Directions" legislation which was initiated in connection with passage of the 1973 Foreign Aid Bill, but continues to be implemented today. (11) "New Directions" embodies a major shift from large public works type projects based on a "trickle down" philosophy to a more direct attempt to assist the poorest of the poor to meet their basic needs. (12) Two areas of potential relevance to U.S. universities are the continued stress within the "New Directions" framework on improving educational resources in LDCs and provision of funds in the FAA of 1975 totaling $20 million for fiscal years 1976, 1977, and 1978 for activities in the field of intermediate technology. (13) (Rep. Clarence Long has offered several amendments to legislation dealing with international organizations and foreign assistance, requiring the U.S. to emphasize the promotion and use of light capital technologies. Included is an amendment which is part of PL 95-105 which requires that the U.S. place important emphasis on light capital technologies in preparing for and participating in UNCSTED.) However, implementation of the intermediate or appropriate technology activity has not yet involved U.S. universities to any extent.

The Legislative Mandate

Name and Public Law No.	Significance
Foreign Assistance	
Foreign Assistance Act of 1961 P.L. 87-195	Reorganization of aid programs; creating of AID Title II-Section 211 pertaining to technical assistance.
Foreign Assistance Act of 1966 P.L. 89-583	Amendment to Title II. Section 211(d) added which provides for strengthening international capabilities of U.S. universities.
International Development and Humanitarian Assistance Act of 1971 S. 1656, S. 1657, never passed	Proposed reorganization of AID, creation of International Development Institute and International Development Corporation.
Foreign Assistance Act of 1973 P.L. 93-189	"New Directions" in policy. Emphasis on basic needs, poorest of the poor.
International Development and Food Assistance Act of 1975 P.L. 94-161	Amended FAA of 1961 with Title XII, involving U.S. universities in decision making on agricultural projects. Also, Section 107 makes funds available for new programs in "Intermediate Technology."
International Development Cooperation Act of 1978 S. 2420-not acted upon	Proposed reorganization of foreign aid programs.
International Mandates for Domestic Agencies	
P.L. 90-407; 1968	Authorizes National Science Foundation in participating in international cooperative scientific activities.
Anti-Nuclear Proliferation Act of 1978	Authorizes Department of Energy involvement in international small-scale energy activity.
Food for Peace Act of 1966-Section 406	Authorizes U.S. Department of Agriculture to enter into contracts or agreements with U.S. universities to conduct research on agriculture related to developing countries (funded in 1974).
Food and Agricultural Act of 1977 P.L. 95-113	Authorizes U.S. Department of Agriculture to strengthen U.S. universities for working in international development (not funded as of 1978).
Education	
International Education Act of 1966 P.L. 89-698	Provided broad mandate for involvement of U.S. universities in international education (never funded).
National Defense Education Act of 1958 P.L. 85-864 amended by International Education Act of 1966	Created international institutes for secondary school teachers. NDEA also authorized U.S. Office of Education to support foreign area studies and language programs.
Immigration	
Immigration and Nationality Act of 1952 P.L. 414 as amended	Created special categories for non-U.S. students and professors.

Fig. 1.1. Some legislation pertinent to the role of U.S. universities in science and technology for development.

Another important policy thrust came in 1975 when foreign aid legislation emphasized international food and agricultural development. This legislation with its emphasis on prevention of hunger and famine led to the creation of Title XII of the International Development and Food Assistance Act of 1975, (15) a very important piece of legislation as far as land-grant colleges and state universities are concerned.

In late 1977, AID was reorganized. The Technical Assistance Bureau (TAB) was absorbed within a new Development Support Bureau. The Office of Science and Technology (OST) within the old TAB retained its identity but the office has been hampered by low, fixed staffing levels. Furthermore, as part of the decentralization thrust of the organization, OST lost some of its decision-making ability to the overseas missions. OST's program does not include all science and technology within AID; for example, major agricultural activity is excluded. Energy activity was split off from OST but appropriate technology activity was added. The name change of the bureau and related developments may have signaled a de-emphasis on science and technology, although the change from "technical assistance" to "development support" may simply reflect a growing sensitivity to the attitudes of the LDCs.

Another significant piece of legislation is Title V of Public Law 95-426, the Foreign Relations Authorization Act, Fiscal Year 1979. This title provides a mandate for a science and technology foreign policy for the United States and requires the State Department to strengthen its science and technology policy activities.

CREATION OF AGENCIES WITH INTERNATIONAL OBJECTIVES

Agencies With International Missions

The Foreign Assistance Act of 1961 created the Agency for International Development (AID) as a consolidation of the International Cooperation Administration and the Development Loan Fund. (16) Dissatisfaction with its structure came soon after its creation. (17) Through the late sixties and early seventies, continued calls for reorganization of foreign assistance programs generated a task force study known as the Peterson Report, (18) and eventually, in 1971, legislation was introduced into the Senate proposing the creation of two new aid agencies - an International Development Institute to handle all technical assistance programs, and an International Development Corporation to process loans and economic aid to LDCs. (19) These agencies were meant to replace AID. While extensive hearings in the Senate considered the two bills - S 1656 and S 1957 - jointly called the International Development and Humanitarian Assistance Act of 1971, it seems that issues surrounding the war in Indochina, including the controversial end-the-war amendment, clouded the issue of foreign assistance reorganization. Bitter debate held up even basic appropriations. In fact, for the first time in 24 years, appropriations for foreign assistance programs of 1972-1973 were not passed by Congress until two-thirds of the way into the fiscal year. (20)

The International Development Cooperation Act of 1978 was introduced in the Ninety-fifth Congress on January 25, 1978, by Senators Sparkman and Case. (21) It was authored by the late Senator Hubert Humphrey. The act proposed to consolidate a wide variety of activities within a new International Development Cooperation Administration (IDCA), including AID, AID's Title XII and 211(d) programs, and support for international financial institutions such as the World Bank. It also would have brought the Peace Corps into the new agency by establishing an International Development Institute, incorporating the Peace Corps and private and voluntary organization programs. In the form the bill was introduced in the Senate, it is not clear what impact, if any, this legislation would have on U.S. university involvement. It appears not to break new ground in this regard. Furthermore, there appears to be neither emphasis on building an indigenous LDC science and technology base nor focus on science and technology. This legislation did not win passage in 1978. However, a revised version of IDCA was under active consideration in 1979.

In October 1977, an interim report was presented by the Brookings Institution to the Department of State, recommending the establishment of two organizations to replace AID. (22) These two organizations resemble to some extent those proposed in the Senate in 1971 but never implemented. One of these, an International Development Foundation, is of particular relevance to universities and is described in Appendix B. It places strong and explicit emphasis on research and development, and on science and technology which is generally missing from other legislation.

During his April 1978 trip to Venezuela, President Carter proposed a new Foundation for International Technological Collaboration (FITC). The objectives and functions of the new foundation are summarized in figure 1.2 and are similar to those proposed in the Brookings study. One difference which may emerge is that the Brookings study envisioned the IDF as being independent of AID whereas the administration now proposes including FITC along with AID in a new International Development Cooperative Administration as two separate but closely related entities.

1) Strengthen S&T capabilities of selected LDC institutions through collaborative relationships with U.S. institutions.

2) Create "centers of excellence" - projects supporting the generators and users of technology in LDCs.

3) Support collaborative R&D projects between U.S. and LDC institutions, e.g., energy, natural resources, transportation, communication, small-scale industries, and traditional AID areas, e.g., agriculture, contraception.

4) Support collaborative assessment of global problems of mutual concern, e.g., ocean, atmospheric degradation, tropical diseases, urban poor.

5) Study past failures and successes of bilateral and multilateral aid.

(6) Perform policy evaluations of R&D priorities for U.S., LDCs, and FITC.

7) Support research on process of technology acquisition, innovation, and international industrial trends.

8) Orient programs of U.S. and LDC universities, government agencies, corporations, professional and trade associations towards development problems.

9) Strengthen LDC access to U.S. and worldwide S&T information.

Fig. 1.2. Objectives and functions of proposed Foundation for International Technological Cooperation (FITC).

The Legislative Mandate

International Mandates for Domestic Agencies

The designation of international mandates for federal domestic agencies has also been an important trend. For example, the National Science Foundation (NSF) is authorized and directed "to initiate and support basic scientific research and programs to strengthen scientific research potential and science education at all levels and to appraise the impact of research upon industrial development and upon the general welfare." (23) There was no reference to international science in the original mandate. However, in July 1968, PL 90-407 was passed which authorized the foundation "to initiate and support specific scientific activities in connection with matters relating to <u>international</u> cooperation [emphasis added], national security, and the effects of scientific application upon society by making contracts or other arrangements (including grants, loans, and other forms of assistance) for the conduct of such activity. When initiated or supported pursuant to requests made by any other federal department or agency, . . . such activities shall be financed whenever feasible from funds transferred to the Foundation by the requesting official . . . " (24) Subsequent amendments substituted "scientific and educational activities" for "educational activities," and authorized NSF to initiate and support specific scientific activities in connection with matters related to the effects of scientific applications upon society. (25)

In the past, the National Science Foundation has never had strong emphasis in its programs on international cooperation. When the RANN (Research Applied to National Needs) program was initiated, there was practically no international component, reflecting a philosophy that international development concerns were not part of national needs. However, the 1977 NSF Authorization Act directs NSF "to assist in the resolution of critical and emerging problems with significant scientific and technological components, such as the world food and population problems." (26) A current focal point of international activity in NSF is the Division of International Programs in the Directorate of Scientific, Technological, and International Affairs.

The United States Department of Agriculture has a history of legislative mandates to become involved in international agricultural development. The Agricultural Trade Development and Assistance Act of 1954 (PL 480), as amended in 1958 and 1959, authorized the Special Foreign Currency Research Program (SFCRP). This program is designed to make use of local currencies paid to the United States for sale of surplus U. S. food. Research undertaken is both relevant to USDA programs and has the potential for producing beneficial results for the foreign country as well. Furthermore, under Section 406(4) of the Food for Peace Act of 1966, as amended,

USDA is authorized to enter into contracts or agreements with land-grant universities and colleges and other institutions for conducting research on tropical and subtropical agriculture for the improvement and development of food production and distribution techniques in developing countries. However, the first funds under this authority did not becoming available until FY 1974. (27)

More recently, Congress passed section 1458 in PL 95-113 of the Food and Agricultural Act of 1977. Section 1458 authorizes USDA to strengthen U.S. universities for work in international development. According to George Waldman (Assistant Director, Interagency Relations, Office of International Cooperation and Development, USDA), it is "believe [d] that this situation provides USDA the needed authority to permit it to become much more actively involved in international programs." (28) Although in 1978 funding had not yet been authorized for this activity, USDA has apparently begun planning to develop the program. The wording of this legislative mandate calls for a program that is not duplicative of Title XII.

Many other federal agencies are involved in international activities through varying mandates. The Office of Education of the Department of Health, Education and Welfare (HEW) was given a major mandate in 1966 through passage of the International Education Act to support international education, but no funds were provided to implement programs. Funds were provided, however, through the National Defense Education Act for area studies and language programs. Support for these programs has decreased in recent years, as has support for the Fulbright scholar exchange program now funded by the International Communication Agency.

The Anti-Nuclear Proliferation Act of 1978 calls for the new Department of Energy to initiate international activites on small-scale, alternative energy sources. The National Aeronautics and Space Administration (NASA), has an Office of International Affairs to assist in bringing the benefits of space technology to all mankind. NASA participated in the SITE experiment which brought instructional television via satellite to Indian villages. Such diverse organizations as the Department of Commerce, the Department of Labor, the Department of the Interior, and the Environmental Protection Agency are involved in a variety of projects through international mandates which we have not examined in detail. Often "pass-through" funds for such activity are provided by AID, which remains the predominant funding source for international involvements on the part of federal agencies.

LEGISLATION PERTAINING SPECIFICALLY TO EDUCATION OR UNIVERSITIES

Foreign Assistance Act

Section 211

Section 211 of the Foreign Assistance Act of 1961 authorizes the president to furnish assistance in order to promote the economic development of less developed countries, with emphasis on developing human resources through technical cooperation. (29) It also emphasizes providing assistance to educational and other institutions contributing to social progress, the integration of new with existing efforts coupled with recipient country commitment to the welfare of its people, an attitude of self-help and willingness to share costs, and recognition of situations possibly adverse to the U.S. economy.

Subsection (d) was added to Section 211 by the Foreign Assistance Act of 1966. It provided for not more than $10 million total to be made available to be "used for assistance... to research and educational institutions in the U.S. for strengthening their capacity to develop and carry out programs concerned with the economic and social development of the less developed countries." (30) The Foreign Assistance Act of 1968 increased the amount to an annual maximum expenditure of $10 million for this purpose. (31)

The first 211(d) grants were made in 1968. As of December 31, 1975, AID had made 54 grants to 45 institutions, totaling $42.9 million; this compares with $442 million in all AID awards to U.S. colleges and universities for such services as training, technical assistance, or research in fiscal years 1967-1975. (32)

In its 1976 analysis of the 211(d) program, the General Accounting Office found that, "not all such grants had been made in priority areas of interest, which limited use of the capabilities being developed. GAO recommends that such grants be made only when clearly necessary to develop capabilities the Agency needs." Also: "Contracts and grants have usually been awarded noncompetitively, often on the basis of work proposals the universities developed. GAO recommends that the Agency procure services from the universities only in response to Agency programming needs and consider all potential sources when awarding contracts for such services." (33) The last paragraph would seem to call for open, competitive bidding for grants and contracts in response to needs and tasks defined by AID.

AID provided a lengthy rebuttal to the GAO report. (34) Furthermore, in a letter dated October 21, 1977, from Jean P. Lewis, Assis-

tant Administrator of AID for Legislative Affairs, to Senator Thomas
Eagleton, it is indicated that following an internal review of the 211(d)
program in 1973, a new policy determination was issued by the AID administration in 1974 which reaffirms the importance and rationale of
the 211(d) program but calls for certain operational changes. Since
1974, according to Ms. Lewis, new grant and grant extensions have increasingly emphasized, among other things, output oriented to solving
LDC development problems and closer professional interaction between
staff of the institution and AID technical offices. Certain 211(d) programs are discussed in more detail in chapters 2, 3, and 4.

It is our impression that the 211(d) program has now been deemphasized at AID. Mounting criticism, a relatively low budget level,
and the new Title XII authorization have all been factors in this development. The 211(d) grants averaged about $5 million annually from 1967
to 1973, and about $2 million annually since that time. (35) According
to a 1978 letter from an AID official:

> It seems likely that a number of new grants will stay very small
> and that the mechanism will be used to strengthen selected minority institutions for greater participation in our overseas development assistance programs. For example, Africa Bureau is currently in process of initiating four 211(d) grants in aspects of
> health services to four leading minority institutions for eventual
> services in African LDCs. (36)

The 211(d) program is specifically focused on enabling U.S. universities to build up their own capacities for international involvement,
a process we have termed "resource base development." The same
process is called "institution strengthening" in the Title XII context.
Although currently in eclipse at AID, expansion of the program has
been called for by two National Research Council reports, including
one prepared in connection with UNCSTED preparations. (37) As the program has been constituted, it has been neither big enough financially,
nor well publicized enough, nor open enough to enable a broad segment
of the university community to participate. Only four 211(d) grants
were made by AID's Office of Science and Technology. One such grant,
the Georgia Tech Small Industries Program is described in chapter 2.
The U.S. university agricultural community has now turned its attention to the Title XII program.

Title XII of the International Development and
Food Assistance Act of 1975

This 1975 amendment to the Foreign Assistance Act seeks to
assist in preventing famine and establishing freedom from hunger

The Legislative Mandate

through strengthening the capacities of U.S. land-grant and other eligible universities in agricultural institutional development and to aid in U.S. government efforts to increase world food production. The title also provides for increased and long-term support for scientific approaches to LDC food and nutrition problems, with emphasis on the combination of teaching, research, and extension. Assistance is directed to be furnished to cooperative programs involving U.S. and LDC institutions, as well as international organizations such as FAO (Food and Agricultural Organization of the United Nations) and UNDP (United Nations Development Program). This title authorizes the permanent establishment of the Board of International Food and Agricultural Development, and outlines its general duties and responsibilities.

A major source of information on Title XII is <u>The First Year - A Progress Report</u> of the Board for International Food and Agricultural Development, November 1977. According to this report, "Broadly defined, the main provision of Title XII, and its central intent, is to provide an expanded role for U.S. agricultural colleges and universities in helping to solve the critical food problems of the developing world.

The legislation is based upon the fact that much of U.S. agriculture's success is due to the combined approach of teaching, research and extension in our agricultural colleges and universities as well as on the proven effectiveness of these institutions in agricultural development activities abroad." (38)

The 1978 BIFAD consists of two presidents and two agriculture deans of major land-grant universities, and three private sector representatives. The chairmen of two key committees, the Joint Committee on Agricultural Development (JCAD) and the Joint Research Committee (JRC), are also university administrators. The board has a new and unique relationship with AID which gives universities much more control and direction of major AID programs than in the past. According to the report:

> The Board pursued three basic objectives in its early stages. First, the Board assumed that the objective of Title XII was the development of a sound long-term program of involvement by U.S. universities. Second, the Board sought to achieve the fullest possible measure of participation by the Board, its subordinate units and the universities in ongoing AID policy, program, and procedure formulation and implementation - a goal which called for integration rather than separation. Third, the Board sought to achieve a relationship with AID which would maximize the strengths and comparative advantages of the universities and of AID in achieving the goals of Title XII. (39)

Title XII is considered by some to be landmark legislation for U.S. universities in that when fully implemented, it would establish assured, long-term involvements in international agricultural assistance and, perhaps more importantly, it gives U.S. universities, through BIFAD, significant inputs into definition and design of certain AID programs. Title XII, which became law on December 20, 1975, was championed by Congressman Paul H. Findley of Illinois and the late Senator Hubert H. Humphrey of Minnesota. Its passage was the result of much effort on the part of the state universities and land-grant colleges which mobilized strong political support from congressmen in almost every state while overcoming objections from the Agency for International Development. Implementation of Title XII, since its passage, has been slower than desired by Congress and the universities, as AID and BIFAD have tackled many issues, some of which had still not been resolved in 1978.

Among the achievements of the first year were the identification of a "core" program of activities which meet the criteria of university participation under Title XII. These are listed in figure 1.3. Estimated Title XII budget levels are presented in table 1.4. Although it is not entirely clear which funds will be directly spent on U.S. university involvement, a 1978 AID progress report to Congress regarding Title XII indicates:

> ...Approximately $259 million of the Food and Nutrition activities under Section 103 of the Foreign Assistance Act are within the Title XII definition and offer opportunities for eligible Title XII university implementation; however, this does not mean that all of these activities will be reviewed by the Title XII Board or that the universities have all of the expertise required by A.I.D. The U.S. Department of Agriculture, private firms, voluntary agencies and others will also have an opportunity to be considered for implementing these projects. (40)

Title XII activities in FY 1979 represent 38 percent of the total $673 million Food and Nutrition Program requested of Congress. Thus, activities under the Title XII rubric have increased substantially since FY 1976 when only $100 million, or 17 percent, of the Food and Nutrition Program could be defined under Title XII. According to BIFAD's definition, Title XII activities cover all of AID's technical assistance funded from Section 103 with two exceptions: resources specifically earmarked for support and development of programs administered by private and voluntary organizations and use of the 211(d) authority. Capital costs directly connected with research, training and extension are covered. (41)

The Legislative Mandate   13

1. Research which includes: (a) support to International Agricultural Research Centers and similar organizations; (b) food production and nutrition components of AID's centrally-funded contract research program; (c) a new Collaborative Research Support Program; and country- or regional-specific research falling within the Title XII mandate.

2. The balance of the centrally-funded technical assistance program, concerned with the adaptation and application of agricultural and nutrition technology.

3. Strengthening developing country institutions in research, teaching, extension and other services essential to agricultural development.

4. Advisory services to developing country governments and private sectors on such food and nutritional development activities as agricultural production and marketing, credit, irrigation and water management, general nutrition projects, and technical assistance for rural development, in which developing or strengthening of research, educational or extension capabilities, though often an important by-product, is not the central purpose.

Fig. 1.3. Core program activities under Title XII that meet criteria for university participation.

SOURCE: BIFAD, The First Year - A Progress Report, AID, November 1977.

Table 1.1. Estimated Title XII Levels for FY 1976-FY 1979 as Submitted by the President to the Congress (in millions).

Title XII	FY 1976 $	%	FY 1977 $	%	FY 1978 $	%	FY 1979 $	%
Centrally Funded Research and Research Support (a)	100 (30)	17	118 (40)	21	195 (43)	33	259 (49)	38
Adaptation and Application of Technology (a)	(4)		(10)		(23)		(25)	
Strengthening Developing Country Institutional Research, Teaching, and Extension (a)	(28)		(18)		(42)		(90)	
Advisory Services to Developing Countries (b)	(38)		(50)		(87)		(96)	
Residual (c)	482	83	422	79	391	67	414	62
Total Food and Nutrition (Section 103 AID Request)	582	100	540(d)	100	586(d)	100	673	100

(a) Includes related capital costs.

(b) Does not include related capital costs.

(c) Includes all non-Title XII Section 103 activities (e.g., rural road construction, fertilizer production or procurement, etc.). Includes also activities closely related to Title XII such as support to build capacity at U.S. agricultural universities under Section 211(d), capital costs of advisory services to developing countries, and activities of voluntary agencies.

(d) These figures are the amounts requested by Congress by the Administration and are used for the purpose of comparison with the FY 1979 request. Actual Food and Nutrition obligations amounted to $474 million in FY 1977 and are expected to amount to $549 million in FY 1978.

SOURCE: Report to the Congress on Title XII—Famine Prevention and Freedom from Hunger of the Foreign Assistance Act of 1961 as amended, AID, April 1, 1978; and BIFAD, The First Year: A Progress Report, AID, November 1977.

The Legislative Mandate

An initial program of four elements has been defined, following BIFAD's ".... intensive studies of the needs of universities in this area" in order to comply with the Title XII provision "to strengthen the capabilities of (U.S.) universities in teaching, research and extension work to enable them to implement current programs authorized" in other parts of the act. The four elements are:

1. Formula-based, recurrent annual funding for Title XII eligible institutions included on the BIFAD-approved roster.

2. A special Title XII grant program designed to strengthen eligible minority institutions.

3. Appropriate revision of AID policies, practices and operating procedures to facilitate university participation in Title XII.

4. A central program of activities designed to facilitate university involvement in Title XII. (42)

A number of activities are outlined in the BIFAD first-year report, including baseline studies of the capabilities of agricultural education, research, and extension systems in LDCs.

As of February 1978, some $3 million in funding to plan some initial research projects had been committed; these are referred to as baseline studies. Considerable attention has also been given to definition of which universities are eligible to participate in the collaborative research support grants and the institution-strengthening grants programs. All land-grant and sea-grant colleges and universities are included. Other colleges and universities are eligible if they have proven capability for agricultural research, teaching, and extension. Noneligible institutions can participate as subcontractors to eligible institutions. As of August 1978, five collaborative research areas have been identified by BIFAD and are in various stages of planning for collaborative research support grants. Each grant will involve several U.S. universities in research focused on one agricultural problem area, with one of the universities serving as the primary coordinating agent. Relationships will be developed with LDC universities, international agricultural research centers, and other international organizations. It is expected that two of these research projects, sorghum and millet, and small ruminants, will be initiated before October 1, 1978, using FY 1978 funds. BIFAD feels that the $9.5 million budgeted for collaborative research activity is inadequate. (43)

To our knowledge, no institution-strengthening grants have yet been made. The latter appear to be somewhat similar to 211(d) grants in providing resource base support for U.S. universities.

Title XII represents a significant step toward greater involvement of a segment of U.S. universities in international assistance and cooperative programs directed toward food and agriculture. Through BIFAD, universities are given much more input to policy and program definition than in the past. Although it is not yet clear how funding involving U.S. universities under Title XII will compare with previous funding, it appears that there is potential for expansion.

There are mixed expectations about Title XII and BIFAD at AID. One individual referred to it as a "university raid on the federal treasury." Others feel that it could serve as a useful model for all university involvements in technical assistance and cooperation activities. The year 1979 will be a crucial one in the development of the Title XII program. For further information, see chapter 3 and the thesis by Perez. (44)

International Education Act of 1966

Congress passed the International Education Act of 1966 but never appropriated funds for its implementation. (45) The act provides for:

1. Funding for establishment, strengthening, and operation of graduate centers at various U.S. institutions of higher education to function as national and international resources for research and training in international studies and international aspects of other fields, with concentration on area studies, multicountry issues or concerns, or both. Expenditures may be made for teaching, research, visiting faculty, staff training... and travel, and stipends for students to study and travel within the U.S. and abroad. Grants are authorized also for private and public nonprofit organizations, including professional and scholarly associations.

2. Funding of undergraduate education in international studies. Expenditures may be made for faculty training overseas, expansion of foreign language courses, student work-study-travel programs, visiting faculty and scholars, and English language training for foreign faculty, scholars, and students. Grants for undergraduate education may also be made to public and private nonprofit organizations.

The act authorized funding of $40 million for the FY ending June 30, 1968 and $90 million for the FY ending June 30, 1969 to be used for Title I (graduate and undergraduate education) provisions. It provides also for the establishment of a National Advisory Committee on International Studies, to be chaired by the assistant secretary

The Legislative Mandate

of HEW to advise the secretary of HEW on executing provisions of the act. The Advisory Committee is authorized $1 million for a report to the president and the Congress recommending modes of carrying out the purpose of Title I. The act also amends other laws pertaining to its purposes, such as authorizing $9.5 million for international affairs institutes for secondary school teachers through amending Title XI of the NDEA of 1958. (46) Title II directs the secretary of HEW to conduct a study on the numbers of foreign students remaining in the U.S. after completing their education, including determination of reasons for failure to return to their home countries.

Many U.S. universities had high expectations for the International Education Act in anticipation of obtaining financial support and developed various educational programs, some of which have fallen by the wayside. Some funds did become available through the Office of Education and the National Defense Education Act for geographic area programs and language programs, but support has lagged in recent years. It is not clear why the International Education Act was never funded, although 1966 coincided with the growth of involvement in Vietnam, which may have been a major factor. The act stressed education, including broadening the international perspective of U.S. universities, rather than international development or national defense or technical assistance. There was little if any focus on science and technology.

LEGISLATION PERTAINING TO VISAS, PASSPORTS, IMMIGRATION, AND EXCHANGE OF PERSONS

The International Communication Agency

Precursor Legislation

The U.S. has a long history of educational and cultural exchanges. The U.S. Information and Educational Exchange Act of 1948 (47) and the Mutual Educational and Cultural Exchange Act of 1961 (48) were designed to increase understanding between the peoples of the world and to improve and strengthen the international relations of the U.S. The first of these acts emphasized information services with exchange of books, publications, and translation of such writings. The latter act also stressed information, but went further to encourage educational exchanges and financing of studies, research, instruction, and other educational activities of, or for, Americans in foreign countries and citizens in American schools abroad.

Another piece of legislation dealing with exchanges has been the Mutual Security Act of 1960 which established the Center for Cultural and Technical Interchange in Hawaii. (49) Known as the East-West Center, it is located in Honolulu at the University of Hawaii and has been quite active in numerous exchanges with developing countries in Asia and the South Pacific.

Creation of the ICA

All of the legislation previously cited, legislation affecting informational exchange, and small parts of other acts, were incorporated into the Reorganization Plan No. 2 of 1977 which proposed creation of the International Communication Agency (ICA). (50) A principal objective of legislation proposed by the Carter administration was to reduce the degree to which misperceptions and misunderstandings complicate relations between the U.S. and other nations. This proposal was made into law on April 1, 1978. The U.S. Information Agency, the U.S. Advisory Commission on International Education, the Advisory Committee on the Arts, and the Department of State's Bureau of Educational and Cultural Affairs were combined as were their respective functions. (51) The Fulbright scholar exchange programs are now supported out of the ICA and administered by the Institute for International Education and the Committee for International Exchange of Scholars, both private organizations. While the primary thrust of the ICA is now toward cultural exchange and communication, if its emphasis were shifted more toward technical interchange, this agency might serve as a focal point for greater involvement of U.S. university personnel in S&T for development. However, these exchange programs have not fared well of late and such a reorientation might further drain the limited resources available for educational and cultural exchange that is not of an S&T nature. (52)

Immigration, Visa, and Passport Regulations

It is beyond the scope of this project to delve into all the legislation affecting this category. However, this is an important area for the free movement and exchange of faculty and students between the U.S. and LDCs. Certain immigration and visa regulations affect the length of time a foreigner may remain in the U.S., which potentially affects the program of study, research, teaching, etc. pursued. Also, clauses regulating the obtainment of employment by non-U.S. citizens in the U.S. affect both the types of experience possible for LDC stu-

The Legislative Mandate

dents and faculty, and the arrangements for compensation of LDC visiting consultants, researchers, and the like.

The importance of changes in U.S. immigration laws and regulations has been pointed out by Kidd in a recent study, "Manpower Policies for the Use of Science and Technology for Development." (53) Kidd attributes a sharp decline from 1971 through 1975 in permanent immigration of scientists and engineers to the U.S. from developing countries to the elimination of scientists and engineers in 1972 from the list of shortage occupations which permitted their entry in large numbers prior to that date. Previous to that, in 1965, preferential treatment of immigrants from northern Europe was eliminated resulting in a rapid rise in migration from Asia. The reader is referred to the study by Kidd for further analysis of the migration and the "brain drain."

OTHER ASPECTS OF THE MANDATE

This chapter has focused on principal legislation which bears on U.S. international involvement, with particular emphasis on U.S. universities and on science and technology. Policy and programs sometimes evolve in this field without the passage of major new laws. Often, policy statements are included through the insertion of clauses in annual foreign aid legislation. Congress can add or eliminate programs by adding or removing items from budgets or by increasing or decreasing funds for those items. The executive branch also has some flexibility in initiating specific programs - for example, bilateral agreements for international scientific cooperation - and in creating new agencies or units by reorganization proposals, as was the case with the International Communication Agency. However, Congress has been scrutinizing the executive branch very carefully in recent years and can often influence the outcome through the appropriations process.

We have not been able to consider all of these elements in detail in this study. For example, an important area we have not touched upon is U.S. participation and funding for international organizations. We hope, however, that we have managed to include many of the important trends and developments that relate to the role of U.S. universities in S&T for development.

ANALYSIS

Foreign Assistance Legislation

Our examination of the past legislative mandate and current legislative proposals reveals several threads that are relevant to the role of U.S. universities in helping to build an indigenous science and technology base. First, foreign assistance legislation has been of considerable importance in shaping that role, through specific programs, by setting policy directions, and by providing the principal share of funds for such activity. Currently, that legislation is heavily focused on fulfilling the policy of "New Directions" legislation, meeting basic needs, helping the poorest of the poor, and, in the language of the proposed Humphrey legislation, growth with equity. This focus is considerably different from the earlier emphasis on long-range economic development which underpinned large-scale projects.

The implications of AID's policies for universities are significant. U.S. university involvement in foreign assistance programs needs to be responsive to the intent of those policies. However, as Segal (54) and others have pointed out, the basic-needs policy may be incompatible with much of U.S. science and technology as well as much of what universities have to offer. Kidd has analyzed the situation and found a general moving away from direct support for LDC universities by the U.S. as the new policies were implemented in the 1970s, presumably because the basic-needs strategy required emphasis on other educational sectors and nontraditional approaches. (55)

A related issue concerns "AID-graduate" countries. The AID mandate, with few exceptions, does not currently allow involvement with countries whose gross national products exceed AID limits. The countries not eligible for support include many that are growing in political importance and that have various relationships with U.S. universities. Although other federal agencies have been approached to develop programs with AID-graduate countries, not much has been done as yet. According to Segal, countries that can pay full costs for NSF assistance can be accommodated but many others that can only pay for part of the costs are excluded. (56) Basic needs relates mainly to the poorest countries. OPEC countries can pay for help; it's the middle income countries where the gap lies. (57)

The ability of U.S. science and technology, in general, and U.S. universities, in particular, to relate to the current AID mandate can be debated on several grounds. First, although universities may not have much experience in directly relating science and technology to basic needs, there seems to be growing interest in such activity.

The Legislative Mandate

Second, it can be argued that the long-run contributions to development needs through S&T infrastructure building should be an essential element of a basic-needs strategy, along with the shorter-run, more direct kinds of action and extension programs. Indeed, it is probably true that the LDCs themselves look to the long-term rather than the short-term when they seek assistance and cooperation from the U.S. Thus much hinges on the manner in which AID interprets its basic-needs strategy, as set forth by the Congress. Many participants at our workshop felt that a broader interpretation by AID of the basic-needs strategy was essential if universities were to make the contributions they are capable of making in helping to build an indigenous S&T base in LDCs. This broader mandate would seem to be particularly important for the proposed Foundation for International Technological Cooperation.

Funding authorizations for foreign assistance programs involving U.S. universities have not been analyzed in any detail. However, as F.A. Long has pointed out, there have been fluctuations in both amount and expectation of support for university programs which are likely to continue into the future. (58) AID has been the principal source of funding for programs specifically targeted toward U.S. universities, like 211(d) and through other activities such as its central research and technical assistance programs. Furthermore, many of the other federal agencies that have international programs that can involve universities do so by receiving "pass-through" money from AID and are hence governed in this activity by AID policies. Segal feels that they need their own funding authority. (59)

Within the foreign assistance legislation, Title XII provides an important, potentially well-funded opportunity for a segment of U.S. colleges and universities to participate in international food and agriculture programs. However, parallel opportunities do not seem to exist in engineering and science. We examine the experience of U.S. universities in international agriculture and Title XII further in chapter 3.

International Education

It is clear that while the U.S. has had a commitment to international education, the emphasis has been on low-key, highly visible exchanges to promote good international relations for the U.S., rather than primarily to serve the educational needs of the LDCs. Furthermore, the education and exchange thread seems to be oriented neither toward international development nor toward building an indigenous LDC S&T base. A major increase in international education activity was contemplated in the International Education Act of 1966 which was passed but never funded. Perhaps the act was too ambitious and out of tune with the spirit of the time. Recently, with the formation of the International Communication Agency, it may be that some commitment to international education will be revived. However, the trend in recent years in area programs, language studies, and Fulbright exchanges has been one of loss of momentum and support. (60) Thus although science and technology are currently in the limelight in connection with UNCSTED, it may not be wise to emphasize S&T heavily in programs that have had primarily an educational and cultural focus and are hurting for support for those activities.

International Mandates for Domestic Agencies

There appears to be a trend in which primarily domestic agencies are acquiring new mandates for international activity. For example, the Department of Agriculture, through the 1977 Food and Agricultural Act, is authorized (but not funded as of 1978) to strengthen U.S. universities for work in international development. The Department of Energy, through the Anti-Nuclear Proliferation Act of 1978, is to become involved in various aspects of small-scale alternative energy sources for international development. This trend, if it continues, could have impact on U.S. university involvement, particularly if direct funding authorizations are received rather than "pass-through" funds from AID. The impact of specific agency involvements on building an indigenous S&T base needs further examination. A survey of activities of government agencies that might relate to S&T for development is being carried out in connection with preparations for UNCSTED.

Is There A Mandate for Helping to Build an Indigenous S&T Base?

To what extent have science and technology been and will they continue to be a focus for U.S. policy toward LDCs? An examination of the history of foreign assistance legislative efforts reveals a fluctuating trend in this regard. From the early sixties to the early seventies, emphasis on technical assistance seems to increase, culminating in 1971 with proposed legislation to create a separate International Development Institute to coordinate research and technical assistance programs. This legislation was unsuccessful and interest in science and technology appears to have waned since that time, as reflected

The Legislative Mandate

by the 1977 AID reorganization. However, a recent report by the Brookings Institution proposes a new organization, an International Development Foundation, (61) which appears somewhat similar to the 1971 International Development Institute. Through the initiative of the Office of Science and Technology Policy in the White House, a new Foundation for International Technological Cooperation (FITC) is being planned as a major thrust of United States UNCSTED preparations. "The new Foundation's programs will include assisting developing countries in strengthening their indigenous scientific and technological education and manpower training programs directed toward developing countries' needs, encouraging both public and private institutions in scientific and technological efforts relevant to developing countries, and supporting collaborative research and development between U.S. and developing country institutions." (62)

Our examination of the past legislative mandate reveals no clearcut focus on involving U.S. universities in helping to build an indigenous S&T base in LDCs. There have been a variety of different emphases, for example, basic needs, long-term economic development, and some specific mission orientations (e.g., alternate energy sources and food), but we fail to find a sharp, integrated emphasis on the central focus of UNCSTED, namely science and technology for development. If the U.S. government wishes to provide such an emphasis and mandate as part of its foreign policy, new legislation would seem to be required.

The FITC (now called the Institute for Scientific and Technological Cooperation - ISTC) can provide this focus and mandate. However, it must win approval and funding from Congress; related proposals have failed to do so in the past. Furthermore, there is serious question as to whether the FITC can function effectively if it is incorporated within the existing AID policy and operating framework rather than as an independent agency. We discuss the FITC further in chapters 5 and 6.

2 Engineering

TYPES OF INVOLVEMENT

U.S. engineering schools have a long history of involvement in international activity. A 1976 report of the National Academy of Sciences and National Academy of Engineering summarizes that involvement and makes recommendations for future activity. (1) Table 2.1 summarizes present and future opportunities for U.S. engineering schools to render development assistance, as taken from the NAS/NAE report, while figure 2.2 summarizes its principal recommendations.

Institution Building

Institution building has been a major focus of U.S. engineering school activity since the 1950s. The University of Wisconsin has been active for more than 20 years in helping to build engineering colleges in India, Singapore, and Indonesia. Consortia of U.S. universities linked to a nonprofit private organization, the Educational Development Center (EDC), were involved in programs to assist in building the Indian Institute of Technology at Kanput (India) and the College of Engineering at Kabul (Afghanistan). More recently a consortium of universities has assisted and is continuing to assist in building the University of Petroleum and Minerals, Dhahran, Saudi Arabia, which awarded its first engineering degrees in 1971. The EDC-consortium approach is being employed in a new $129 million, ten-year project to build a newly founded Algerian institute, INELEC, with major focus

Table 2.1. Summary of Present and Future Opportunities for U. S. Engineering Schools to Render Development Assistance

	Program	Conducted in: U.S.	Conducted in: LDCs
1.	Undergraduate Engineering	Basically sound; continue to offer broad, balanced flexible programs. Suitable for students interested in development	Continue to upgrade quality
2.	Engineering Technology	Well-established; suitable for students interested development	Promising applicability; need to establish in LDCs
3.	Graduate Engineering	Continuing influx of LDC students; good, long-term impact	Create more regional graduate centers of high quality
4.	LDC-oriented Graduate Education	New courses, more experimentation to match specific LDC needs; cooperative efforts with LDC universities	U.S. provide assistance in developing advanced programs; develop applied research; provide faculty training at regional centers
5.	Industrial Extension Service	Develop additional models	Develop new university-based centers for transfer of technology
6.	Contract Special Training	Potential for expansion, both for schools of engineering and engineering technology	Potential activity at regional centers
7.	LDC Faculty Upgrading	Workshops and courses at proximate U.S. universities	Short courses at regional centers and LDC universities
8.	Seminars on Special Subjects	Provide professors, experts	Conduct for students/faculty at LDC universities and regional centers
9.	Short Courses for LDC Engineers	Develop courses; provide visiting professors	Conduct at LDC schools and regional centers
10.	Continuing Education	Technical assistance in developing courses for use in the LDCs	Facilitate development and application. Need permanent activity in continuing education
11.	Research (wherever possible cooperative between LDC and U.S. universities)	Development methodology and processes; development-oriented technical research; management techniques; energy; natural resources; etc.	LDC faculty work on products and processes; marketing; technology applications; housing; energy; natural resources

SOURCE: NAS/NAE Report, The Role of U.S. Engineering Schools in Development Assistance.

1. <u>Research</u>. Calls for AID and other technical assistance agencies to find ways, especially by new funding, to increase research on LDC development problems at both U.S. and LDC institutions, particularly at regional graduate/research centers. Also calls for adoption of impartial award procedures by funding agencies, such as peer review used by the National Institutes of Health. Wants strengthening of institutional capability of U.S. institutions to work on LDC problems by strengthening the AID 211(d) program.

2. <u>Curriculum Improvement</u>. Calls for funding to improve courses and programs at selected U.S., LDC, and regional graduate centers for graduate students interested in economic development. Also calls for innovative work-study programs at both LDC and U.S. institutions. Wants U.S. colleges of engineering technology funded to test the potential for this type of technical education in LDCs.

3. <u>Technology Transfer</u>. Emphasizes examining concept of establishing engineering/industrial extension services for LDC rural areas; expansion of contract services by U.S. schools to offer special education in development-related subjects to LDC government and industrial managers. Calls for support of short-term consulting and advising assignments for U.S. engineering faculty abroad, and for the development of short courses, workshops, and seminars given in LDCs by visiting U.S. faculty. Calls for U.S. schools to assist LDC universities to develop continuing education courses for engineers.

4. <u>Evaluation</u>. Recommends evaluation against original objectives of past programs involving "sister" schools and consortia of U.S. schools that assist LDC institutions. Calls for clearly stated objectives and an evaluation phase for future programs. Wants criteria and means developed for evaluation of U.S. faculty performance.

Fig. 2.1. Recommendations of NAS/NAE Report on role of U.S. engineering schools in development assistance.

Engineering

on engineering and technology programs in electricity and electronics. A significant shift over the years has been from involvement in projects funded by AID to projects funded by OPEC country governments.

We know of no good, single source of information on the overall extent of and funding for past involvements, which include, among others, Colorado State University aiding the Asian Institute of Technology, MIT assisting the Birla Institute of Technology in India, and the University of Kentucky working with an engineering college in Indonesia. U. S. engineering faculty, through UNESCO, UNDP, World Bank, and NSF-SEED program support, as well as with support from their own universities, have gone on both short- and long-term assignments to help build LDC institutions. Later in this chapter, we describe several institution-building cases.

Cooperative R & D

We found very few examples of cooperative research and development programs between U. S. and LDC institutions in the literature. They may exist but rarely get reported. Our impression is that they tend to be somewhat one-sided with the U. S. serving in more of an institution-building capacity. More information might dispel that impression.

In the 1960s, MIT had a program of cooperative research on civil engineering programs generic to both Latin America and the United States. Projects were worked on in the areas of structures, soils, systems, materials, and water resources in Colombia, Mexico, Venezuela, Argentina, Brazil, Chile, and Peru. (2) The Rural Industrialization Technical Assistance (RITA) Program and the Georgia Tech Small Industries Program have elements of cooperative R&D and are described later in this chapter. There are also cases of smaller collaborative research projects arranged by two faculty members such as the summer program on appropriate chemical technologies at the National Autonomous University of Mexico (3) and a program on materials science between Argentina and the U. S. supported by NSF's Cooperative Science Program.

U. S. Resource Base Development

A series of five-year grants made by AID's Office of Science and Technology (OST) in the late 1960s under the 211(d) program enabled engineering schools at three universities to become more involved in international activity. Georgia Tech's Small Industries

Program, centered in their engineering experiment station, and their M.S. degree program with emphasis on industrialization benefited from this grant program, as did MIT's Technology Adaptation Program and Cornell's program in Science and Technology Policies for Developing Nations. A fourth grant was made to the University of Arizona; we are not clear about the extent of engineering school involvement. Other AID 211(d) grants made outside of OST that have involved engineering schools include Utah State University and Colorado State University in Water Resources.

Education and Training

During the past 20 years, between 20 percent and 25 percent of all foreign students enrolled in degree programs in the U.S. have chosen to study engineering. (4) Table 2.2 shows the region of origin of foreign engineering students in 1973-1974. There has been a major increase in the percentage of foreign students coming from OPEC countries. In 1977, Iran headed the list of countries with 11.5 percent of all foreign students; Nigeria was in third place with 5.8 percent, Venezuela was in eleventh place with 2.8 percent, and Saudi Arabia was in twelfth place with 2.3 percent. Engineering and technology programs are heavily selected by this group of students who are supported by their home-country governments.

Foreign student enrollments in U.S. engineering programs rose from 31,187 in 1973-1974 (5) to 42,000 in 1975-1976 to 48,990 in 1976-1977.(6) Although we have no data on sources of funds for these students, we do have data on sources of funds for all foreign students in 1976-1977 (see table 2.3). The largest category is personal funds from foreign sources (44.4 percent), with home government (13.5 percent), and private sponsor/foreign (3.9 percent), bringing the total foreign sources to 61.8 percent. We are not sure that this latter figure is for engineering students only. Graduate students in engineering are often supported by research and teaching assistantships from the U.S. institution, which would tend to lower this figure. On the other hand, heavy emphasis on technical skills by OPEC country governments will tend to raise it.

Foreign engineering enrollments are particularly significant at the graduate level. There were reportedly 18,230 foreign graduate students in engineering in the U.S. in 1975-1976. (7) According to Engineers Joint Council statistics, total graduate engineering enrollments for that period were 36,479 full-time and 26,842 part-time students. (8) Thus, the percentage of engineering graduate students fell somewhere between $18,230/(26,842 + 36,479) = 29$ percent and

Table 2.2. Foreign Students in Engineering in the U.S.: Home Region and Academic Level, 1973-1974

Region	Undergraduate		Graduate		Total	
	Number of Students	Percent of Total	Number of Students	Percent of Total	Number of Students	Percent of Total
Africa	978	6.1	727	5.2	1,767	5.7
Europe	774	4.8	1,257	8.9	2,108	6.8
Far East	4,318	27.0	8,242	58.7	13,022	41.7
Latin America	3,064	19.2	1,230	8.7	4,430	14.2
Near and Middle East	6,132	38.5	2,029	14.4	8,530	27.3
North America	379	2.4	222	1.6	616	2.0
Oceania	63	0.4	54	0.4	124	0.4
Stateless	43	0.3	28	0.2	73	0.2
Country Unknown	214	1.3	271	1.9	517	1.7
Total	15,965	100.0	14,060	100.0	31,187	100.0

Note: Totals do not equal sum of Undergraduate and Graduate because of "other" student category.

SOURCE: Institute of International Education, Open Doors (New York: IIE, 1974), pp. 20-27.

18,230/(36,479) = 50 percent, depending upon whether part-time students are included in the totals. For corroboration, the 1976 Digest of Education Statistics reports that in 1974-1975, 41.3 percent of doctoral degrees in engineering were awarded to individuals of foreign citizenship. (9) From table 2.3, it can be seen that about 80-90 percent of foreign students studying engineering in the U.S. are from developing countries.

Data on the percentage of foreign engineering students that return to their own countries do not appear to be available. Data compiled by the National Science Foundation indicate immigration of engineers to the U.S. increased from about 5,000 in 1966 to 9,000 in 1970 and then dropped to 4,000 in 1974. (10) The fall-off in the early 1970s was due to changes in U.S. immigration laws which eliminated scientists and engineers from the category of shortage occupations. More than half of immigrant scientists and engineers over the last ten years have come from Asia. We guess that they are from countries with "surpluses" of engineers, i.e., India and Taiwan.

The years 1976 and 1977 saw sharp increases in the number of foreign students studying in the U.S. Whether this will lead to an increase in immigration of engineers remains to be seen. One final piece of data is that the high percentages of foreign graduate enrollments are in marked contrast to the overall enrollment of foreign students in all programs in the U.S. According to 1974 and and 1975 UNESCO statistics, the U.S. was in twenty-first place with 1.6 percent foreign students compared with Canada (first at 16.9 percent), Switzerland (second at 15.6 percent), and France (third at 12.4 percent). (11)

Most foreign engineering students come to the U.S. to study traditional engineering specialties such as civil, electrical, mechanical, and chemical engineering. For many years, there has been an ongoing debate within the engineering education community concerning the relevance of the type of education received here to LDC needs. (12) It seems important to consider the subject again, in view of the large percentage of foreign engineering graduate students in the U.S.

Several programs have been developed that attempt to relate to specific international education and development needs. Programs in this category include the engineering student exchange between the University of Wisconsin and Monterrey Tec in Mexico. (13) We also include degree curricula of specific relevance to LDCs like the M.S. degree emphasizing industrialization at Georgia Tech, the Technology Adaptation Program at MIT, and the technology and international development area within the M.S. and D.Sc. programs in the Department of Technology and Human Affairs at Washington

Table 2.3. Primary Sources of Funds for Foreign Students Studying in the U.S., 1976-1977

Primary Source of Funds	Percentage
Personal/U.S.	7.6
Personal/foreign	44.4
College or university	12.7
Home government	13.5
U.S. government	5.0
Private sponsor/U.S.	2.6
Private sponsor/foreign	3.9
On-campus employment/part-time	4.2
Off-campus employment	3.9
Employment of spouse	0.4
Other	1.9

SOURCE: Institute of International Education, Open Doors (New York: IIE, 1978), p. 27, with permission.

University. Many foreign nationals have come to the U.S. for AID-sponsored, nondegree participant training programs, but we do not know how many have been in engineering.

Engineering faculty have worked overseas as individuals in a variety of programs. Organizations such as the World Bank and United Nations Development Programs hire them as expert consultants to advise on institution building. The NSF-SEED program has supported individual U.S. faculty members for teaching and research activity in LDCs. There appears to be a demand at the moment for faculty to serve in certain OPEC countries where new institutions are being built.

MECHANISMS

A primary mechanism for involvement of U.S. engineering schools in institution building abroad has been AID contracts to individual U.S. universities and consortia of universities. Increasingly, AID support is being replaced by contracts with OPEC governments. In the Kabul and IIT-Kanpur cases, a private organization, the Education Development Center (EDC) was a useful means of fostering cooperation among U.S. universities. EDC is now managing a U.S. consortium of engineering colleges, engineering technology programs, and private industry which is helping to build INELEC in Algeria. The consortium appears to be a necessary mechanism for funding enough U.S. faculty with the right combination of skills to support large institution-building projects.

The 211(d) program grants in the late 1960s supported U.S. resource base development involving a small number of engineering schools. Although this program is not very active, there is interest on the part of some U.S. engineering educators in seeing it revived to provide the kind of support they feel is needed to permit expanded international involvement.

Another mechanism involves linkages between U.S. and LDC institutions in which cooperative programs of various kinds (including R&D) are undertaken. Linkages are the subject of reports by Harrington (14) and a detailed analysis by Glyde. (15) Specific details of U.S.-LDC engineering links are not readily available.

Mechanisms supporting individual U.S. faculty initiatives include the program for Scientists and Engineers for Economic Development (SEED), administered by the National Science Foundation and the older, Fulbright program which involves only a small number of U.S. engineering faculty. Regional and international consultancies play a role as does support from U.S. engineering institutions.

Mechanisms facilitating the entry of foreign students to engineering programs in the U.S. include financial support from home governments (e.g., the Gran Mariscal de Ayacucho Foundation Fellowship Program, Venezuela), and research and teaching assistantships from graduate engineering departments. The Institute for International Education and regional fellowship programs such as LASPAU and AFGRAD seek to match foreign students with U.S. institutions.

ENGINEERING CASES

The RITA Program

The Rural Industrialization Technical Assistance (RITA) program, implemented between 1962 and 1968, aimed for small private-industry development in commercial centers and rural intermediate towns in northeastern Brazil. The program was funded by USAID and the government of Brazil, and initially involved cooperative work between students and faculty at the University of Ceara in northeast Brazil and the Latin American Center at UCLA.

RITA relied on local equity capital to fund target industries, substantially involved the community in the development process, relied on local markets and factors of production, and employed university group members in catalyzing community mobilization of human and material resources and local capital for the creation of new local industries.

General procedures for project planning and implementation operated as follows:

1. Identification of available small and medium industry niches.

2. Organization of local companies for exploitation of industrial opportunities.

3. Across-the-board assistance in implementation of sustained new company operation (including basic research, product design, local construction, technology adoption, equipment testing, etc.).

4. Technical assistance to counterpart university faculty and students.

5. Involvement of the local population in industry formation, ownership, and management.

Asimow and McNown describe a pilot RITA project which resulted in the establishment of a corn-processing industry. At completion, the processing plant was expected to store and process 1,000 metric tons of corn per annum, realize about $1 million annually, and employ 60 to 70 persons. About 300 local small and medium sized farmers owned stock in the company. According to the authors, this RITA project apparently fulfilled program objectives, including economic advancement of a local area, expansion of agricultural production, training of local personnel for management, the use of U.S. dollars to initiate, rather than subsidize, LDC industry, and improved education of U.S. and LDC students. (16)

The history of the RITA project is interesting and needs to be written. It was initially largely propelled by one U.S. engineering professor, Morris Asimow from UCLA, whose efforts resulted in a substantial development effort linking U.S. and Brazilian universities and businesses in rural areas. It is our impression that Asimow's efforts were not fully supported at UCLA, particularly after a change in engineering school administration.

In 1976, the World Bank sponsored an evaluation of the RITA program by Neil Boyle, formerly associated with the RITA project. (17) This bank document draws a number of conclusions concerning the economic impact of the project for the period 1962-1968. Of interest to our study are the conclusions concerning university participation. Boyle concludes that though none of the Brazilian universities are continuing their role as catalysts in rural industry formation, universities can nevertheless play an effective role in "grassroots rural development" under certain circumstances.

> Several of the NE [Brazilian] universities now have projects which combine education and development in the interior of their States, something that did not exist in the early sixties. Brazilian universities are suitable because they carry a great deal of prestige... and because they, more likely than any other Northeast institution, can design and develop appropriate technology needed for a rural/urban industry program and at the same time meet their own objective of education. For example, the Federal University of Paraiba is presently gearing up to introduce the concept of the "engineering clinic" where engineering students, professors and consultants are engaged in solving real problems for real clients. The Brazilian university is being assisted in this regard by an American professor from Claremont University through an NSF-SEED grant from the National Science Foundation. The American professor is the former coordinator for the RITA project in the State of Paraiba.

The engineering and economic curricula of Northeast universities is improved because of RITA. Some of this improvement was directly provided by the American university. For example, this is the case with the Federal University of Paraiba. The curricula for the master's degree in mechanical engineering was developed by professors from California State University at Los Angeles. Curricula is also reported to be improved indirectly by the faculty members trained by the RITA program. (18)

What emerges from the above is a case in which U.S. university involvement appears to have contributed to strengthening the indigenous S&T base in Brazil enabling that country to work on rural industry and other problems. Unanimity does not exist, however, concerning the results of RITA. We received one letter that stated,

> I have asked professional engineers in Brazil about the RITA project and have had very few complimentary responses to that project. Their attitude has been "who cares, if the U.S. wants to spend their money that way, it's their business." It is not surprising that the evaluation of such entrepreneurial type programs is being done by academics and not those who know about such things as small businessmen, bankers, etc. The evaluations turn out great. The businesses themselves have in almost all cases gone bankrupt or collapsed. (19)

Boyle's evaluation for the World Bank does not agree with the latter pessimistic appraisal. He visited 4 of 6 field sites and collected data on 9 of 19 known RITA firms. He indicates that knowledgeable Brazilian officials feel there are a good many more than 19. His study concludes that the RITA project has benefited the development of the northeast.

The Boyle-World Bank evaluation indicates that some strengthening of LDC educational institutions took place through U.S. university involvement. One could conceive of similar U.S. university involvement in other countries that are less far along than Brazil, or of a continued linkage with the Brazilian university, focused on development problems. The engineers who commented negatively on RITA might or might not have been genuinely interested in rural development. On the other hand, the World Bank evaluation was performed by someone who had been involved in the RITA program. In any event, it may be that the objective of U.S. involvement should be to strengthen the development orientation and S&T base of a local counterpart organization, i.e., the university, rather than to interact directly with local developers. An indigenous capability means having local institutions cope with local problems.

Georgia Tech Small Industries Program

Georgia Institute of Technology was granted $800,000 from AID under section 211(d) of the Foreign Assistance Act for a five-year period (1973-1978) to work in the area of employment generation through stimulation of small-scale industry. (20) The 211(d) grant provided core support for strengthening the ability of Georgia Tech's Economic Development Laboratory to work on international development problems. Grant objectives were: consolidation of methodologies for small-scale industrialization and job creation; development of methodologies such that generally applicable and transferable principles and procedures may be derived and applied; development of new technological approaches to industrialization and employment generation. (21)

Certain program elements were emphasized in pursuit of these objectives. They were:

1. Establishment and maintenance of linkages with four counterpart institutions to form a real-world laboratory for cooperative investigation and testing of approaches.

2. Data base compilation, analysis, and codification.

3. Analysis of methods and technologies for solving small-scale industry problems and encouragement of small-scale industry expansion and diversification to enable isolation of factors leading to success or failure under various circumstances.

4. Evolution and field-testing of new approaches and alternative methodologies.

5. A graduate degree program.

Work based on these objectives has resulted in a range of activities, including:

1. Formation of 11 formal cooperative agreements with educational institutions, government organizations, and research institutes in the developing world.

2. Establishment of the International Development Data Center at Georgia Tech in 1973.

3. Publication of industry studies, economic analyses, international market potential analyses, comparative analysis of

counterpart country financing programs, small industry case studies, and other publications on research findings.

4. Development of training programs in Korea, Nigeria, and the Philippines, with subsequent field-testing.

5. Design, establishment, and development of a new M.S. curriculum in the School of Industrial and Systems Engineering at Georgia Tech with concentration on industrialization.

Georgia Tech 211(d) activities were reviewed for AID in January 1977. (22) The review team regarded implementation of small-scale industry development programs as a critical need in LDC development. They stated that

> ...the Georgia Tech capability in small scale industry development represents one valuable resource for AID use in achieving this goal. . . . We are appreciative of the excellent, well organized presentation of the Georgia Tech staff. The review was a rewarding experience...because of the frank and open discussion of all the issues.... The positive tone of the meeting facilitated an in-depth exploration of the numerous potentials for utilizing the capability which has evolved at Georgia Tech under this program... This report is a result of...careful review of progress to date and the appropriate utilization of this capability in the future. . . . (23)

The Tech final report lists funded activities that 211(d) has directly or indirectly played a part in generating. These activities are listed in figure 2.2. The dollar impact on Georgia Tech of the 211(d) grant is shown in table 2.4.

The review team's delineation of Tech's accomplishments very nearly coincides with those listed in the Tech project reports, which indicates their general agreement that the grant objectives had been met. They go on to raise several issues concerned with how effective use might be made by AID of the 211(d) capability that had been developed. The issues include: What should AID and Tech do to insure recognition and utilization of capability by various AID elements? Who are other potential users and how should they be approached? Should Tech, with one grant-year left, alter their pattern of choosing well-motivated counterparts in LDCs with an existing industrial base, to work with counterparts in the poorest LDCs with small industrial bases where the probability of success may be much less? What do counterparts expect from the Tech relationship and what external support do they need to maintain their developed capabilities? What mechanisms would support continued foreign student involvement in Tech's program, given the students' need for financial support? How can 211(d) grant restrictions in technical assistance to counterparts be minimized?

If we assume that grant objectives were being met, was Tech contributing to building an indigenous S&T base in LDCs? Based upon the information available to us, it appears as though jobs were created, new technologies devised, and existing technologies adapted to a variety of different local conditions. Further, Tech made substantial efforts to <u>transfer</u> the knowledge and modes of application to developing world institutions in a variety of sectors (public, private, and academic) that could be expected to diffuse it further; training was a significant part of this transfer. It is not unimportant that knowledge/application modes were actively transferred rather than passively diffused (i.e., limited to technical journals).

The Georgia Tech 211(d) grant would appear to be a reasonably successful one. Factors for this outcome probably include: (1) strong support from the top by the president of Tech, an active participant in international development activity; (2) the linking of the international activity to ongoing, related domestic small industry activity; (3) the focus of the program within the Engineering Experiment Station rather than in a traditional university department, which meant that program personnel had less traditional academic responsibilities and were free to travel and to pursue this activity as a full-time pursuit. It is hypothesized that these factors may not all have been prevalent in other 211(d) grants which were less successful.

It should be kept in mind, however, that our analysis of Georgia Tech's program is based solely upon their own reports and the report of one internal AID review team. We have not solicited any feedback from any of Tech's counterpart institutions, nor have we come across any independent evaluations. We are also interested in the extent to which the Georgia Tech program, with its focus on local entrepreneurs, was able to respond to the basic needs thrust of the "New Directions" policy. The case is a fruitful one to explore further.

The Kabul Afghan-American Program (KAAP)

Background and Objectives

An AID-funded consortium of 12 U.S. institutions, (24) administered by the Education Development Center (EDC), assisted in development of the College of Engineering at Kabul University (KU) in Kabul, Afghanistan, between 1963 and 1973. (25) The American faculty members at Kabul were known as the U.S. Engineering Team (USET).

Engineering 39

1) **Small Industry Grant Contract.** A Tech contract with AID to administer grants to five LDC organizations to assist in problem solving for about 300 small industries in Korea, Brazil, the Philippines, Nigeria, and Ghana. Tech and the East-West Center also provided small industry-related training consultation to the organizations to supplement their small industry extension services. The author states that these activities have "resulted in strengthening the counterpart organizations and have been instrumental in saving or creating employment in the assisted countries."

2) **OAS Ecuador Survey.** A small industry specialist worked in OAS's survey of Ecuador as part of that country's development planning.

3) **AID Small Industry Loan to the Government of Colombia.** Tech staff assisted AID/Bogota in development of small industry loan package and in review of the package by Colombian officials. The loan was approved. Seven persons from Colombia visited Atlanta to work on this with the Office of International Programs.

4) **Basic Ordering Agreement.** This technology transfer agreement concerned provision of problem-solving assistance for IRRI-designed machinery in Ecuador.

5) **Training for CONACYT (Mexico) Staff.** Five staff members trained five weeks in information systems, in Atlanta; designed to increase staff capability to respond to inquiries by industry, and other requests for technical information.

6) **International Development Conference/Seminar.** About 120 persons attended the conference on "Techniques and Methodologies for Stimulating Small-Scale Labor-Intensive Industries in Developing Countries," and 20 persons, mainly from LDCs (invitation only), attended the seminar, both in Atlanta.

7) **Science and Technology Symposium.** An annual Bureau of Technical Assistance, Office of Science and Technology (AID/TA/OST) Symposium. About 50 persons, attending by invitation, discussed small industry development and related topics.

8) **Small Industry Development Network.** Quarterly newsletter. First supported for two years by AID. Continued under a three-year extension. About 3,000 subscribers in 131 countries.

9) **Rice Machinery Industrial Extension Program.** Tech provided on-site marketing and other forms of assistance to IRRI in that organization's efforts to stimulate manufacture of IRRI-designed rice machinery.

10) **Pyrolytic Conversion of Wood Wastes - Ghana.** Tech found that raw material supply was sufficient for converting sawdust to charcoal, oil, and gas for use as alternative energy sources, and did preliminary design and economic analysis of a six-ton/day pyrolytic converter.

11) **Pyrolytic Conversion of Agricultural Wastes - Indonesia.** Investigation of potential for small pyrolytic converters utilizing rice hulls.

12) **Solar Energy - Korea.** Design of a solar system for heating a typical Korean residence. Funded by the Korean Institute of Science and Technology.

13) **AID Manually Operated Water Pump Field Test.** Manufacture of the pump in Nicaragua and Costa Rica and purchase of competing in-country pumps; pumps were installed in village water wells. Testing, evaluation and comparison were underway.

14) **Chile - Small-Scale Rural Enterprises and Intermediate Technology.** Team visited Chile to develop background information on two proposed AID loans to the government of Chile.

15) **Training of Industrial Extension Personnel.** Tech conducted a training course, focused on industrial extension practices and techniques, for industrial extension personnel from the Institute Centroamericano de Investigacion y Technologia Industrial (ICAITI).

16) **Rural Development Potentials for the Siliana Districts of Tunisia.** Evaluation and recommendations of potential for and approaches to development.

17) **Desalination Feasibility Study.** Tech team visited the Cape Verde Islands to review the feasibility of a proposed desalination/power plant.

18) **Technology Transfer Training: Indonesia.** Design and presentation of a training program for 18 Indonesian librarians for a period of three months.

Fig. 2.2. Other international activities related to Georgia Tech's 211(d) grant.

SOURCE: Georgia Tech 211(d) Fifth Annual (Final) Report.

Table 2.4. 211(d) Grant Impact on Georgia Tech International Development Programs

Fiscal Year	211(d) Grant Expenditures	Other Intl. Program Expenditures	Direct Tech Support for Intl. Programs(2)	Total Intl. Program Expenditures	211(d) % of Total Intl. Exp.	No. of Intl. Projects in Fiscal Year
1973	$ 26,900	$ 78,100	$ 13,000	$ 118,000	22.8	3
1974	170,900	5,600	16,000	192,500	88.8	4
1975	180,600	115,400	29,000	325,000	55.6	8
1976	172,700	187,300	39,350	399,350	43.2	13
1977	167,000	263,000	29,000	459,000	36.4	18
1978(1)	81,900	450,000	81,000	612,900	13.0	22
Totals	$800,000	$1,099,400	$207,350	$2,106,750		

(1)Estimated.

(2)Other IPO (International Programs Office) and Tech programs provide substantial indirect support.

SOURCE: Georgia Tech 211(d) Fifth Annual (Final) Report, with permission.

Engineering 41

The initial USET objective was to establish a college that would form a base for its own expansion with external assistance; USET's longer-term goal was to strengthen the college so that it would become a self-sustaining institution appropriate to national needs. To these ends, the KAAP assigned 42 consortium faculty and staff to the college, provided special consultants and services, and trained 55 prospective Afghan instructors and staff at consortium schools.

The nature of the program was overtly influenced by competition with the Soviet-assisted Polytechnic in Kabul (which became a part of KU during the KAAP):

> ...we had been told that one of the reasons for increased American assistance to the College was to forestall the establishment of a Russian engineering school. Two engineering institutions seemed more than Afghanistan would need.... Therefore, in order to survive in competition with the larger Polytechnic for students, staff and University support, the College would have to produce a different type of engineer. Our new planning strategy called for limitation in size of the student body, prevention of duplication of courses, and emphasis on quality to produce a more sophisticated, versatile and management-oriented graduate. (26)

The consortium effected a variety of innovations in the first five years which supports this approach and which perhaps reflects the nature of the mostly private U.S. institutions that made up the consortium. These innovations also illustrate the strong differences in culture and tradition that had to be contended with in the Afghan case. The innovations include:

1. Outreach recruiting of secondary school students.

2. Preengineering work in English and mathematics for those students needing it.

3. Extrauniversity practical training.

4. University activities conducted off-campus, such as research.

5. Regular intraterm exams.

6. Passing courses rather than entire semesters; elimination of second-chance exams.

7. A woman teaching male students.

8. A modern stores system.

9. Establishment of a research institute.

10. Establishment of an external (nonuniversity) advisory committee.

Participant Training

USET's key interest was in strengthening the Afghan faculty. Therefore, the program included substantial participant training in U.S. institutions. During the first three years of the program, selected trainees, almost without exception, had never been overseas. They were handicapped by inadequacies in English and technical competence. Also, their advisors were typically unfamiliar with Afghan conditions. Programs were, thus, not always well tied to prior academic performance, and participants were sometimes started at degree program levels for which they were not entirely prepared. Nine of the 23 participants who began training between 1963 and 1965 did not complete planned degrees at that time.

The middle period of the program, 1966-1967, was characterized by instability. There were fears about participant success rate, and fears about competing with the Soviet-assisted Polytechnic at KU. Even without competition, KAAP had trouble finding qualified participants among prior college graduates. In late 1966, the decision was made to choose participants from among first-year college undergraduates. These students were largely unclear about their academic interests; this caused a change from matching undergraduate work to planned graduate work to increased emphasis on fundamentals in undergraduate work.

Return of former USET staff members to their home institutions, and further EDC coordination of college/consortium school communication helped support the programs. Also, the quality of the college had begun attracting better students, some of whom had spent their senior year in high school in the U.S., and older Afghans with teaching or engineering experience, some of whom had previously studied overseas.

Twelve of 13 participants who started training in 1966 and 1967 achieved degree objectives. There were problems in that many trainees excelled and returned to KU with high expectations, causing some conflicts. Yet, of a total of 55 trainees, only 2 did not return to Afghanistan. By 1968, the viability of generating enough staff through the participant program was recognized. Several faculty members obtained graduate assistantships overseas, and others looked for support for doctorates.

Engineering

While arrangements were made for six participants to pursue doctorates, about half were unsuccessful. As a result, the Visiting Instructorship alternative was devised and implemented. It provided a year abroad for M.S. degree holders and other senior faculty to take selected courses, teach, do research, and participate in departmental operations without being in a degree program. (Barry points out that a similar shift to nondegree study took place in connection with University of Wisconsin programs in the 1950s and 1960s.) (27)

Table 2.5, taken from the final report, (28) summarizes the participant training program.

Analysis

The final reports written by departing USET staff pointed out that institution building would not be completed by the planned contract expiration date of 1973. A 1970 steering committee assessment came to the same conclusion.

Essentially, many USET staff felt that the faculty was less than optimally developed. Yet when the college was assigned 180 new students in 1973, they included a "large number" of the top 1 percent of lycee graduates. It is pointed out that this reflects favorably on the college's development over the prior ten years. In addition, about 10 percent of college graduates had competed successfully overseas with graduates of other institutions, those Afghan graduates having received 90 percent of their training from Afghan faculty.

Most of the members of entering classes had had minimal or no lab experience or experience with technology materials or operations. Graduates had little on-the-job guidance. Good programs in practical lab and shop work, design courses, and a supervised practical training semester were lacking at the college (one author regarded this as the college's most serious deficiency), and while ample opportunity for practical training existed, it was not effectively utilized.

The final report of this ten-year, institution-building project indicates that KAAP contributed significantly to development of an indigenous scientific and technological base in Afghanistan by expanding and strengthening indigenous numbers and quality of Afghan engineering faculty and, in turn, Afghan engineering graduates. However, it was the consensus of the USET that further development assistance for the college was needed. The overthrow of the royal government of Afghanistan in July 1973 precluded immediate future USET assistance.

There are a number of aspects of this case that are interesting, including the strong influence of cold war politics, the big gulf that

Table 2.5. Kabul Afghan-American Program - Summary of Participant Programs

Total Number of Candidates	Candidates in Various Degree Programs	Average Duration of Training	% Successful
23 Candidates starting prior to 1 January 1966	6* for BS degrees 15 for last year BS degrees and/or regular MS program 2 for Ph.D degrees	18 months 34 months 35 months	50 66 0

(*) Many of whom had completed several years training under the old College of Engineering curriculum.

13 Candidates starting after 1 January 1966 and prior to 1 January 1968	7** for BS degrees 4 for MS degrees 2 for Ph.D degrees	43 months 30 months 41 months	84 100 100

(**) One received a BS and MS degree simultaneously; the majority had completed only the freshman year at the College of Engineering.

19 Candidates starting after 1 January 1968 and completing training prior to 30 June 1973	1 for BS degree 12 for MS degree 3 for Ph.D degrees 3 for nondegree training	31 months 21 months 27 months 9 months	100 90 33 100

10 Candidates in training at the end of the project	5 for MS degrees 4*** for Ph.D degrees 1 Visiting Instructorship		

(***) One of these was changed into an on-the-job engineering assignment.

SOURCE: Education Development Center, Final Report, KAAP Program.

Engineering

needed to be bridged between the consortium's conception of an engineering school and the environment in which it was to be placed, and the dedicated efforts of many U.S. engineering faculty who went to Afghanistan to help build an engineering institution.

There are also some questions that arise. Was a U.S.-style engineering college what was really needed? How did the Afghans perceive the college? What were their views? What has happened since 1973? How do the Russian and U.S.-supported efforts compare today? We do not have any data or independent evaluations that would enable us to make valid judgments on whether the Afghan faculty is now doing quality work without benefit of external assistance, nor are we able to judge whether the KAAP realized both its short- and long-term objectives.

The Kanpur Indo-American Program (KIAP)

Background and Objectives

Development of the Indian Institute of Technology at Kanpur (29) was assisted by a consortium of the Education Development Center (EDC) and nine U.S. institutions. (30) The program was supported by an AID contract with EDC, which in turn set up agreements with the other consortium members. EDC administered the program; a steering committee, to which each university head appointed a representative, formed program policy.

The authors of the report regard two aspects of the program, which distinguish it from other institution-building efforts, as particularly noteworthy: 1) U.S. field staff operated and were administered separately from Indian staff (the former reported to the KIAP leader); and 2) Indians were, throughout the program, responsible for Indian Institute of Technology, Kanpur (IIT/K) administration. IIT/K staff were thus more autonomous than they would have been had KIAP staff been integrated into the IIT/K structure. This structuring mode precluded both the use of program staff as visiting professors and of any possible "we-make-it-you-take-it . . . intellectual imperialism that inhibits collaboration, or even worse, fosters dependence." (31)

At the beginning, AID and the government of India clarified assistance program characteristics and agreed that AID should select a U.S. institution with which to contract. AID approached MIT, which set up a committee to advise MIT's president. The committee found that Indian undergraduate engineering education was similar to that of the U.S., and considered that the Indian government might want a graduate-level-only school. The committee visited India in January 1961 with support from the Ford Foundation; their experience there led them to the conclusion that a science-oriented engineering institution, including an undergraduate program which would define graduate admission standards and influence Indian undergraduate education, was needed.

The committee found that their thinking was in accord with IIT/K's director, Dr. P. K. Kelkar, who felt that technical education should be tied to national needs, and that the traditional educational system in India was too rigid to do this. He regarded U.S. technical education as concordant with his objectives. Both Dr. Kelkar and the committee assigned high value to an assistance pattern that would include a genuinely collaborative high quality approach.

Later, the consortium approach was considered and adopted. The universities began looking for top faculty and endeavoring to provide for project work at IIT/K to be regarded as regular duty as far as consideration for promotion, salary increases, and other concerns. (Merton R. Barry feels that these changes in the reward system were a key to the success of the Kanpur project. They had been negotiated originally by the University of Wisconsin in earlier AID contracts in India. (32) Finally, it was agreed that the consortium and IIT/K discuss assistance program content. To this end, AID and EDC signed a preliminary contract (the first of this kind for AID) which funded the consortium to hold discussions in the U.S., and for the steering committee and the EDC staff to set up a work plan with Indian officials and AID/Delhi.

Recruitment and curriculum development were early concerns. The director concluded that much of the faculty would have to be obtained from Indians abroad, because there were few resident engineering faculty who would integrate well, on the basis of training and attitude, with IIT/K's goals. The director was able to offer competitive salaries because of his plan to keep numbers of lecturers to a minimum; this plan grew out of his wish to emphasize independent work over classroom contact.

During the curriculum planning/faculty recruitment phase of operations, the director appointed a Committee on Committees, selected from IIT/K and KIAP faculties, to propose the structure and organization of the IIT/K senate committees and procedures for the senate itself. The committee succeeded in structuring the senate such that academic matters would be handled through faculty interaction, displacing the conduct of academic affairs through administrative procedure. The IIT/K senate mode of organization is (or, perhaps, was) unique in India, rendering an unusual degree of autonomy to individuals and senate committees as compared to power held

Engineering

by the hierarchic structure. Innovation in both undergraduate and graduate programs are seen by the author as following from this unique senate structure.

Curriculum and Program Development Results

Early plans called for use of quality texts, reduced class hours/week, more frequent examinations with decreased emphasis on the final exam, and experimental lab work with emphasis on creativity. In addition to the "strong interdisciplinary orientation" of the traditional engineering departments, there were to be departments of chemistry, physics, and mathematics, each with a graduate program and a "substantial component in the humanities and social sciences." (33)

While American technological education in 1960 was based on utilization of science through engineering methods, and technological leadership was based on social sciences and the humanities as well as engineering, Indian technological education was geared to emphasis on "...narrow competence in traditional state-of-the-art engineering." (34) KIAP staff endeavored to promote the U.S.-type approach.

Table 2.6 shows data for placement of IIT/Kanpur's first graduates through the then latest available data (in the 1970-1971 annual report).

Table 2.6

Placement of IIT/K Graduates

	Higher Education		Employment		
	India	Abroad	India	Abroad	Not Known
B. Tech* (840 students)	15%	26%	48%	5%	6%
M. Tech., M.Sc., and Ph.D. (576 students)	15%	17%	60%	2%	6%

* The B. Tech and M. Tech Degrees are equivalent to bachelor's and master's degrees in engineering in the U.S.

SOURCE: KIAP Final Report, p. 114.

The report contends that this does not support arguments for a high level of "brain drain," though it was then too early for conclusive findings, since 42 percent of undergraduates and 32 percent of graduate students were still in school, including 20 percent of those overseas. However, of 111 Ph.D's graduated, all were employed in India.

It had been traditional for an Indian university to require candidates for faculty positions to be interviewed by screening committees at the university. IIT/K was the first institution in India to drop this recruiting element, and its example has been followed by other Indian schools. Subsidy of new faculty from overseas for personal and family travel and transport of household goods to IIT/Kanpur has also helped to alleviate "brain drain" and to obtain top recruits.

Analysis

Professor Normal Dahl of MIT, the first KIAP leader, notes that his viewpoint is based not only on his experience with KIAP, but also on his experience of three years with the Ford Foundation in Delhi following KIAP work. (35) He states that no other combination graduate/undergraduate institution in India is as open and intellectually stimulating as IIT/K. The undergraduate program was modeled after the characteristic teaching methods and curricula of some of the best U.S. institutions, and the performance of IIT/K graduates in U.S. graduate engineering programs at the same institutions attests to its quality.

He points out that questions have been raised about the appropriateness of theoretical aspects of the undergraduate program in Indian development and responds to these questions by arguing that the program is more accurately described as "science-based" and that both theoretical and empirical aspects of science are certainly practical components of an engineer's knowledge. Whether IIT/K graduates are contributing to Indian development, he says, will be answerable only in 10 or 20 years. He expresses disappointment that (then) current IIT/K faculty were not addressing themselves to national problems and suggests that the absence or lack of work on national needs at IIT/K only tends to push students to do graduate work abroad because there is little opportunity at IIT/K to work on either national needs or their own careers. Dahl contends that IIT/K:

> Has been an irrelevant factor in the industrial and social progress of India, . . . a kind of isolated island of academic excellence, but not a part of the main stream of India's development. The faculty, as a group, has not yet recognized the necessity for social relevance to coexist with academic excellence. (36)

Dahl wrote these last words in 1972 and they are similar to words that one of us (Morgan) wrote about IIT/Kanpur in 1967. (37) Yet much can happen in five or ten years. It is our impression that IIT/Kanpur graduates have assumed significant roles in India's science and technology efforts. With the push for relevance in many engineering institutions today, it seems likely to expect that IIT/Kanpur has begun to shift its emphasis to the developmental needs of the country, although we have no data or contacts to indicate whether or not this is the case. However, we also speculate that IIT/Kanpur, if it is becoming relevant, is doing so in relation to modern, industrial science and technology needs. It would be interesting to know how it is responding, if at all, to the recent Desai government emphasis on appropriate technology. That the indigenous science and technology base in India has been strengthened through KIAP, and those programs of the four other IITs which were supported by four other countries, seems clear. The nature of that base and the uses to which it is being put are less clear.

Aside from the usual aspects of institution building, KIAP's main thrust was toward effecting changes in the way engineering education was to be conducted at IIT/K. In essence, these changes were geared to "westernizing" Indian engineering education, at least at IIT/K, primarily through development of a broader, more theoretical, science-based curriculum that was typically found in India, and through changing attitudes of IIT/K faculty and students toward education and toward one another.

Intervention of developed countries in Third World affairs is a sensitive subject and, if done, should be conducted with extreme caution. KIAP operated at the invitation, and with the approval, of the Indian government. Their approach was mutually agreed upon by the director of IIT/K and officials of the Ministry of Education. The cooling of U.S.-Indian relations in the early 1970s precluded continuing relationships and also resulted in the canceling of a joint U.S.-Indian conference to evaluate the Kanpur experience. So once again we are left with a lack of independent evaluation. It may be possible now, however, to get Indian perceptions of what was accomplished and what the U.S. role was in these accomplishments.

Other Involvements

The University of Wisconsin

The College of Engineering at the University of Wisconsin-Madison has involved itself in a number of institution-building and international education projects.

Wisconsin in India

From 1953 to 1967, Wisconsin assisted in institution building in
India under two consecutive AID contracts. The earlier work helped
pave the way for subsequent activity, like the IIT/Kanpur effort. The
second contract supported Wisconsin work in developing graduate
programs and engineering educator training at the Bengal Engineering
College, Howrah, and the University of Roorkee, in Roorkee.

The Wisconsin final report on the second contract indicates that
much of Wisconsin's work outside of mainline institution building involved, like KIAP's, the promotion of attitude change among their
counterparts. (38) Wisconsin's efforts in this direction included reduced emphasis on external examiners, less reliance on rote memory,
punctuality of instructors in meeting classes, sufficient preparation
of lectures, and reduction of low status associated with demonstrat-
ing the use of equipment to students. In addition, Wisconsin set up
an extensive participant training program, under which Indian faculty,
administrators, and students did graduate work or received specially
designed training at the University of Wisconsin and other U.S.
schools.

The Wisconsin-Singapore Degree Development Program

The University of Wisconsin College of Engineering also assisted
in the development of an engineering degree program at the University of Singapore between 1966 and 1976, under a grant from the Ford
Foundation. (39) Wisconsin provided visiting professors, library
consulting, scholarships for Singapore staff to study in the U.S., lab
instruments, and research and travel funds. Training of indigenous
personnel facilitated utilization of indigenous engineers on the faculty.
Visiting staff and Madison campus support staff assisted the Singapore
staff in working with funding agencies.

Wisconsin has continued its linkages with Singapore. The United
Nations Development Program is supporting the use of the facilities
of Wisconsin University's Instrumentation Systems Center to help the
Singapore Institute of Industrial Research develop its metrology
capability. (40)

The Wisconsin-Monterrey Tec Exchange Program

Wisconsin began offering third-year engineering students the
opportunity to study at El Instituto Tecnologico y de Estudios
Superiores (Monterrey Tec) in Monterrey, Neuva Leon, Mexico, in
1961. (41) The program became one of exchange in 1963 when Wiscon-

Engineering

sin began making tuition-remission scholarships to a limited number of Tec students.

The program is still continuing. Some Tec participants have returned to Madison or gone to other U.S. schools to obtain advanced engineering degrees, and Tec has sent some of its younger faculty members to Madison for graduate work. By the end of the 1975-76 academic year, 84 Wisconsin students and 51 Tec students had participated in the program.

The MIT Technology Adaptation Program (TAP)

TAP was associated with an MIT 211(d) grant. It aims at adaptation or creation of technologies geared to removing obstacles to LDC development. It focuses on: development of a knowledge base; identification of criteria for technological creation and/or adaptation; knowledge of introduction, dissemination, and utilization processes; and socio-economic costs and benefits of choices to import technologies rather than improve indigenous ones. (42)

MIT approaches these foci through linkage with LDC institutions and government agencies, collaborative research, and education at MIT. The education component is represented in the interdisciplinary master's degree program, through which graduate research assistantships and fellowships are provided.

Work on MIT's 211(d) grant, "The Adaptation of Industrial and Public Works Technology to the Conditions of Developing Countries," was reviewed by AID in 1974, and again in August 1975. (43) The review concentrated primarily on identifying issues, though the review team did state that the quality of MIT's linkage with LDC institutions was less than desirable, and that funding of some projects was insufficient to achieve project goals.

Other Cases

There are many other cases of involvement of U.S. engineering institutions and individuals in international S&T activity. Some of these that we are aware of include:

1. Activities of the American Universities Consortium (44) in assisting in the development of the University of Petroleum and Minerals, Dhahran, Saudi Arabia.

2. Colorado State University's assistance to the Asian Institution of Technology and the former's On-Farm Management Program in Pakistan.

3. The work of the East-West Center and its Technology and Development Institute.

4. Programs of Stanford and UCLA in Latin America.

5. Cornell's program on Science and Technology Policies for Developing Nations.

6. Washington University's teaching and research programs in technology and international development.

7. University of Kentucky's assistance to the Bandung Institute of Technology, Indonesia.

We know of no current catalog of, and no comprehensive effort to analyze, past and present programs. We hope that this study provides a beginning for compiling such information.

ENGINEERING TECHNOLOGY ACTIVITY

In the U.S., the field of engineering technology is defined as "that part of the technological field which requires the application of scientific and engineering knowledge and methods combined with technical skills in support of engineering activities; it lies in the occupational spectrum between the craftsman and the engineer at the end of the spectrum closest to the engineer." (45) Engineering, in turn, is defined as "the profession in which a knowledge of the mathematical and natural sciences gained by study, experience, and practice is applied with judgment to develop ways to utilize, economically, the materials and forces of nature for the benefit of mankind." (46)

Engineering technology programs in the U.S. have received formal status in recent years and are now accredited by the Engineers Council for Professional Development (ECPD). In general, such programs must be technological in nature, and in higher education. Instruction is in the broad area between engineering and vocational education - industrial technology. The following schools offer engineering technology programs: technical institutes, junior/community colleges, colleges of technology, polytechnic colleges, divisions of colleges and universities, proprietary schools. ECPD calls graduates of baccalaureate (four-year) programs in engineering technology, "engineering technologists," and graduates of associate (two-year) programs, "engineering technicians."

Engineering

There are now 124 technical colleges and affiliated technical colleges offering engineering technology programs that are institutional members of the American Society for Engineering Education. These programs have had some difficulty in the past in defining their relationship to the more traditional bechelor of engineering programs. In general the technology programs are more practically and vocationally oriented. However, the amount of mathematics and science preparation can be considerably greater than other more vocation-oriented programs offered at junior colleges.

Because of their close link to engineering and their more practical orientation, engineering technology programs are of potential interest to LDCs where "middle-level" manpower and technicians are often an important need. Rao wrote several articles during the 1960s on technical education in developing countries which focus on this sector and the efforts to address this need by the programs of the Ford Foundation. (47) For example, in Pakistan according to a 1965 article, polytechnic institutes at Karachi, Rawalpindi, and Dacca received $4.4 million in Ford Foundation support to develop as teacher-training centers for a complex of 50 other polytechnics then being planned with help from the World Bank and other agencies. Consultants from Oklahoma State University helped in devising the institutes' curricula, including three-year courses in mechanical, electrical, and automotive engineering and radio-electric technology. (48) According to a recent letter from Rao, Oklahoma State University, the University of Houston, and the Dunwoody Industrial Institute have all been active in this area. (49) The Ford Foundation program for the most part seemed focused on training technical education teachers who would function at the precollege level.

In 1978, the Wentworth Institute of Technology in Boston was heavily involved in assisting the development of the Shiraz Technical Institute, under contract with the Imperial Organization for Social Services in Iran. (50) The institute's engineering technology programs seem patterned after U.S. engineering technology programs, with Wentworth providing engineering technicians-laboratory assistants in the field as well as advisors. MIT was providing some consulting services in an interesting example of university-technical institute collaboration.

One of the largest institution-building projects to date is currently underway in Algeria, where both electrical and electronics technology and engineering programs are being developed in a new institution, INELEC. The magnitude and duration of this project are significant: a total of $129 million over ten years from the government of Algeria to support a consortium of U.S. universities under the management of the Education Development Center (EDC). Consortium members include Case Western Reserve University, University of Houston, University of Missouri-Rolla, Oklahoma State University, Stevens Institute of Technology, GTE-Sylvania Training Operations, Wentworth Institute, and the University of Wisconsin-Stout. Not only is there a mix of technology and engineering but industrial participation as well. (51)

Although the number of LDC students studying engineering technology in the U.S. is small, it is increasing. The percentage of foreign students enrolled at two-year colleges increased from 10.6 percent in 1970 to 15.6 percent in 1976. A major development was the decision by the government of Nigeria to send large numbers of high school graduates to the U.S. under the Middle Level Technical Manpower Training portion of the Nigeria Manpower Project. The program is being administered by AID but paid for by the Nigerian government. The students are located in 119 American community and junior colleges and technical schools in 30 states and the District of Columbia. (52)

ASEE INTERNATIONAL ACTIVITIES SURVEY

The Joint Committee on International Activities of the American Society for Engineering Education (ASEE) developed a questionnaire to survey current involvement in international engineering program activity of U.S. and Canadian institutions offering engineering and engineering technology programs. The questionnaire was mailed May 19, 1978 to deans of the 385 U.S. and 24 Canadian ASEE member institutions, which include engineering colleges and technical colleges - the latter offering engineering technology degrees. About 50 percent of all institutions responded.

The principal findings were as follows: (53) 42 institutions, or about 21 percent of all respondents, reported 56 formal international programs, 39 of which involve LDCs. Of the latter, 18 involve OPEC and 21 non-OPEC countries. Twenty of the engineering and technical colleges report having a coordinator or director of international programs; 14 report special arrangements for international grant or contract activity which differs from U.S. work. Almost all institutions enroll foreign graduate students; although the data are not reliable, there may be some indication that large engineering foreign student graduate enrollments are beginning to fall off. Annual funding for formal programs averages about $110,000; of 28 U.S. programs involving LDCs in which sources were identified, 7 were funded by the U.S. government, 17 by foreign sources, 4 by combined U.S. govern-

Engineering

ment and foreign sources. OPEC country funding was prominent. Table 2.7 indicates the nature of the programs in which ASEE member institutions were involved.

ISSUES

Introduction

In this section we identify some of the issues that arise in the engineering education field with regard to international involvements. Some are peculiar to engineering while others are shared with agriculture and science. We identify issues based upon the cases we have examined, the literature, and upon current thinking.

Lack of Funding

Lack of funding is a perennial cry in most academic endeavors. In international engineering activity this issue arises in several different ways. First, there would seem to be a dearth of U.S. government funding in general for efforts to involve engineering colleges and individuals in international activity. Engineering schools do not have a Title XII program to look toward, and the program that had begun to show some interest in them, 211(d), seems to have been de-emphasized. Lack of funding is most significant with respect to resource poor countries that are not able to pay. If it is desired to increase U.S. engineering school involvement in helping to build engineering education infrastructure in those countries, then U.S. government funding needs to be increased.

OPEC country governments have become a significant source of funding for some U.S. engineering schools. Funding is provided to pay tuition for OPEC students, to support U.S. university involvement in consortia, to assist in developing INELEC in Algeria and the College of Petroleum and Minerals in Saudi Arabia, and even to endow chairs at U.S. universities.

A related issue concerns the nature of programs being funded. The institution-building efforts in Algeria and Saudi Arabia, if they follow the pattern of previous institution-building efforts, could have little impact on aspects of U.S. engineering education. This activity has elements of an international service or contract function. On the other hand, funds for international collaborative research and development between U.S. and LDC institutions, funds to develop U.S.

curricula focused on international development problems, and funds to provide more relevant experiences for foreign engineering students in the U.S. do not seem to be available. These latter activities may prove to be important for aiding in strengthening the indigenous LDC S&T base after the initial institution-building period is over. They also would serve to involve larger numbers of engineering faculty in teaching and research related to development problems, just as have funds for aerospace and energy activity in the 1960s and 1970s, respectively.

Lack of Continuity

Lack of continuity is most problematic in institution building and resource base development. In institution building, as noted in the GAO report, (54) some assisted LDC institutions have regressed due to termination of assistance from U.S. schools. A major reason for premature termination can be a change in the political climate between the two countries, something over which there may be little control. Another reason may be that the LDC becomes an AID-graduate country and is no longer eligible for AID support. The latter situation would be ameliorated by new programs of support for cooperative relationships with AID-graduate countries.

Lack of continuity also arises in connection with U.S. resource base development. Under the 211(d) program, AID has provided a series of five-year grants to develop capability for doing international work. After the five-year expiration, some grants have been extended, others terminated, and others supplemented after the grant by contracts in the same area. By the time grants run out, the university is expected to have developed other funding sources to maintain the capability, so that it will be available to AID should the agency wish to utilize it later. In some cases, the support to sustain this capability may not be available.

One longtime participant and observer of international engineering programs remarked that AID administrators and planners suffer from what he calls the pre-1961 "bridge-building" syndrome. You build a resource base. It's there to stand on its own, to be traveled on, to be used. In his opinion, this attitude shows a lack of understanding of the nature of a university and of the long-term nature of human resource development.

A related issue concerns the possibility that AID may feel obligated to sustain a resource base effort to justify its selection in the first place. Such a practice can have the effect of excluding others from competing for contracts and grants. It is here that independent

Table 2.7. ASEE International Activities Survey Program Elements

Element	Number times element was cited		
	Engineering Colleges and E.C. Affiliates	Technical Colleges and T.C. Affiliates	Total
a. Organized study for U.S. students abroad (undergraduate level)	8	0	8
b. Organized study for U.S. students abroad (graduate level)	5	0	5
c. Student exchange program (undergraduate level)	8	1	9
d. Student exchange program (graduate level)	13	1	14
e. Foreign students to study in U.S. (undergraduate level)	14	8	22
f. Foreign students to study in U.S. (graduate level)	22	0	22
g. Special curricula for foreign students in U.S.	8	1	9
h. Special training programs (nondegree) in U.S.	7	4	11
i. Institution building abroad	18	5	23
j. Faculty exchange	18	2	20
k. U.S. faculty abroad	19	4	23
l. Foreign faculty to U.S.	16	2	18
m. Cooperative research and development between U.S. and foreign institutions	20	2	22
n. Formal linkage with overseas educational or research institutions	12	2	14
o. Membership in a consortium of universities working abroad	5	2	7

SOURCE: Soule, T.N., and R.P. Morgan, "Summary of ASEE International Activities Survey," Draft Report, August, 1978.

evaluation becomes especially important to ascertain whether in fact the resource base institution is performing well. There would also appear to be the need for programs to encourage and enable individuals from a larger spectrum of colleges to participate, rather than to focus on a very limited number of institutions.

Lack of Clout in Washington

We have been struck by the inability of engineering schools to marshal their resources to win support for international program initiatives. This is in spite of the fact that there are many more engineering schools than there are agricultural colleges and many more foreign students studying engineering than agriculture in the U.S. By contrast, the National Association of State Universities and Land-Grant Colleges (NASULGC) appears to be a highly effective force for international agriculture programs, as passage of the Title XII program would indicate. This organization has as its constituency most if not all of the colleges of agriculture in the U.S. By contrast, there are many engineering schools outside of NASULGC so that this organization may not be a good vehicle for such an effort.

New mechanisms within the engineering community would seem to be called for. The American Society for Engineering Education needs a full-time office of international programs. Other professional engineering societies might be enlisted in such an effort. A focus on energy might give engineering schools a cohesive effort, just as food and agriculture provide the focus for agricultural problems.

Red Tape

Most reports we have reviewed in the areas of institution building and resource base development comment on difficulties encountered in interacting with bureaucracies, be they agencies of the U.S. or foreign governments or institutions. Substantial attention to red tape is beyond the scope of this study; we suggest, however, that loss of efficiency and effectiveness due to red tape might be reduced by a trend toward decentralization and scale-down of technical assistance programs - for example, relying more on small-scale cooperative programs managed largely by the primary actors themselves (universities and their departments or individual faculty members) or by relatively independent elements of the U.S. foreign assistance bureaucracy. On the other hand, large-scale institution-building projects may only be able to succeed if sufficient faculty can be mobilized through consortia. Furthermore, it can be argued that resistance by academic personnel to bureaucratic structure prevents them from using it effectively.

Evaluation

We have been unable to locate any independent evaluations of programs. In essence, the lack of independent evaluation means that there is no work toward establishment of a baseline, no set of criteria that lends itself to a concept of what is of adequate or inadequate quality. The absence of a baseline precludes comparability of one program to another in terms of effect, or of comparing a program to an ideal concept of what that kind of program should do.

The notion of "intuitive" evaluation, of collecting together all the effects of a program and subjectively judging the set of them is not without value; yet it is doubtful that it will satisfy the demands of government sponsoring agencies, nor does it provide researchers or workers with objective, external value judgments of what is good, better, or best.

Finally, we have located no evaluations by recipient country institutions or governments. These kinds of evaluations, were they done, would provide added perspective on the programs and stimulate thought on how they might be improved.

LDC Engineering Education

Institution-building efforts of U.S. engineering schools have generally sought to transplant U.S. engineering education to LDCs. The philosophy seems to be that education of LDC engineers should be similar to that of many U.S. students: broad, science based, and theoretical as well as practical and applied. An alternative approach holds that LDC students should be educated almost entirely along the lines of practical knowledge and application of technical competence. The latter approach is based on the notion that graduates will be trained in the kinds of problems they will face in their local environments, while the former notion reflects the claim that a broad knowledge base and the ability to innovate are essential characteristics of a good engineer, no matter what kind of situation he/she practices in.

These conflicting views extend to the LDCs in terms of what kinds of institutions they want built, or what kind of results they want institutional assistance to provide. Some parties want centers of excellence in their country; others object strenuously, claiming that such institutions simply try to copy the models of elite, developed country institutions, and that investing financially and psychologically in these copies is in fact contrary to national development goals geared to national identity, accomplishment, and self-reliance. Hazeltine has provided a good analysis of the options available to LDC engineering schools. (55)

At the Project Workshop, Barry and Hazeltine stressed the relevance of the "software" side of engineering to LDC education, particularly industrial engineering and science planning. Barry quoted from an LDC government official at the 1978 World Congress on Engineering Education who said that "what I need in my ministry is people who can tell if we are being sold a bill of goods, who can understand contracts, etc." (56) Hazeltine stressed the need for science and technology planners and planner-managers who understand broad issues such as choice of technology, as well as for industrial engineers who have a systems orientation. (57)

It should be pointed out that a body of literature does exist on LDC engineering education. UNESCO has published reports on such topics as Continuing Education for Engineers and the Design of Engineering Curricula. (58) In 1976, UNESCO sponsored an International Conference on the Education and Training of Engineers and Higher Technicians which produced a report with many recommendations. (59) Regional organizations such as the Pan American Federation of Engineering Associations (UPADI) hold annual conventions in conjunction with Latin American Congresses on Engineering Education. International networks of engineering professional organizations appear to be strong, with participation and leadership by LDC representatives. We have not had sufficient time to examine this literature or these organizations in this study. Such an examination should yield useful insights.

Political Issues

The U.S. move to set up an engineering school at Kabul University, partly in order to forestall establishment of a Soviet-sponsored institution at Kabul, underlines the political root and implications of much of institution building, as well as assistance overall, and obscures and complicates efforts at defining criteria to be used in evaluating the success of such projects.

Some parties associate the notion of external assistance in institution building with control by the assisting country. However arguable this might be, it should be considered as a factor to look for and consider in any future U.S. university-assisted institution-building plans. According to Kidd, at the Middle East Technical Institute, Ankara, Turkey, at the University of del Valle, Cali, Colombia, and at the Indian Institute of Technology, Kanpur, there was suspicion of U.S. domination. (60)

Currently, the U.S. seems to be paying increased attention to emerging, resource-rich countries such as Nigeria, Brazil, and Saudi Arabia, as well as to Egypt because of its pivotal role in the Middle East. From this interest flow commitments to institution building and other efforts in engineering education. This situation is one which highlights the need for new mechanisms to support activity with AID-graduate countries that are not fully reimbursable.

U.S. Education for LDC Engineering Students

Our analysis has shown that in 1965-1976, somewhere between 29 and 50 percent of all graduate engineering students in the U.S. were foreign nationals. Although the corresponding undergraduate percentages are nowhere near these numbers, many undergraduate engineering students from LDCs continue to come to study in the U.S. (e.g., in 1978 from Iran, Venezuela, Nigeria, and Saudi Arabia). The question of special curriculum and other activity for these students has been debated for several years, (61) and no strong consensus for such curricula has yet emerged.

At the Project Workshop, Gomez stated that the U.S. engineering education system is the best model for LDCs, (62) a sentiment echoed by Hazeltine who stressed the system's flexibility, ability to deal with a wide set of students, and good pedagogy compared with other models. (63) Gomez does not want the current emphasis on science-based education changed, stating that LDCs need the flexibility and versatility it provides. (64) Hazeltine sees the need for this type of education for LDC students who return to work in modern industries in LDCs, but he also feels that planner-managers are needed. (65)

Several nontraditional activities have emerged in the U.S. that may be suitable for LDC students. These include engineering-based graduate programs focused on technology and international development (e.g., Georgia Tech, MIT, Washington University), and problem-oriented research institutes which couple the university and the student to industry. Science policy programs may provide the opportunity for LDC students in the U.S. to work on international problems in collaboration with LDC institutions. However, the bulk of the foreign students will continue to enter traditional engineering programs. Summer programs and courses oriented toward LDC problems might be useful for these students as well as thesis problems oriented toward development problems in their home countries. The large number of foreign graduate students would seem to make the case for some such activity compelling.

A related issue concerns the appropriateness of large numbers of foreign engineering graduate students coming to the U.S. to study in the first place. If a significant number of these students do not go back, the effect on building an indigenous S&T base could be very negative. Furthermore, state legislatures might well ask why one-third to one-half of graduate engineering degree candidates at schools in their states are taken up by foreign students. Even if they pay their own way, the cost of tuition covers only 20-25 percent of the total cost of educating a student at a public university.

On the other hand, data indicate that immigration of engineers to the U.S. has slowed in recent years, with immigrants coming from countries in which there is unemployment of engineers. (66) Foreign graduate engineering students are needed by U.S. engineering faculty and institutions to maintain programs and sustain research. Many countries need to send their students for graduate engineering education abroad. Involvement is likely to continue for some time, given what appears to be the mutual interest of both parties. At the same time, one measure of success in building an indigenous LDC S&T base should be the increasing adequacy of and reliance on LDC institutions to educate their own students, a capability that some may be overlooking today.

Both Gomez and Hazeltine agreed at the Project Workshop that as seen from overseas, U.S. university efforts look disorganized; a central clearinghouse for information about where to send students in the U.S. and also for U.S. people interested in development is badly needed. This function does not seem to be well served currently by existing mechanisms.

CURRENT THINKING

In September 1977, we wrote to 50 individuals and asked the following questions: "Is there a role for U.S. engineering schools in international development and development assistance, given the changing world situation and priorities? What precisely should that role be? How does it differ, if it does, from five or ten years ago? Where will the financial and institutional support for such an effort come from?" We also asked for information and evaluation, either formal or informal, on successful or unsuccessful programs involving U.S. engineering schools in development assistance.

Engineering

The people written to have all been involved in one way or another in activity related to engineering schools and international development. Almost all are or have been associated with U.S. universities, either as faculty members or administrators. Thus, we were soliciting opinions from a group with strong interest and involvement in the topic.

We have not undertaken a detailed analysis of the responses but the following general results emerge. Most respondents felt that there was a role for U.S. engineering schools. Of those that were not in this category, most sent back responses that did not address the question. Only one was strongly negative. Although many felt that there was a role, opinions differed greatly as to just what the role should be, and no single activity or view appeared to be mentioned by a majority of the respondents.

Few people addressed the question of how things may differ from five or ten years ago. Some responses indicated: (1) more concern today for environment, food, and energy problems; (2) more maturity now on the part of LDC institutions, which means U.S. schools must play less of a dominant and more of a supportive role; and (3) not much has changed. Of the small number of people who responded about financing of U.S. engineering school activity, there were comments on: (1) the need for and lack of U.S. government and international organization funding possibilities, and (2) the prospect of certain LDCs such as the OPEC countries paying for services.

Some respondents stressed the importance of involving U.S. engineering educators who are sensitive to people and problems in other cultures. The dearth of such educators was mentioned, as was the need to find ways to sensitize U.S. faculty and students. Others we wrote to felt that a small number, say one-half dozen, of U.S. engineering schools should become heavily involved because that was all that available funds could support. (These responses came from universities that had substantial funding!) One respondent felt that appropriate technology was needed and that U.S. engineering schools would have a role if they could respond to that need.

In the remainder of this section, we quote without comment from some of the replies we received to our inquiry:

> Most contracts I have observed involve a U.S. institution trying to duplicate what they do at home in a foreign country, and most of those . . . are true academicians or researchers with less than desirable professional expertise.

> . . . I found that all too many [LDC] universities want to be mini-Cambridges . . . or mini-grandes ecoles . . . What they really need is some sort of association or twinning with a good American A and M school, or with a school of technology.

I believe the need for the . . . large scale U.S. government- and foundation-supported efforts [in institution building] will always be with us. Not so much because there is a desperate need . . . as because there exists a group of what might be called professional internationalists who need them to justify their existence.

First, we must educate the educators themselves in the real needs of developing countries and get them to realize that different emphasis must be made for students from developing countries . . .

. . . indigenous competency [as a base for education and training] cannot be overemphasized, [and] educational programs must be responsive to current needs.

Many foreign graduate students contribute as much to U.S. academic and fundamental research as American graduate students, often at less cost. They return sometimes as professors in their countries to contribute to research and teaching students and they or their students may work with or for American companies, and are likely to be involved in purchasing American equipment.

Future projects may be in difficulty if they do not contribute to America's own self-interest.

Not nearly enough emphasis is placed on the fact that both training and research in engineering may - indeed, will - be a two-way street. As we become more conscious ecologically, we may learn from LDC's both through their attitudes and their experiments. This emphasis on return benefits may make any increased U.S. support more palatable at home and at the same time make cooperation more palatable to the receiving countries. Let us avoid the air of being generous gift bearers as much as possible. There should, by now, be genuine, specific examples of return benefits that can be pointed to.

[I have worked in projects] almost always on my own or, occasionally as part of someone else's larger team. In part, this reflects a personal bias. I work in a one-to-one situation in either one of the above cases with only a few others involved. I just do not believe that what might be called the "residual effect per dollar spent" is as great from large-scale government-to-government and institution-to-institution projects as from one-to-one projects with small groups I have been engaged in . . . .

Technical books for students could improve most programs considerably at little expense, as could technical libraries available to those needing them . . . . It seems to me the first effective way an individual can work is as a faculty member . . . without the overlapping administration of international agency, developing country agency, local university administration, and home institution administration. However, this simple type of project never seems to get any appreciable support funds for books or equipment.

There is a growing sense of self-reliance among developing countries . . . [with] emphasis on technical [cooperation] <u>among</u> developing countries. . . .

. . . I am convinced that real and lasting development in a country must be generated internally by the people of that country. . . .

[An] objective for a university in a developing country is related to the need for such a country to achieve an identity and self-respect. A university which is clearly of high quality might very well assist in the psychological decolonization which is essential for many developing nations.

One thing is certain: We cannot, and should not, regard the goal to be molding of other societies into our own image.

The reality is that developed as well as developing societies are in need of major changes to meet new parameters and new interrelationships in their economies, but to be successful the initiatives must come from within each. It seems to me that if the UNCSTED is to be more than an empty "command performance" it must address itself to the issue of <u>how to encourage and support initiative from within</u> each nation and each university.

CONCLUDING REMARKS

U.S. engineering colleges and faculty members appear to be an underutilized resource for strengthening the LDC S&T base. Table 2.1 summarizes the collective thinking of experienced U.S. engineering educators on present and future roles for U.S. engineering schools in development assistance and table 2.2. summarizes their principal recommendations. Rao has singled out for particular attention at this time efforts in engineering technology, industrial extension services, and cooperative research. (67) Continuing education is another area receiving attention.

During the 1950s and 1960s engineering faculty associated with a significant number of universities gained experience in major institution-building efforts. Large-scale efforts of this kind are continuing at the University of Petroleum and Minerals, Saudi Arabia, and INELEC in Algeria. An ASEE survey of member institutions reveals a shift in funding from the U.S. government to foreign governments, particularly OPEC countries. We suspect that U.S. faculty are involved in international cooperative research projects but documentation is lacking. Although some of the expertise developed in the 1960s is currently being utilized, there are also cases in which faculty have returned to their home institutions and have not been able to utilize effectively their international experience.

Engineering education in the U.S. is primarily concerned with supporting the modern industrial sector and rests on a sound scientific base. As such, engineering education should be transferable to other countries and be able to serve this sector. Institution-building efforts have followed this pattern and, in this sense, appear to have been successful in helping to build an indigenous S&T base. Whether these institutions in LDCs are relevant to AID's basic-needs strategy is another matter. U.S. engineering education is less likely to be helpful in this regard unless its own orientation moves in this direction, a trend which does not seem to be well established. There are isolated instances, as in the Georgia Tech 211(d) Small Industries Program and the RITA program, in which this sort of orientation has been attempted in a cooperative international mode, apparently with some success. These judgments must be qualified by the dearth of independent evaluation which characterizes the field.

Engineering is a somewhat fragmented profession. Allegiances are to specific disciplines (e.g., chemical engineering, electrical engineering, etc.) rather than to engineering per se. This may be why U.S. engineering education does not seem to be able to marshal support for international efforts. Another may be the general good health of the domestic engineering enterprise. Organizations such as ASEE are active in international affairs and professional contacts among practitioners are extensive, at least among a limited number of individuals. However, the U.S. engineering education profession does not seem to be able to get the kind of government program support results that the land-grant agricultural colleges have gotten with the Title XII program (see chapter 3), in spite of the fact that there are many more colleges, faculty, and students in the field of engineering than in agriculture. Recommendations by some engineering educators to revive the 211(d) program are understandable, but given the current condition of that program, other approaches may be more feasible. The Foundation for International Technological Cooperation may be an important development in this regard.

Potentially the most important information in this section of the study is that about 25 percent of all foreign students in the U.S. are studying engineering and that an estimated 30-50 percent of all graduate engineering students in the U.S. in 1975-1976 were foreign students, mostly from LDCs. These numbers show the important present and potential impact of U.S. engineering education on LDC science and technology infrastructure. They suggest that new efforts to make U.S. engineering education relevant to specific LDC needs might receive more attention than it has in the past. They also suggest that increased emphasis on efforts to strengthen indigenous LDC institutions might have as a long-term objective the reduction of such heavy dependence of LDCs on U.S. engineering programs.

3 Agriculture

United States universities, especially the land-grant institutions, have been and continue to be actively and extensively involved in international development activities related to agriculture. A major contribution to the agricultural development of LDCs has been through the training of foreign students in U.S. universities. (1) In addition, U.S. universities have been involved throughout the past 25 years in: (A) attempting to transfer U.S. agricultural technology (the early 1950s); (B) developing an extensive institution-building program (the late 1950s and 1960s); (C) developing and/or strengthening their own capabilities in international development (the late 1960s and 1970s); (D) carrying out basic agricultural research relevant to developing country needs (the early 1970s); and (E) contributing to the formation of a global network of international agricultural research centers (the 1970s).

University involvement in agricultural development was and is assured because of heavy emphasis on food and nutrition in technical assistance and cooperation legislation and programs. An important rationale behind past and present involvement is the belief that the U.S. land-grant university which was highly effective in developing agriculture in the United States is a relevant and transferable concept for developing country institutions. University involvement has been and continues to be highly dependent on federal support, primarily through the Agency for International Development and its various predecessor organizations. U.S. universities are presently getting geared up under the new Title XII program for a major renewed involvement by strengthening and/or rebuilding their capacity to effectively deal with agricultural problems and institutions of developing countries, as well as through long-term collaborative research efforts

Agriculture

with developing country scientists to find answers to practical problems via a mission-oriented approach.

OVERVIEW OF U.S. UNIVERSITY INVOLVEMENT

Historical Background

Involvement of U.S. colleges in technical assistance programs and in the training of foreign students before World War II was very small and limited to individual efforts of a few universities. (2) It was not until after World War II that significant programs were developed and supported by official U.S. government policy.

The participation of colleges of agriculture in assisting agriculturally developing countries had its beginning over 100 years ago. In 1876, the Massachusetts State College of Agriculture, now the University of Massachusetts, assisted in developing the Sapporo Agricultural School in Japan. In 1924, Cornell University began its Cornell-in-China program. Under this program, Chinese agriculturalists were trained in plant breeding and general crop improvement. Enrollment of foreign students in U.S. colleges of agriculture began in a small way in the late 1800s, and the number has increased following both world wars.

On January 20, 1949, President Harry S. Truman, in his inaugural address, pledged to set forth a program (now commonly called the Point IV declaration) by which the U.S. would help developing countries help themselves. Truman identified inadequate food production as one of the underdeveloped world's critical problem areas. The Point IV program, which the land-grant universities helped to plan and design, officially marked the beginning of cooperation for international agricultural development. The successful transfer of materials and techniques from the United States to Western Europe and Japan after World War II gave encouragement to the idea that agricultural science and technology could be transferred to the underdeveloped countries as well.

By 1950, three agencies administered government technical assistance: the Institute of Inter-American Affairs (IIAA) in Latin America, the Economic Cooperation Administration (ECA) in Europe and the Far East, and the new Technical Cooperation Administration (TCA) in Africa, the Middle East, and South Asia. In the early stages, the United States Department of Agriculture (USDA) carried on most aspects of technical assistance that dealt with agriculture under these agencies. After 1954, the USDA, IIAA, ECA, and TCA turned over all direct technical assistance contracts with universities to the Foreign Operations Administration (FOA), the first year in which technical assistance was unified under one agency. FOA later became the International Cooperation Administration (ICA), which was eventually succeeded by the Agency for International Development (AID) in 1961. From the very beginning, the USDA turned to the land-grant colleges for help in its overseas programs. Soon IIAA, ECA, and TCA all began to make direct contracts with various agricultural colleges for their respective geographical regions.

IIAA engaged Michigan State University in Colombia (1951) and the University of Arkansas in Panama (1951), TCA signed contracts with Oklahoma State University to assist in the building of a new agricultural college in Ethiopia (1952) and with the University of Arizona for similar assistance in Iraq (1952), and the ECA worked with Cornell University in the Philippines (1952). Other institutions that were early participants in international agriculture development were: Purdue University in cooperation with the Rural University of the State of Minas Gerais, Brazil, Texas A&M University with Pakistan and Mexico, and the University of Wyoming in Afghanistan.

American foundations such as the Ford Foundation and the Rockefeller Foundation also increased their technical assistance activities during the 1950s and, like the government, they turned to American universities for collaboration in their projects. During this early period, foundation work in agriculturally related projects was small. It is not until the late fifties and early sixties that the foundations began to play a significant role in agriculture projects.

This early period was clearly an experimental time both for the U.S. government, the U.S. universities, as well as for the foreign governments themselves. The first efforts by U.S. universities emphasized policies that attempted to transfer U.S. technology; the concept of "institution building" abroad also began during this period.

Early Involvement (Pre-1958)

In January 1957, the Institute of Research on Overseas Programs (IROP) began a three-year study of international programs of American universities, their impact overseas, and their impact on the universities in the United States. Funding for this research was provided by the Carnegie Corporation of New York. This study was the first attempt at assessing these university-related, technical assistance programs, which were quickly becoming a major factor in U.S. policies. Various publications resulted from this study, many of which, at least in part, dealt with the early agriculturally related programs.

Agriculture

One such publication, The International Programs of American Universities (1958), is an inventory and descriptive analysis of programs in all fields. (3) According to this publication, in the field of agriculture the number of universities that reported international agriculture activities were 26; some universities were involved in more than one program making the total number of activities 31. Reported funding sources displayed a wide variance; in 28 cases, by far the majority, at least partial funding was provided by the International Cooperation Administration; in 14 cases host-country contributions were used and in 3 cases the U.S. university itself provided funds. Other sources such as U.S. private foundations provided funding for 2 programs (see table 3.1). The types of programs, as reported by the universities themselves, included: sending U.S. faculty abroad in almost all of the programs (29 cases); sending host students to the U.S. in 22 of the programs; and the providing of equipment and materials to host-country institutions in 18 of the programs. Other aspects of programs reported were student exchange and sending host-country faculty to the U.S.(see table 3.1). The 31 programs reported in agriculture represented 8.1 percent of all subject-matter fields of programs reported. (4)

Later Involvement (Pre-1966)

In 1966, a second edition of The International Programs of American Universities was published, this time, by the East-West Center in Hawaii and Michigan State University, jointly. (5) In the eight years since the publication of the first inventory, the number of universities that reported involvements in international agriculture programs rose to 43; the number of actual programs rose even more dramatically to 90. Funding sources continued to be dominated by the U.S. government; the Agency for International Development (AID) accounted for funding in 48 of the reported cases. During this period, both the Peace Corps and Ford Foundation became important sources of funding, each accounting for 10 programs. The Peace Corps funds were used primarily to train Americans to become overseas volunteers. The Ford Foundation supported various activities including the creation of international research centers. The number of different sources of funding increased as well to 15. Despite changes in emphasis of program directions, the predominant activity under the "type of program" category continued to be sending U.S. faculty abroad; 53 of the programs reported this type of effort. Other major areas of activity included: U.S. training for foreign professionals in 41 programs, sending foreign students to the U.S. in 34 of the programs,

Table 3.1. U.S. University Programs in International Agriculture (reported in 1958)

A) Number of Universities with International Agricultural Programs: 26

B) Number of Programs: 31

C) Funding Sources:

   1) International Cooperation Administration 28
   2) Host Country 14
   3) Servicio Cooperativo Interamerican de Salud 1
   4) Council on the Cultural and Economic Affairs 1
   5) Servicio Tecnico Interamerican de Cooperacion Agricola 1
   6) U.S. University 3
   7) Private Business 1
   8) American Foundations 2

D) Type of Programs:

   1) United States Faculty Abroad 29
   2) Cooperating Country Students to U.S. 22
   3) Cooperating Country Faculty to U.S. 4
   4) Provision of Equipment and Materials 18
   5) Student Exchange 1

SOURCE: Institute of Research on Overseas Programs, <u>The International Programs of American Universities</u>, Michigan State University, 1958.

Agriculture

and the provision of equipment and materials in 39 of the programs. Other types of programs reported included: faculty exchange, Peace Corps training, professional collaboration, research training, student exchange, training Americans to work abroad, and sending U.S. students abroad. One of the major programs during this period was the Agricultural University Development Program in India. The U.S. universities that participated were the University of Illinois, Kansas State University, the University of Missouri, Ohio State University, Pennsylvania State University, and the University of Tennessee. This program is described in more detail later in this chapter.

Rural Development Contracts

Of particular importance during these first two decades of U.S. university involvement in international agriculture were the rural development contracts. Between 1951 and 1966, 35 U.S. land-grant universities undertook 68 rural development contracts in 39 countries with the Agency for International Development. Programs active in mid-1966 involved 24 U.S. universities with 43 contracts in 24 countries. Most of the projects focused on institution building in the broad sense of assisting one or more host-country, agricultural educational research or extension institutions to develop a suitable philosophy and organization, along with adequate resources and facilities. In only 4 of the 68 contracts was there an "action program" component where the immediate goal was an increase in agricultural production. Even these action programs implied the need to develop host-country institutions or programs that would continue to carry out the contract activities. (6)

Total expenditure for the 60 rural development projects during the years 1951-1966 was $99,276,087. (7) Where participant training or commodity assistance were not provided through contract funds, they were sometimes financed directly by the U.S. AID mission in the host country. (8)

Rural development AID contracts were distributed among 35 U.S. land-grant universities. By 1966, 6 universities each held 3 or more contracts for a total of 25, while 29 other universities divided the remaining 43 contracts. (9)

In mid-1966, all contracts in Africa were active while all contracts in the Far East region had expired. In Latin America, two-thirds of the contracts were active, all but two of which were four years old or less. The Near East-South Asia region contained the greatest number of long-term contracts of any region. Two-thirds of the region's contracts were still active in 1966, all of which were nine years old or more. (10)

These university-AID contract projects had three principal components: 1) provision of U.S. personnel; 2) providing of opportunities for host-country nationals to study in the U.S., commonly referred to as participant training; and 3) supplying of equipment, books, and supplies, usually called commodity assistance. (11) The U.S. personnel engaged in overseas assignments with these projects totaled 1,399. Among the contract personnel studied in this rural development research project, animal scientists and agronomists were most numerous. Agricultural economists, including a few rural sociologists, ranked third. Agricultural extension workers, agricultural engineers, and agricultural education specialists and vocational agriculture teachers were next most frequent. (12)

Individuals serving under rural development contract projects served from a few days to as long as 15 years. The most frequent length of assignment was approximately 2 years. About 20 percent served for longer than 30 months and another 20 percent served on assignments of less than 6 months. (13)

University AID contracts provided for financing of training programs of host-country nationals. The objectives of these programs were to improve technical, professional, and managerial knowledge and abilities and to introduce attitudes and values essential to development activities. Following the training, participants returned to their countries to contribute to attaining country objectives under the AID contract project. In 1966 U.S. universities reported that there were 2,360 training participants under their AID rural development contracts distributed by regions as follows: Africa, 179; Far East, 520, Latin America, 417; and Near East-South Asia, 1,244. The large number of participants from the Near East-South Asia region was due to more long-term contracts there than in any other region. Africa had the smallest number, partly because nearly all contracts were relatively new at the time the survey was made and partly because in Africa many participants were funded directly by the AID missions rather than by the contracts. (14)

Most university AID contracts provided for the supplying of commodities to the host institution in addition to staff members serving as advisors or technicians, and provisions for the training of host-country nationals. Commodities included a wide variety of equipment and supplies that were needed in support of contract objectives. At educational institutions, the following items were commonly included: library books, teaching and research laboratory equipment and supplies, visual aids, experimental farm equipment, office equipment, and vehicles for use by contract personnel. In the 60 rural development projects for which data were available six contracts had no budget for the purchase of commodities. Any commodities provided

Agriculture

were purchased directly by the AID mission. Of the total amount budgeted, $99,276,087, the allocation for commodity purchases was $12,469,200. Thus the average amount allocated for commodities was 13 percent of the total contract budget; however, the amount actually budgeted per contract ranged from 1 to 42 percent of the total budget. (15)

Other Types of Involvements

While the land-grant universities/AID rural development contracts clearly were the major sources of U.S. university involvement in international agriculture-related projects, during this period other types of projects with other sources of sponsors were being carried out as well. One idea which began to mature during the 1960s was the cooperation of institutional resources in consortia. Many universities discovered that involvement in international projects often drained their already scarce resources. The formation of consortia served to combine the strength of several institutions. An example of a consortium that involved agricultural universities, established in 1974, is the Midwest Universities Consortium for International Activities, Inc. (MUCIA), which has been involved in other fields of activity as well as agriculture.

U.S. universities also became involved in international agriculture through programs funded by the private foundations. In addition, the foundations began establishing international agricultural research centers in the developing countries, many of which were staffed initially from the U.S. academic community. The best example of this is Norman Borlaug, who left the University of Minnesota to win a Nobel prize for his work at the International Maize and Wheat Improvement Center (CIMMYT) in Mexico.

An area of indirect participation for U.S. universities involved private consulting of individual professors. These assignments ranged from short term (one week, one month, etc.) to long term (two years) and consisted of various activities including advising ministries of agriculture, consulting with private agribusinesses, and assignments with international organizations.

RECENT INVOLVEMENT IN AGRICULTURAL DEVELOPMENT

Overview

The preceding has been a description of what can be termed first (1950-ca. 1958) and second (1958-ca. 1968) generation involvements by U.S. universities in international agricultural development projects. Broadly speaking, the first generation can be categorized as an attempt to transfer American agricultural technology to developing countries. Many problems were encountered during this early period; (16) insufficient effort was made to adapt the American technology to specific developing country needs and conditions, and, more critically, there was a lack of adequate indigenous LDC S&T capability and institutions that could use and benefit from the agricultural technologies being transferred to the LDCs. The second generation of activities, then, was largely an institution-building effort, which entailed large-scale projects to assist LDC agricultural educational institutions in developing. These programs consisted primarily of: (A) sending U.S. university faculty members to LDC institutions to teach, do research, and advise; (B) training students in agricultural programs in the U.S.; and (C) providing materials and equipment needed for the development of the LDC institutions.

The last ten years have seen an ever-growing, worldwide concern for the world's critical food situation, highlighted by the UN World Food and Nutrition Conference held in 1974. The U.S. universities have maintained their involvement in assisting agriculturally developing countries. The institution-building efforts of the second generation were largely reduced, but the third generation of U.S. university activities has emphasized research. U.S. universities, as part of an expanding international agriculture research network, turned their energies, in large part, to scientific research that could benefit developing countries. This research is carried out both in the United States and the LDCs.

No comprehensive survey of U.S. university international agricultural development programs since about 1966 has been located; therefore, it is difficult to analyze activity during the last ten years quantitatively. Nevertheless, some information was obtained from certain written materials and from interviews with knowledgeable individuals who have been involved in this area over the past ten years. It appears that U.S. university involvement in activities related to international agriculture peaked in the early seventies and has since declined slightly, primarily as a result of past AID-funded projects reaching completion with relatively few new contracts to replace them.

Agriculture 77

A 1977 National Research Council study supports these observations. (17) The last two years have witnessed renewed optimism for extensive involvements by U.S. universities as they gear up for participation in Title XII activities. It is still too early to evaluate this fourth generation of involvement but it may be designated as an institution-strengthening phase (both in the U.S. and LDCs), achieved primarily through collaborative research. A distinction is made here between "institution building" and "institution strengthening." The presumption is that the agricultural institutions in the LDCs were, for the most part, "built" during the 1950s and 1960s through earlier AID programs. This latest program, then, seeks to strengthen existing institutions. (The phrase institution strengthening is also used in connection with strengthening capabilities in the U.S. to carry out international research).

Research and Teaching Activities

In January 1978, the American Council on Education carried out a survey of international scientific activities at U.S. universities. In 1977, of the 144 universities that responded to the survey, 41 reported formally organized research programs in the field of agricultural and natural resources dealing with foreign regions. Of these 41 universities, 78 percent reported programs in Latin America and the Caribbean region, 24 percent reported programs in the Far East, 39 percent reported programs in South Asia, 34 percent reported programs in black Africa (south of the Sahara), and 44 percent reported programs in the Near and Middle East and North Africa. Thirty-three universities reported formally organized teaching programs in the field of agricultural and natural resources dealing with foreign regions. Of these 33 institutions, 70 percent reported programs in the Latin American and Caribbean region, 36 percent reported programs in the Far East (excluding Japan), 42 percent reported programs in South Asia, 52 percent reported programs in black Africa (south of the Sahara), and 30 percent reported programs in the Near and Middle East and North Africa. (18)

U.S. University Research Role

With the winding down of an earlier generation of projects, U.S. agricultural universities have increasingly concerned themselves with helping to build research capabilities in the developing countries and with orienting U.S. research activities more to global concerns about hunger and malnutrition. Certain U.S. universities already have well-established research programs of this kind underway. Many universities are getting ready for such activity under the Title XII program which is specifically devoted to agricultural research for developing country needs. The USDA's Tropical and Subtropical Research and Training Program (TSRTP) is also oriented toward this type of research.

The basic assumptions that have led U.S. universities to research-oriented activities are, first, that a strong research base in LDCs is essential to all activities needed to increase the food supply, reduce poverty, and moderate the instability of supplies and prices, and second, that the capacity in the developing countries to conduct research and development on food and nutrution, and to apply the results, is very limited at present. (19) It is the U.S. university's intent, under Title XII, to become one link in the emerging global agricultural research system concerned with increasing food production for domestic use in the developing countries. This system, which is presently far from fully organized and established, is described by a National Research Council study as being composed of five major components: 1) research organizations in the developing countries, 2) regional or intercountry research programs in the LDCs, 3) international agricultural research institutes principally located in the LDCs, 4) research organizations in the high-income countries (e.g., U.S. universities), and 5) cross-country research networks generally concerned with specific commodities (e.g., soy beans). (20) Figure 3.1 outlines these major components.

Within this context of an international agricultural research network, U.S. universities are one of many possible avenues for work on world food problems. Thus, the U.S. academic community is concentrating on research activities that they feel they can do best. Broadly speaking, the U.S. universities are assisting developing countries in building their food and nutrution research capabilities by helping them to establish and operate research institutions, and by helping to train scientists and other research workers. Research capabilities are being strengthened in both the U.S. and LDCs by collaborating on joint research projects, exchanges of research information and materials, exchanges of scientists and students, technical conferences, and advisory services. The International Soybean program (INTSOY) based at the University of Illinois is a good example of this type of activity, as is the International Sorghum Research Network, of which Purdue University, the University of Nebraska, Texas A&M University, and the University of Puerto Rico are all a part.

The international agricultural research centers were, for the most part, pioneered by the Ford and Rockefeller Foundations with

Agriculture

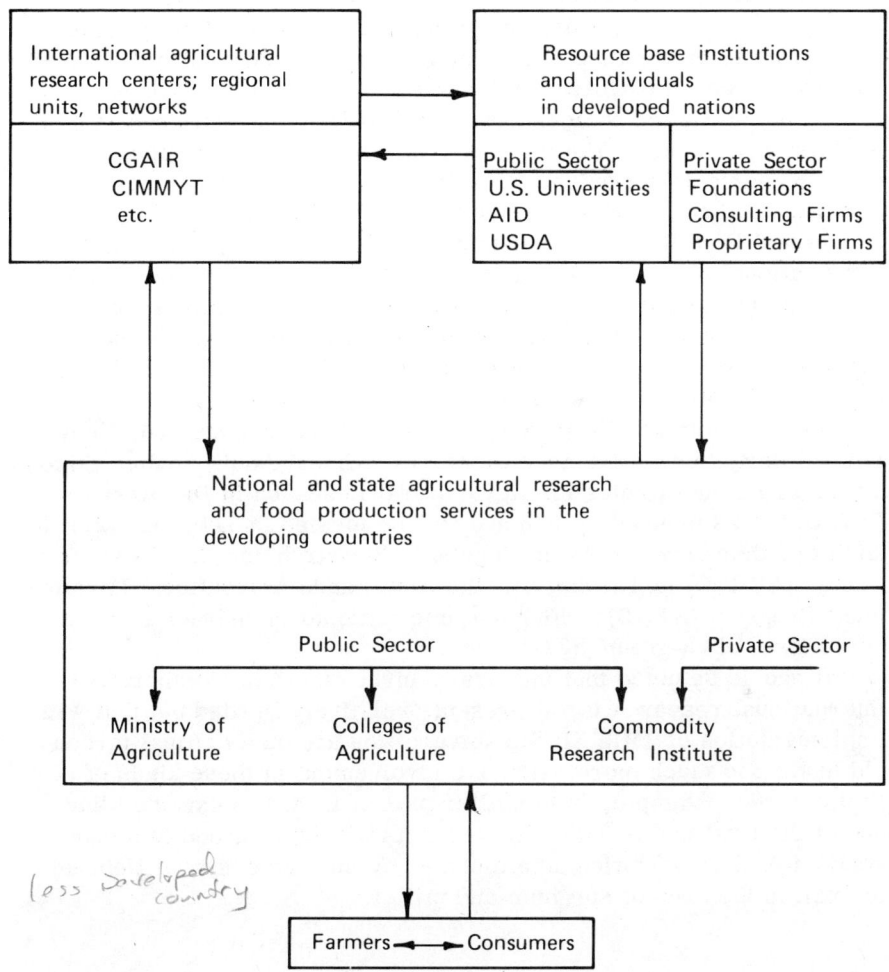

Fig. 3.1. Generalized structure of international agricultural research for developing nations.

SOURCE: Supporting Papers: World Food and Nutrition Study.

U.S. universities playing an important role by providing personnel for the initial staff of these centers. Today, these centers play a major role in international agriculture development. Financing is now arranged through the Consultative Group for International Agricultural Research (CGIAR), which was organized by the World Bank, FAO, and the United Nations Development Program; AID provides 25 percent of the funds. The centers are important linkage mechanisms for U.S. agricultural universities both in coordination of priority activities and collaboration in joint efforts. They include IRRI (International Rice Research Institute) Los Banos, Philippines; CIMMYT (International Maize and Wheat Improvement Center) Mexico; IITA (International Institute for Tropical Agriculture) Ibadan, Nigeria, CIAT (International Center for Tropical Agriculture) Cali, Colombia; CIP (International Potato Center) Peru; ICRISAT (International Crop Research Institute for the Semi-Arid Tropics) India; ILRAD (International Laboratory for Research on Animal Diseases) Nairobi, Kenya; ILCA (International Livestock Center for Africa) Addis Ababa, Ethiopia. The International Center for Agricultural Research in Dry Areas (ICARDA) was planned in 1976 and will be located in Lebanon. (21) In addition, there are the Asian Vegetable Research and Development Center (AVRDC) in Taiwan, and the Arid Lands Agricultural Development Program (ALAD), which are international in character but are not in the CGIAR group. (22)

It should be noted that university involvement in collaborative international research is, at present, relatively limited but that with implementation of Title XII the universities are on a critical threshold leading to much more extensive involvement in these kinds of activities. For example, in the latter part of 1978 it is expected that one of the first major Title XII contracts will be awarded to a consortium of 11 universities administered by the University of Nebraska, to work in the area of sorghum and millet.

Foreign Agricultural Students at U.S. Universities

One important area of continued university involvement today is that of teaching agriculture to foreign graduate students at U.S. universities. The National Center for Education Statistics lists a total of 14,674 students (U.S. and foreign) who were enrolled for master's and doctor's degrees in the field of agriculture and natural resources during the 1975 academic year. (23) The number of these graduate students who were foreign is given by the Open Doors publication as 3,240. (24) This indicates a foreign graduate student enrollment of 22 percent of all agricultural graduate students in 1975.

Agriculture   81

While this number may not be as large as the percentage of foreign engineering graduate students (estimated at roughly 40 percent), it is still a significant portion of graduate students. Furthermore, discussions with university directors of international agricultural programs indicated that this percentage may have been as high as 33 percent at their institutions during the 1977-1978 academic year.

211(d) Grants

In a 1975 AID directory of 211(d) grants, 31 of the 53 grants were awarded predominantly to land-grant universities in areas directly related to food and nutrition. (25) These grants were typically for five years and included funding of $200,000 for each of six universities participating in agricultural development in India, $1 million for the University of California, Riverside, to work on soil and water development in arid and subhumid areas, and $1.5 million for Land Tenure research at the University of Wisconsin. Between 1968 and June 30, 1974, Section 211(d) grants awarded in food and nutrition related areas totaled $17,225,000. These totals accounted for approximately 55 percent of the 211(d) grants awarded in all categories during that time. Contracts from other offices in AID to use the capacity built with 211(d) grant funds provided additional financial support roughly equal to the grants themselves. The grants averaged about $5 million annually from 1968 to 1974, but were greatly reduced in number and amount since 1975. Specific examples will be given later in this chapter.

Title XII

In 1975 the Congress enacted the Title XII amendment to the Foreign Assistance Act of 1961 (see chapter 1). (26) Title XII is an attempt to bring American agricultural research, education, and production expertise to bear on the world food problems. It recognizes the contributions that the land-grant universities have made to America's agricultural efficiency and offers this educational research and extension resource to help develop agriculture in the less developed countries.

The legislation mandates an expanded involvement by U.S. land-grant universities in U.S. efforts to aid the poor countries in providing a better way of life for their people. To facilitate this involvement, the legislation provides for three important changes in the relationship between the university community and the Agency for International Development. First, it provides a mandate for the land-grant and other eligible institutions to involve themselves in efforts to assist poor countries of the world in improving the economic and nutritional well-being of their population. Second, it recognizes the importance of long-run involvement of U.S. institutions and individual faculty members, by providing for long-term funding to allow institutions to plan and staff for such involvement. Third, it provides for the development of a true partnership between the land-grant university and AID by creating a mechanism for university involvement in the total process of problem identification and program conceptualization, organization, and implementation.

To bring U.S. universities into a closer relationship with AID, the Congress created the Board for International Food and Agricultural Development (BIFAD). The board is presidentially appointed, has an obligation to report annually to Congress, and acts as the policy-setting body for the Title XII program. This is an unusual organizational arrangement - a presidentially appointed board that reports to Congress but is within an existing government agency. The board is a seven-member board, four of whom by legislative mandate must be representatives of state universities and land-grant colleges. (27) The first BIFAD chairman is Clifford Wharton, former president of Michigan State University and currently head of The State University of New York (S.U.N.Y.) system.

Furthermore, the Title XII legislation permitted the creation of two standing committees under the direction of BIFAD - the Joint Research Committee (JRC), and the Joint Committee on Agricultural Development (JCAD). These committees are composed principally of university and AID representatives. The JRC is responsible for developing a collaborative research program; the JCAD is charged with preparing baseline studies that should help organize the assistance programs to the LDCs.

After a slow start, BIFAD and its committees have been increasingly active since the second half of 1977. Among their most important accomplishments are: (1) the preparation of operating guidelines by both joint committees; (2) the preparation by the JCAD of a methodology for assessing the capabilities of existing research, teaching, and extension systems in developing countries in order to (A) establish a benchmark (baseline) for future Title XII programs, and (B) jointly develop with LDCs a long-term program for participation of U.S. universities in helping LDCs to strengthen the institutional framework of their agricultural sector; (3) the development of a program aimed at strengthening the capabilities of U.S. universities to participate in Title XII programs; (4) the identification of a preliminary list of collaborative research planning grant priorities by the JRC;

Agriculture

and (5) the initiation of six collaborative research planning grants by the JRC.

Four of the research programs are in the planning and design stage and are in the areas of: (1) sorghum and millet, (2) small ruminants, (3) fisheries and aquaculture, and (4) human nutrition. All four of these categories are oriented toward the "poorest of the poor," which is consistent with the legislative mandate. As of August 1978, the sorghum and millet, and small ruminants projects have been approved as planned by the JRC and are being recommended to AID for funding. The sorghum and millet research project will be carried out primarily by a consortium of 12 U.S. universities. The functions of this project will include collaborative research, technical assistance, and training for LDC students. The consortium has requested a $3.5 million one-year budget to carry out 45 projects. As of August 1978, the JRC and BIFAD had recommended to AID a fifth collaborative research planning project on edible beans and cowpeas, which will be administered by Michigan State University. The next such project recommended will be in tropical soils.

The inclusion of Title XII in the Foreign Assistance Act is financially significant to universities in that: (1) a congressionally imposed limitation on all centrally funded contract research was eliminated, (2) U.S. universities are being asked to provide money of their own for international development activity in the form of matching contributions and, most important, (3) funding for development activities is long term. In the 1960s and 1970s the U.S. Congress imposed a monetary ceiling on all AID centrally funded contract research with the exception of research on population. The level was $9 million in FY 1970 and was raised to $12 million in FY 1974. A National Research Council study suggests that this ceiling effectively restricted the amount of available funds, therefore limiting, or in some cases, preventing further research in important areas of agricultural development. (28) According to this study, "highly desirable complementary activities such as training, dissemination of information, and outreach were either not done or were at best picked up under other programs." (29) With the passage of Title XII this ceiling was eliminated.

State universities and land-grant colleges receive most of their funds from their respective states. These funds are specifically earmarked for domestic activities. Funds used for international activities have been predominantly from federal sources. Little money has come from the universities' own budgets. According to an April 1, 1978, AID report to Congress, funding to institutions for Title XII university-strengthening grants will be given, "providing that AID's support be a matching contribution to the universities' own efforts." (30) Since AID stipulates that the matching funds be nonfederal, many universities are encountering serious difficulties in coming up with the money. However, only the university-strengthening activities come under this stipulation. (31)

Finally, of most significance to universities, Title XII gives hopes of providing stable long-term funding. The expectation is that there will be guaranteed funding for two years and a planning horizon of five years, with the prospect that most of the projects will run at least ten years. This assured long-term funding is a response to continuing university frustrations with the uncertainty that was customarily associated with AID contracts. These contracts were typically written for periods of one or two years and universities were never quite sure when the program would be phased out; this presented many problems, the most serious of all being their inability to hire qualified personnel for long periods of time.

The Title XII program is still very much in a developmental stage. There are many problems, both within the universities and within AID, which still need to be ironed out. Many of the new projects are of an experimental nature; it is expected that all concerned with Title XII will learn much from the two projects that were to be initiated later in 1978.

USDA-Supported Activities

The U.S. Department of Agriculture (USDA) supports university activities in international agriculture, although to a much lesser extent than AID. USDA is primarily a domestic-oriented agency, but it has been given various mandates by Congress authorizing it to collaborate with U.S. universities in contributing to the development of international agriculture. The pertinent legislation includes the Agricultural Trade Development and Assistance Act of 1954 (PL 480), as amended in 1958 and 1959, Section 406 of the Food for Peace Act of 1966, and Section 1458 of the Food and Agriculture Act of 1977 (PL 95-113) (see chapter 1).

The first two of these laws authorize USDA to enter into research contracts or agreements with U.S. state universities, land-grant colleges, and other institutions, and make the results available to LDCs. (32) The programs resulting from these two laws are the Special Foreign Currency Research Program (SFCRP) and the Tropical and Subtropical Research and Training Program (TSRTP), respectively. The third law (Section 1458) authorizes USDA to assist U.S. colleges and universities in strengthening their capabilities for agricultural research and extension relevant to agricultural development activities overseas. As of August 1978, funding authorization for the implementation of Section 1458 had not been provided.

Agriculture

The Special Foreign Currency Research Program (SFCRP) makes use of local currencies paid to the United States for the sales of surplus American food (PL 480 sales). Under this program, research is undertaken by U.S. universities and other institutions that is of potential benefit both to the U.S. and LDCs. The research is carried out primarily in the developing country. Expenditures for PL 480 sales in the LDCs totaled $2.16 million in FY 1970 and $3.1 million in FY 1975. (33)

Under Section 406(4) of the Food for Peace Act of 1966, USDA is authorized to assist developing countries by entering into research contracts or agreements with land-grant universities, and colleges, and other institutions, and by conducting research on food products and making the results available to developing countries. For these and other activities the department was authorized to spend up to $33 million per year. Actual funding would not come from the Food for Peace Act but through regular USDA channels. From this mandate, the Tropical and Subtropical Research and Training Program (TSRTP) was created. Funding for TSRTP was first provided in FY 1974. In that year, $500,000 was appropriated with the following general guidelines:

-- Two research and training centers, one in Hawaii and one in Puerto Rico, would be established.

-- Land-grant colleges and state universities would be an integral part of the activities.

-- Research and training centers would be considered networks rather than fixed facilities.

Two primary objectives emerged: providing tropical training and experience for USDA and land-grant college personnel by working on specific tropical research problems under tropical research conditions, and providing foreign nationals a place to learn techniques and methodology under tropical conditions with U.S. specialists supplying the training. (34)

As it has evolved, TSRTP now centers on the University of Hawaii and the federal experiment station at Mayaguez, Puerto Rico, designated as the Mayaguez Institute for Tropical Agriculture. In addition, some universities have projects underway through the International Programs Division of the Agricultural Research Service (USDA) financed under Section 406. Funding for TSRTP was $500,000 in FY 1975, $529,000 in FY 1976, and $681,000 in FY 1977. (35)

Sources of Funding Available to U.S. Universities

Funding is of central importance to U.S. universities that wish to become involved in international development activities. Many universities have been restricted in or prevented from participating in international agricultural development activities because of lack of sufficient funds. This has proved especially true for non-land-grant universities. There are multiple sources available for funding these types of activities but analyzing what these sources are and how much money is actually available has proved complicated and quite difficult. It is clear, though, that the largest source of funding for international programs in agriculture is from the U.S. federal government. Within the federal government the Agency for International Development (AID) supports the largest number of programs and provides the greatest financial support. Although to a much lesser extent compared to AID, the USDA also provides support to U.S. universities for international agricultural activities. Figure 3.2 illustrates the wide variety of funding sources that support U.S. university activity in international agricultural development.

As the chief administrative agency for U.S. foreign assistance, AID has been, and continues to be today, the predominant source of funds for activities in international agricultural development. The broad category of U.S. university projects that AID has supported include: research potentially beneficial to developing countries (both basic and applied), technical assistance, technical collaboration, institution building (U.S. and LDC), education (both here and abroad), and others. A complete picture of AID-funding totals and patterns was difficult to obtain, but, for example, for the fiscal year 1975 alone, the dollar value of all university contracts, grants, and amendments for food and nutrition awarded was $23,485,000. An additional $9,440,000 went to education (including participant training) of which some portion can be assumed to be agriculturally related. These amounts represent a very large portion of the total budget of almost $43 million for all AID-supported university activity in FY 1975. (36)

Although most funds for international involvement in agricultural development come from AID with an additional small amount from USDA, there are many other sources that fund various programs today. The mid-1970s have seen certain United Nations organizations engaging U.S. universities in activities overseas. The Food and Agriculture Organization, for example, has funded a three-year, $862,000 contract with the University of Illinois for a soybean development project in Sri Lanka with CARE and UNICEF providing an additional $454,000. (37)

Agriculture

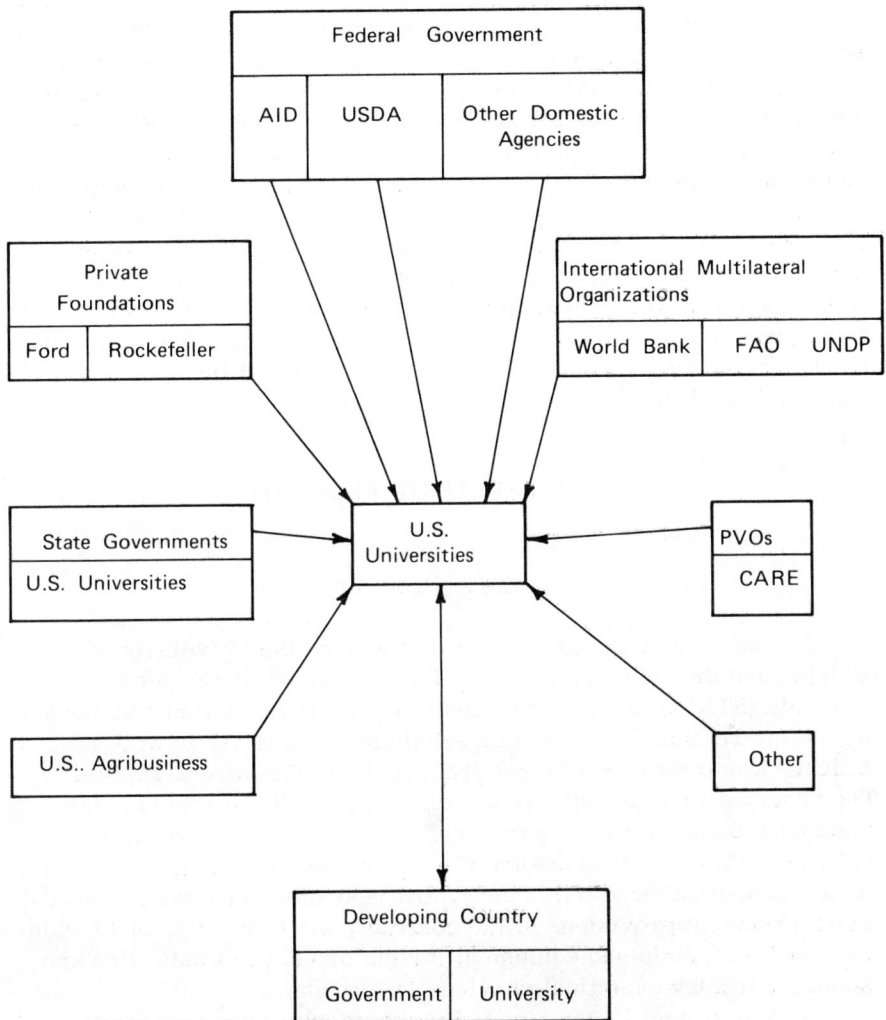

Fig. 3.2. Flow of funds to U.S. universities for involvements in international agriculture activities.

The private foundations, most prominently Ford and Rockefeller, are also a source of funds. Support from the foundations began in the fifties, peaked in the early seventies, and has since seen a major diminution of direct funding to U.S. universities as the foundations routed funds earmarked for agricultural development directly to the international agricultural research centers which they pioneered. Some foundation money, though, still goes directly to U.S. universities; in 1974, for example, the Rockefeller Foundation awarded a $100,000 grant to the University of Puerto Rico to support tropical soybean production research in connection with the International Soybean program. Other funding sources include: the developing countries (especially the new oil-rich countries), very limited state funds, the international financial institutions (i.e., World Bank), and private individual funds.

EXAMPLES OF PROGRAMS

Early Programs

An example of an early program involved the University of Florida and the Servicio Technico Interamerican de Cooperation Agricola (STICA) in San Jose, Costa Rica. The program was funded by the International Cooperation Administration (ICA) as well as STICA. Contract expenditures from 1954 to 1958 were $118,849. The program was largely a research program where Florida staff members assisted counterpart technicians in the Ministry of Agriculture of Costa Rica in design of experiments, equipping of laboratories, importation of trial seeds, and laboratory analysis. Through reimbursement provisions of the contract, the University of Florida provided materials and equipment for the project in Costa Rica and some laboratory analytical services in Florida.

A total of nine University of Florida faculty members visited Costa Rica for varying lengths of time assisting government officials and employees through consultation and by performing research. In addition, two Costa Rican government technicians were brought to Florida as participants in 1956 for four-month, nondegree, in-service programs in animal husbandry and veterinary science. One participant spent most of his time at the Range Cattle Experiment Station in central Florida and the other with the veterinary department on campus. (38)

Another early program involved Montana State University and the Ministry of Agriculture in Jordan. This program, which lasted from

Agriculture

1954 to 1957, was funded by ICA. The objective of the project was to assist the government of Jordan in developing and implementing a program to train Jordanians in (1) agricultural extension methods, (2) home economics, and (3) agricultural marketing.

Two contracts were signed by Montana State University. The first contract called for the contractor to work with the cooperative department. Contract staff members assisted in establishing an extension service in agriculture and home economics, in training Jordanians in subject matter and extension methods and grading and marketing of agricultural products, and in establishing demonstration research programs through assistance in improving teaching methods, curriculum planning, and research.

The second contract called for assistance to the Joint Fund in developing programs in irrigation and forest and range resource management. Contract staff members worked closely with the U.S. Operations Mission and assisted with operation of government forest nurseries, training local government employees in nursery operation, establishing a tree distribution system, and supervising the land development terracing program. Seven Montana State staff members were assigned for two-year tours in Jordan. There were no training participants under either contract. The total expenditure under the first contract was $123,000. The budget for the second contract was $200,000. (39)

Involvement in India

A book by Read, Partners with India, documents the history and involvement of U.S. agricultural universities in assisting the development of agricultural universities in India. Although the book is more promotional and less analytical and evaluative than one might desire, it does provide some useful background information. During the 20-year period from 1952 to 1972, six U.S. universities were involved in programs that had as their objectives the improvement of teaching, research, and extension activities of existing Indian agricultural colleges, the establishment of new agricultural universities, and the reorganization and improvement of the higher education system relevant to agriculture, veterinary medicine, home economics, and related fields. During those 20 years, more than 300 staff members from six U.S. universities went to India on either short- or long-term assignments and more than 1,000 Indian faculty members and graduate students studied in the U.S. Funds were also spent for equipment, libraries, and other materials needed in India. During this period, almost 31 million U.S. dollars and another 11 million U.S. owned

Indian rupees were spent, and about 700 man-years of U.S. staff time was devoted to helping India improve its system of agricultural research and education, including the creation of nine new agricultural universities. (40)

A 1974 case study on the India project was made by Roland R. Renne, who collaborated with Read on Partners with India. Renne obtained and interpreted data on the growth and development of the nine agricultural universities that were established during the partnership years of the six U.S. land-grant universities. Renne concludes that, in general, the nine universities assisted by the U.S. universities and AID have developed competency in those applied fields that are of most significance to India's welfare and progress in her present state of development, i.e., improving agricultural production, technology, and rural living areas which for many years were neglected. Renne found that this was accomplished by applying the land-grant college philosophy of integrating teaching, research, and extension education at the departmental level. He concludes that this approach achieves a "high degree of relevance in helping to solve India's major agricultural and rural problems." (41)

Some of the major areas in which progress was made by the nine U.S.-assisted Indian universities during the period from their first full year of operation through the fiscal year 1973 include:

(1) The number of home campus colleges for the nine universities increased from 21 in their first full year of operation to 37 in 1973, while the number of all-campus or branch campuses decreased from 20 to 18 . . . .

(2) The number of professional staff more than tripled from 2,015 to 6,507 . . . .

(3) Enrollments increased about two and a half times from 9,790 to 23,213.

(4) The number of staff members with Ph.D. degrees increased approximately fivefold, from 251 to 1,234 . . . .

(5) The number with advanced degrees from the U.S. more than tripled from 140 to 486 . . . .

(6) The number with the master's degree as their highest degree increased nearly fourfold, from 1,095 to 4,073 . . . .

(7) Those holding only the bachelor's degree increased less than twofold, from 616 to 1,133.

Agriculture

(8) The number of degrees granted by the nine universities multiplied nearly two and a half times from 1,740 to 4,119 during their first decade of operations. (42)

Renne also points out that in 5 of the 9 universities the basic sciences were strengthened significantly, following closely the action of land-grant universities in the U.S. where there is a gradual process of strengthening the basic sciences in order to strengthen the applied science fields of agriculture, veterinary medicine, engineering, and home science. In addition, Renne collected data on 11 Indian agricultural universities which were not assisted by the United States. He makes certain quantitative comparisons between the 9 U.S.-assisted and 11 nonassisted Indian agricultural universities:

(1) The nine [U.S.-assisted] agricultural universities . . . account for more than 23,000 of the some 33,000 students attending [all twenty] Indian agricultural universities, or approximately seventy percent of the total . . . .

(2) The nine [U.S.-assisted] universities have a total of 721,000 volumes in their libraries, compared with less than 400,000 volumes for the eleven nonassisted agricultural universities.

(3) All but one of the nine [U.S.-assisted] Indian agricultural universities have "earn-as-you-learn" programs, and all but one of the eight unassisted agricultural universities . . . had these programs.

(4) The nine assisted universities have about four-fifths of the professional staffs of all Indian agricultural universities, so that the student-staff member ratio is somewhat lower for the U.S.-assisted than the nonassisted Indian agricultural universities. . . .

(5) The nine assisted universities have 85 percent of the Ph.D. degree staff members of all the agricultural universities, approximately three-fourths of the master's degree staff, but only about half of the total staff members whose highest degree is the bachelor's.

(6) Of those [staff] having advanced degrees (beyond the bachelor's degree) from U.S. universities, the nine U.S.-assisted universities have nearly nine-tenths of the total. (43)

Renne suggests that the quantitative superiority of the 9 U.S.-assisted universities over the 11 nonassisted universities in almost all the categories he looked at reflect, at least in part, the effects of USAID support for such programs as participant training which financed graduate work in the U.S. for staff members of the assisted Indian agricultural universities and for programs that provided considerable sums of money to these same universities for the purchase of books and periodicals to build up their library resources.

While these comparisons are interesting, we feel it would also be helpful, for the specific purpose of this study, to see a comparison between the U.S.-assisted and nonassisted universities in other areas, such as the extent to which the universities have developed their own indigenous capability and the relevance of their curricula to India's basic needs. It would also be interesting to see what impact the U.S. efforts to help a select 9 universities had on Indian initiatives to help the other 11 universities.

The University of Wisconsin Land Tenure Center

The University of Wisconsin's Land Tenure Center (LTC) was initially awarded a five-year 211(d) grant in 1969, which was a sequel to a contractual relationship between AID and the center which began in 1962. The grant has been extended continually with the present contract not due to expire until late 1979. The accumulated expenditures between 1969 and 1978 total close to $2.5 million. (44)

The majority of U.S. universities that carry out research related to food and nutrition concentrate primarily on scientific research that will lead to the increased production of food. The LTC, on the other hand, is unique in that it focuses on the relationship of land tenure to agricultural development, especially vis-à-vis small farmers and landless laborers. The center describes itself as an institute for research and education on social structure, rural institutions, and resource use and development. The center's working hypothesis is that in many less developed countries, investment in agriculture will significantly increase rural employment, improve income distribution, and improve human nutrition only if preceded or accompanied by basic structural reforms in institutions governing the organization, use, and allocation of land. This concept is now receiving much attention in Congress with the "New Directions" emphasis on meeting the basic needs of the poorest of the poor and achieving growth with equity. The Title XII program calls for similar research on non-technological factors as an essential aspect of the overall program.

Agriculture

The center is a unit of the College of Agricultural and Life Sciences and is closely associated with many faculty members in the department of agricultural economics. In addition, faculty from other departments within the university have affiliation with the center, especially in law, sociology and rural sociology, agricultural journalism and communication, political science, economics, business, history, geography, and anthropology. Although the major focus for the first eight years was on countries of Latin America, faculty associated with the center have conducted research in many countries throughout the developing world. Several hundred students (both U.S. and foreign) have been affiliated with the center, although degrees have been mainly awarded within the several academic departments. Within the past seven years, however, a special experimental interdisciplinary development studies doctoral program has been authorized by the graduate school and is administered by the center. The research carried out has been in a variety of areas including economic, political, and social aspects of land tenure and agrarian reform; the economic, social, and political effects of the Green Revolution; migrant labor and economic development; employment policies, the role of women in development; industrialization and the distribution of wealth; the role of agricultural cooperation; and other related areas.

In most developing countries, especially in Latin America, land tenure is a very sensitive political topic. This is most evident in countries that are presently attempting to carry out agrarian reform primarily through a redistribution of land ownership. Clearly, then, land tenure is an area in which it is critical to develop expertise in indigenous personnel. Regardless of how qualified or well intentioned, a nonindigenous professional working or giving advice on land tenure in a foreign country is open to political attacks and is often the subject of accusations regarding outside (either rightist or leftist) subversion. This usually has the impact of sidestepping the real issues. Developing country individuals trained at the LTC, then, would be less vulnerable to these types of nationalistic problems and subsequently more effective.

In May 1977, the Development Studies program had 45 graduate students enrolled - from 22 foreign countries and the U.S. This degree program provides an academic focus for students who wish to pursue advanced study of development issues within an interdisciplinary framework. In 1977, field research was underway in the following countries: Ghana, Tanzania, Turkey, Brazil, Chile, Guatemala, Mexico, Nicaragua, Venezuela, Korea, Malaysia, the Philippines, Japan, and Thailand. As of 1977, there were 15 graduates from this program, all of whom were reported working in the development field, either in universities, international agencies, or government departments concerned with international development.

The Land Tenure Center has had faculty members working with country institutions (usually universities) in ten Latin American and several Asian and African countries. Short-term missions and student research have included the above plus at least twelve other countries. (45)

The program has been primarily funded through the 211(d) program but is supplemented by other sources such as: the University of Wisconsin itself, Midwest Universities Consortium for International Activities, foundations, local AID missions, Latin American universities and government agencies, and other cooperating agencies. Students are financed from a variety of sources; e.g., a student in Chile was financed by the Inter-American Foundation, a Malaysian was financed by his home university, and another student by the Agricultural Development Council. Several students within the past several years have conducted research under a collaborative arrangement between LTC and CIMMYT. The center also has the capability to mount short- and long-term training programs to meet individual program needs. Trainees have come from Turkey, Ethiopia, and the Philippines.

In addition, the Land Tenure Center has sponsored or cosponsored several conferences. In June 1975, the center cosponsored with the Agricultural Development Council Research and Training Network a world conference on group farming.

A recent (1977) major activity by the center includes research to appraise the role of cooperative organizations in rural development efforts. A primary goal of the LTC is making its research results and the knowledge it accumulates widely available to others, through extensive publications, films, and exhibits. A specialized research library is maintained on the University of Wisconsin-Madison campus. The center also receives requests for advice on a variety of problems, ranging from recommendations on policy problems to evaluations of a country's agricultural legislation. These requests are met either with one consultant or a team of consultants.

The University of Illinois, Champaign-Urbana

The University of Illinois has a history of extensive involvement in international programs, with emphasis on agriculture. In 1952, they signed an AID contract to help develop India's Allahabad Agricultural Institute which was the forerunner of the extensive involvement of land-grant colleges in building agricultural institutions in India from 1952 to 1973. Read describes Illinois' assistance to Uttar Pradesh Agricultural University and Jawaharlal Nehru Agricultural University. (46) From 1964 to 1973 assistance was provided in developing Njala University College in Sierra Leone. The University of Illinois is active as a member of MUCIA (Midwest Consortium for International Activities) in other countries. According to a recent letter from the College of Agriculture's International Program Office, the university was recognized for its outstanding international programs by the Institute for International Education and <u>Reader's Digest</u>. In December 1977, the India League of America recognized their outstanding work in establishing two agricultural universities in India.

A major University of Illinois program in international agriculture was an institution-building program with Indian agricultural universities and institutes, as part of the Council of United States Universities for Rural Development in India (CUSURDI); five other U.S. academic institutions collaborated in this effort. Staff members of the University of Illinois provided both teaching and research services in India, while also instructing Indian students at Urbana. Research supported by 211(d) grants centered on soybean diseases and their control. In 1964, soybeans were introduced in India as a potential source for human food through two counterpart Indian institutions related to the Illinois program. In addition, a system of interuniversity cooperation for handling information requests of a special character has been set up with particular overseas institutions.

A recent activity carried out by the University of Illinois was the cosponsoring of a conference in May 1978 on "Agricultural Technology for Developing Nations: Farm Modernization Alternatives for 1-10 Hectare Farms." This conference was cosponsored by the American Society of Agricultural Engineers and the Interfaith Center of Corporate Responsibility. It was funded from a grant from Deere and Company. The purpose of the conference was to discuss the problems and opportunities associated with farm mechanization alternatives in developing nations. Particular attention was given to the socioeconomic issues involved with agricultural technology transfer as well as the engineering and marketing problems associated with small farm technology.

A few of the many international agricultural activities at the University of Illinois have been outlined. However, one specific activity, the INTSOY program, is of such importance that it is deserving of separate attention.

International Soybean Program (INTSOY)

The International Soybean Program (INTSOY) at the University of Illinois is a major element of its international agricultural activities today. (47) The Brookings Institution review of U.S. technical assistance concluded: "AID experience with the 211d program since 1966 has not been successful, but several impressive models for future program development have emerged . . . . One is the INTSOY program based on the University of Illinois and the University of Puerto Rico." (48)

The University of Illinois, supported by the U.S. Agency for International Development, initiated soybean work in India that was coordinated with their agricultural university institution-building programs. The program was designed to demonstrate multidiscipline and interinstitutional research. This early work indicated that there were potentials for soybean production and for its use as human food in areas where the crop was not commonly grown. Support from both AID and the Rockefeller Foundation from July 1, 1971, to March 31, 1973, provided for demonstration and technical service activities on a broader geographical basis. Variety testing expanded to 11 countries from 1969 to 1972 and to 33 countries in 1973. This work laid the foundations for the establishment of INTSOY, with objectives similar to the international agricultural research centers. The INTSOY program officially began in July 1973. The Technical Advisory Committee of the Consultative Group for International Agricultural Research has supported the general concepts underlying INTSOY as a means of bringing developed country research, education, and development capacity to play a role in the world agricultural situation. INTSOY was designed to create an international pattern of problem-oriented institution linkages (commonly called a network). As of 1978, INTSOY work has resulted in cooperation with 105 countries, providing a direct linkage to national organizations and individuals.

Support has been provided from many sources: both the University of Illinois and the University of Puerto Rico, many individual host countries, USAID, the Rockefeller Foundation, and the combination of the United Nations Development Program (UNDP), the Food and Agricultural Organization (FAO), UNICEF, and CARE (Sri Lanka) (see Figure 3.3).

General activities undertaken with INTSOY include: (1) a publication series, (2) a newsletter, (3) regional and specialized soybean conferences, and (4) training courses. The publication series was established to provide a way of getting information to an international audience and supplementing the established journals, bulletins, etc. Between 1974 and 1978 there were 14 additional publications. The INTSOY

Agriculture

1. U.S. AID. Research Contract (AID/CM/ta-C-73-19)
    Total Funding: $980,000
    Contract Period: April 1, 1973-March 31, 1976
    Title: Development of Improved Varieties of Soybeans

2. U.S. AID. Research Contract (AID/ta-C-1294)
    Total Funding: $2,085,855
    Contract Period: April 1, 1976-March 31, 1979
    Title: Development of Improved Varieties of Soybeans and Supporting Cultural and Marketing Practices for Production in the Tropics and Information Delivery Systems

3. U.S. AID 211(d). Grant to U. of Illinois (AID/CM/ta-g-73-49)
    Total Funding: $500,000
    Grant Period: October 1, 1973-September 30, 1978
    Title: A Grant to Strengthen the Institutional Response Capabilities of the University of Illinois in Improvement of Soybeans for Tropical and Subtropical Areas

4. U.S. AID 211(d). Grant to University of Puerto Rico (AID/CM/ta-g-73049)
    Total Funding: $500,000
    Grant Period: October 1, 1973-September 30, 1978
    Title: A Grant to Strengthen the Institutional Response Capabilities of the University of Puerto Rico in Improvement of Soybeans for Tropical and Subtropical Areas

5. Rockefeller Grant
    Total Funding: $100,000
    Grant Period: August 1, 1974-July 31, 1975
    Purpose: To support tropical soybean production research centered at Puerto Rico including seed storage under tropical conditions

6. Food and Agricultural Organization of the United Nations, UNDP, and UNICEF/CARE
    FAO/UNDP Funding: $862,000
    Period: 1975-1978
    UNICEF/CARE Funding: $454,000
    Period: Mid-1977-1978
    Purpose: Soybean development in Sri Lanka

7. Soybean and Maize Development in Peru
    AID Funding: $652,000
    Period: November 1977-October 1979

Fig. 3.3. Major contracts for INTSOY program.

SOURCE: "A Brief Outline Summary of Progress 1973-1978: The International Soybean Program, INTSOY."

Newsletter is intended to provide communication among those interested in soybeans, on a world scale. It was first issued in August 1974 and is released quarterly. The first mailing was to about 500 people. Requests for names to be added have resulted in expansion of the mailing list to over 1,600.

Three regional soybean conferences have been held in Puerto Rico, Ethiopia, and Thailand. These conferences provided a means of encouraging exchange of ideas and information among soybean workers across national lines. The combination of the publication series, INTSOY Newsletter, and conferences has been a means of bringing national organizations and individuals into the international soybean network.

Two nondegree training courses were established in 1975 under the USDA/AID training program format.

1. Soybean Processing for Food Uses – five-week course. By early 1978, 47 trainees from 22 countries had participated.

2. Technical and Economic Aspects of Soybean Production – 18-week course. By early 1978, 46 trainees from 28 countries had participated.

These courses have resulted in educating soybean workers with whom INTSOY is now cooperating in country programs. Special training courses have been conducted for groups from Peru and Iran. Training courses on soybean production and use were held in Ecuador (1976) and in Peru (1977). Plans are underway for offering more specialized courses on a regional or national basis. These courses are supplementary to degree work available at the University of Illinois, University of Puerto Rico, and many other universities with soybean expertise.

INTSOY carries out many individual-country programs, supported by the individual countries and their institutions, AID, and by international organizations within the United Nations family such as UNDP, FAO, UNICEF, and CARE. Finally, INTSOY has developed administrative mechanisms for cooperation with national and international organizations. One effective means is the general Memorandum of Understanding with Letters of Agreement to provide for specific projects and activities. Memoranda of Understanding have been completed with organizations such as: the International Institute of Tropical Agriculture (IITA) in Nigeria, the Asian Vegetable Research and Development Center (AVRDC) in Taiwan, and the Fundacao Instituto Agonomico do Parana (IAPAR) in Brazil.

Agriculture

Michigan State University

The College of Agriculture and Natural Resources at Michigan State University has extensive involvements in international agricultural development activities. A report summarizing the highlights of the college's international activities during the period of June 1977-June 1978 lists a wide variety of involvements in agricultural development by faculty and students. (49)

According to this report, the College of Agriculture and Natural Resources has: established or revised agriculture courses with international content; provided foreign students with academic training; established and/or strengthened linkages with U.S. and foreign institutions and organizations; published works oriented toward international subjects; fostered faculty and student seminars, workshops, and other similar activities on topics of international interest; and engaged in overseas projects oriented toward technical assistance, research institution building, and student study and exchange.

In the area of training, MSU's College of Agriculture and Natural Resources reported almost 300 foreign students representing 59 countries enrolled in the college during the time period between June 1977 and June 1978; almost all (93 percent) were graduate students. In addition, over 200 foreign researchers, educators, government officials, agribusiness persons, and farmers visited with faculty and staff, both on and off campus in Michigan.

During this same time period, according to the summary report, "...the College of [Agriculture and Natural Resources] was involved in 24 international technical assistance, institution development and/or research projects supported by AID; the recipient countries, themselves; ACTION; FAO; the Michigan Partners of the Americas; Heifer Project International; USDA; NSF (Mexico and Washington), and others." (50) An example of one of these projects was a study of cattle nutrition in the tropics undertaken as a research project by the department of dairy science of MSU's College of Agriculture and Natural Resources. The research was carried out in Mexico and was financed by NSF/Washington and its counterpart group in Mexico City and the Mexican National Institute of Livestock Research. Figure 3.4 provides a brief summary of selected international projects involving the College of Agriculture and Natural Resources from June 1977 to June 1978.

There are many other cases of U.S. agricultural university involvement that could have been chosen (Illinois and Wisconsin cited above are fairly accessible to Washington University). For example, Purdue University in March 1978 published volume 1, number 1 of their International Education and Research Newsletter, reflecting

a) Livestock Feeds Research Project; Belize; Departments of Animal Husbandry, Poultry Science (MSU), and colleagues from the University of Minnesota; financially supported by Michigan and National Partners of the Americas, Heifer Project International, MUCIA, local government agencies, CARE, and others.

b) Michigan Partners of the Americas; Dominican Republic and Belize; technical assistance, training, and exchange; Cooperative Extension Service (mostly 4-H) and animal science departments at MSU; supported by the Partners, Michigan, and Belizan institutions and organizations.

c) MUCIA-Indonesia; interdisciplinary; institutional development; AID funding.

d) MSU Brazil-MEC; interdisciplinary; support to graduate education; institution building; supported by AID loan to Brazil and government of Brazil.

e) Peace Corps Intern Program; Philippines, Thailand, Malaysia, and Nepal; financed by ACTION- Peace Corps.

f) MUCIA-Somalia; Department of Crop and Soil Sciences; research and training; financed by FAO.

g) Better Understanding of Poor Rural Households-technical change and income distribution in LDCs; research and technical services; oriented to West Africa; Department of Agricultural Economics; AID financed.

h) Female Participation in West Africa; research; Department of Agricultural Economics; AID financed.

i) Sahel Secretariat and Documentation Center; Department of Agricultural Economics; consulting services, applied research, seminars, translation of research results; AID financed.

j) Master's Degree Training for 26 Students from Sahel, Africa; Department of Agricultural Economics; AID financed.

k) Agricultural Economics Services; African Sahel, Africa; Department of Agricultural Economics; AID financed.

l) Integrated Rural Development, Upper Volta/ORD; technical assistance; Department of Agricultural Economics; AID financed.

m) Agricultural Economic Analysis, Niger: design of agricultural research program as part of Niger's National Institute of Agricultural Research; Department of Agricultural Economics; AID financed.

Fig. 3.4. Selected international projects involving the College of Agriculture and Natural Resources of Michigan State University (June 1977-June 1978).

SOURCE: Highlights-International Activities of College of Agriculture and Natural Resources, June 1977-June 1978, Michigan State University.

their renewed interest in international involvements. The newsletter describes a variety of activities including: studies of the economics of food production and small-scale irrigation in Central West Africa funded by AID; a report by the University Senate Committee on International Education Programs defining future directions for research, education, training, and administration; the establishment of a new Peace Corps office; programs for study abroad in Mexico; the work of Purdue's Laboratory for Remote Sensing (LARS) in using earth observation satellite data for analyzing global resource problems; a program of assistance to Brazil's National Agricultural Research Service concerned with energy uses and costs in agriculture; and a news item about the appointment of Dr. Woods Thomas, former Purdue director of International Agriculture Programs, to be executive director of the Board for International Food and Agricultural Development (BIFAD).

Cornell University has been one of the U.S. universities most actively involved in international agriculture. J.F. Metz, director of International Agriculture, described a variety of their programs in international agricultural development, (51) including: (A) Cornell's collaboration over a period of two decades with the College of Agriculture at the University of the Philippines, Los Baños; (52) (B) work with the International Potato Center in Lima, Peru; and (C) research on highly weathered soils of the tropics.

Recently an ad hoc committee of the National Association of State Universities and Land-Grant Colleges surveyed some of their members for information on successful experiences in international agriculture. As of 1978, this information was not generally available.

EVALUATION OF U.S. UNIVERSITY INVOLVEMENTS

The preceding sections presented an overview of the involvement of U.S. universities in agriculturally related international programs over the past 25 years. Any attempt at making an analysis of this past involvement is made difficult by the general scarcity of independent case studies or evaluations. Lack of time and resources did not permit obtaining views of participants from both ends of such programs. Furthermore, whether or not an individual project or major program can be considered a success depends heavily on the attitudes of the evaluator, unless criteria are carefully defined. Nevertheless, this chapter reviews some of the past evaluations that have been made regarding university involvements in development efforts. Most comments are of a general nature which assess broad programs rather than one individual university project, and while some do not deal specifically with agricultural universities or agricultural projects, all of the evaluations presented are clearly relevant to the agricultural experience.

Institute of Research on Overseas Programs

An early study which did attempt to evaluate case studies independently was carried out under the direction of the Institute of Research on Overseas Programs (IROP), and funded by the Carnegie Corporation. One result of this study was a collection and analysis of some case studies of programs in which U.S. universities were involved in international programs. Richard N. Adams and Charles C. Cumberland, two social scientists, published <u>United States University Cooperation in Latin America</u> (1960) which reviewed 13 university programs in Mexico, Peru, Bolivia and Chile. Two of the case studies dealt with agricultural programs. (53)

The first study examined the project in agriculture of the University of California and the Direccion Tecnica Interamericana Cooperative de Agricultura of Chile (DTICA), which was broadly conceived to aid Chilean agriculture in instruction, research, and extension, through advice to Ministry of Agriculture agencies and Chilean colleges of agriculture. Adams and Cumberland identify the major problem in this early program as a misunderstanding between the University of California and DTICA as to what the objectives of the project were. Despite some basic problems in the program, Cumberland and Adams cited achievements in teaching, research, and extension. Although the achievements are quantitatively reported, there is no analysis of how much of the original contract was accomplished, nor were assessments of long-term impact estimated.

The second case study examined a program between Texas A&M and the Escuela Superior de Agricultura "Antonio Narro" in Mexico. Texas A&M was to assist the School of Agriculture in improving teaching methods and curricula, and was to aid in the organization, administration, and development of research, extension, and demonstration programs. Cumberland and Adams report that this project encountered more grave problems than any other program studied. The many and varied problems reported, with blame placed both on Mexico and the U.S., were of such gravity and such diversity that the program, whose total expenditures reached $94,047 for the two-year period, never actually got underway. Concrete achievements, which included some basic planning, were very few. The entire project was finally canceled at the request of the U.S.

Another study undertaken by IROP was <u>The World Role of Universities</u> written by Edward W. Weidner and published in 1962, which outlined the main findings of the IROP project as a whole and assessed their theoretical significance and policy implications. (54) Weidner describes the basic activities of most programs as that of administration, teaching, consulting, or doing research. The two most common activities were teaching and consulting by Americans overseas. Three kinds of teaching were most frequent: (1) direct teaching while acting as a substitute teacher; (2) direct teaching in demonstration class; and (3) teaching of intermediary groups such as teachers or the teacher of teachers. Priority was given to the latter two; however, according to Weidner, a major problem was a lack of counterparts to train in the field or no personnel who would watch demonstrations. Weidner concludes that the results were, "rather disappointing, particularly in those programs where nearly all teaching in a new field has been carried on by American Professors. . . . Building of institutions has not progressed rapidly." (55)

In demonstration teaching, two main aspects are subject matter and teaching techniques. Weidner concludes strongly that in subject matter, the practices demonstrated by Americans were inappropriate to the host-country conditions. Most courses were American practices without any modification. A case was cited in Bolivia where students complained that courses taught by Americans had too much content that did not apply to the Bolivian situation.

For the most part, though, Weidner found that teaching techniques were modified from those found in American colleges because local conditions, such as language barriers and differences in values, required changes. In particular, agriculture professors encountered great resistance by the local educated elite who did not want to "condescend to engage in such work" as field work and laboratory methods. In a third or more of the programs, Weidner found American professors became involved in teaching and other activities before they could seriously examine the suitability of the content and teaching techniques in their courses.

All the technical assistance contracts provided for advice and consultation by the American university personnel to host-country students, college professors, and administrators and on numerous occasions to host-government agencies such as the ministries of agriculture. One of the more frequent recommendations made was that the host institution adopt "a land-grant philosophy" - i.e., application of knowledge in teaching, research, and extension activities. Americans recommended that the extension and experiment station functions should be placed under the agricultural university's jurisdiction. Weidner found that this recommendation was almost uniformly rejected and that American university personnel usually did not come up with suitable alternatives. Weidner cites this situation as a good example of American professors having only superficial knowledge of host-country practices which inevitably led to unsuitable recommendations.

Other important activities found by Weidner in these early projects included research and obtaining educational equipment and books. Research was aimed primarily at the preparation of teaching materials or the collection of documentation and statistics. Requests for educational equipment and books were more frequent than for any other kinds of assistance. An interesting finding by Weidner was that the " . . . reluctance of financing organizations to underwrite costs of equipment without American faculty members to advise as to its use has led some host institutions to request, reluctantly, visiting American faculty members." (56) Weidner criticizes projects in which equipment bought was merely duplicative of the kind of equipment found in American universities, rather than being related to the proposed work project. Concluding his analysis, Weidner says that, "though the activities of staff members have been primarily teaching and consultation, the underlying role has been one of being agents of change." (57)

Assessment of these early pioneer programs led those involved to conclude that institutional deficiencies were a very large barrier to overcome and greatly hindered any significant large-scale beneficial results to the host country. The domestic experience of the U.S. agricultural colleges was gained through a unique concurrent rise of national commitment to rural life and agricultural innovation. Understandably, as the U.S. agricultural colleges initially planned and executed foreign aid projects, they assumed their own long experience could in some measure be transferred abroad. But, as Weidner points out, the direct transfer of the U.S. agricultural experience itself was far from successful either in reforming agricultural practices around the world or in increasing total agricultural output to any significant extent.

Education and World Affairs

Established in 1962 as a private, nonprofit, educational organization, Education and World Affairs (EWA) was concerned chiefly with the activities of American colleges and universities in international relations. Its principal purpose was to study, analyze, and make recommendations about those activities. EWA drew its basic support from grants of the Ford Foundation and the Carnegie Corporation.

Agriculture

In 1964, EWA, in cooperation with AID, published its first report, AID and the Universities. (58) In it John W. Gardner, then president of the Carnegie Corporation of New York, conducted a study of the whole pattern of relationships between AID and the university community in the United States. While acknowledging a "record of solid accomplishment that reflects great credit on both sides of the partnership," (59) Gardner deliberately focused on problem areas that had hindered the achievements of joint AID-university activities. The report was based on seven months of study; comments were obtained, orally and in writing, from many hundreds of individuals who had first-hand experience with AID-university relations. At the time Gardner undertook this investigation, university activities had already been extensive. As of December 31, 1963, 72 universities in the United States had performed technical assistance tasks under 129 separate contracts with AID or its predecessor agencies. More than $158 million was involved in those contracts.

The report is divided into seven areas: (1) the university role in technical assistance, (2) the AID-university relationship, (3) participant training, (4) research, (5) university contracts and contract administration, (6) personnel and training, and (7) organization. The study supported the concept of using universities as an outside resource for the government in meeting its technical assistance goals but cautioned that:

> (1) The Federal agency involved must have a nucleus of first class people capable of dealing with outside individuals and institutions on terms of professional equality . . . .(2) [T]he relationship between government and the university must be defined in such a way as to preserve to each party independence of action in those functions that it must perform unimpeded. (3) The relationship must be such that each party not only can perform at its best but can gain added strength from its participation. (60)

Gardner suggests that university involvement must be based primarily on long-term goals. The study found that many of the short-term technical assistance tasks carried out by the universities were "not only impaired but rendered meaningless by such a shortened time perspective." (61) Gardner concludes that "With minor exceptions, the university must address itself to the achievement of long-term purposes: educational growth and human resource development, the advancement of knowledge, and the application of knowledge to basic problems." (62)

In looking at the AID-university partnership, Gardner suggests that the relationship was somewhat troublesome and irritating to both sides; he describes the nature of the complaints:

The universities say that AID lags far behind other agencies, such as the National Science Foundation and the Office of Naval Research (to name only two) in its understanding of the universities. They say that AID doesn't grasp the nature and purpose of universities, doesn't know how to use them wisely, doesn't allow them to make the distinctive contribution that only they can make. If AID really understood these things, say the universities, the Agency would take a more generous view of the research component in contracts; would not devise and administer contracts so rigid and detailed as to frustrate the purposes they are designed to further; would take a more generous view of the kind of contract provisions that would strengthen the university itself, and would not insist on measuring contract performance by externals and expecting precise evidence of short-term accomplishment.

AID responds that the universities make no attempt to understand its problems - its constant need to justify its actions to Congress, its inescapable responsibility for program decisions, its accountability to the taxpayer. It points out that universities have often acted irresponsibly - sending third-rate personnel overseas, neglecting the needs of the host country while they concentrate on what *they* want to do, engaging in aggressive tactics to get contracts, taking on tasks they are not equipped to do well, failing to put the full weight and resources of the university behind a contract and so on. Some AID officials add that no United States university ever willingly terminated a contract program, no matter how valid the reasons for doing so. (63)

Among the problems discovered was a "perhaps unresolvable" difference of views between AID officials and university people on the degree of autonomy enjoyed by the university in the field. Another problem was a lack of commitment by the university as a whole in many of the overseas contracts. Gardner recommends that universities recognize overseas activity as, "an integral part of university life and work." (64) If a university fails to meet this responsibility, the study strongly recommends that it get out of overseas activity entirely. In addition, a warning was made that some U.S. universities had their home-base resources weakened as a result of their activity abroad. It was recommended that action be taken, both by AID and the universities themselves, to strengthen the U.S. institutions.

Participant training - the bringing of individuals from developing countries to the U.S. for training - was also reviewed. To begin with the study found a lack of consensus among (A) the participant, (B) the U.S. university, (C) AID, and (D) the host government as to the rationale and specific objectives for the training program. Various problems were highlighted such as lack of coordination on campus and a lack of extracurricular services for the participant but the "most frequently noted deficiency of present university participant programs is that trainees too often receive training that is only partly relevant to their needs on returning home." (65) The study cited as the reasons for this deficiency, the difficulty and expense of providing special training programs and the failure of the university to understand a participant's needs. While acknowledging that an AID participant costs more than the U.S. student to train, the report stressed that the university must accept the responsibility to tailor special programs and render a multitude of services to AID participants.

The study strongly suggested that more research on development was needed. It felt that not enough high level support was given to AID's program of research contracts and grants to universities. Research should be both applied and basic - even research that does not have immediate, visible utility should be supported. In particular, the study endorsed basic research on social processes. In looking at the universities, the report found that more "first-class" minds needed to be involved in international development research. In addition, they found that "one of the characteristics of much research on development is that it is interdisciplinary in nature, and in a good many academic departments the resistance to such boundary-crossing is formidable." (66) The universities were found not to have faced up to their responsibilities of making concerted efforts to bring other sources of funds, governmental and nongovernmental, into the productive support of international development research.

Regarding university contracts and contract administration, it was found that the selection of contractors was often haphazard and based on chance encounters. No adequate philosophy or strategy governing the choices was found. The agency tended to award contracts to a small cluster of the "most obvious institutions . . . As of December 31, 1963 there were 72 universities with 129 contracts. Of these universities, 7 had a total of 37 of the contracts!" (67) Conflicts resulting from contract administration were found to be the most persistent irritants in the AID-university relationships.

> Universities accuse AID of undue rigidity, incomprehensible delays, unsympathetic attitudes, and excessive and costly emphasis on small details. AID points out that universities have at times behaved irresponsibly and with little recognition of the requirements of accountability under which a government agency must function. (68)

The study pointed out that many of the problems stemmed from a "confused definition of appropriate roles and purposes, incomplete commitment of host country leaders, short-run view of the aid program, budgtary and organization difficulties of the agency, and so on." (69) The study also raised the issue of conflict of interest considerations ruling out a potential contractor if he had already done the precontract survey or feasibility study. It found that there were "great advantages" to the contracting university doing the precontract studies.

The study concluded that there was a severe shortage of people qualified for development work. The leading contributor to this situation was a lack of adequate career opportunities. Professionals who accepted two-year assignments abroad had no assurances that opportunities would be available at the end of the term.

In regard to organizational problems, the study

> yielded conclusive evidence that AID's present organizational arrangements for dealing with educational and human resource development [are] far from satisfactory, and in particular its machinery for dealing with the universities and other nongovernmental groups is inadequate. (79)

The study went on to recommend a new unit within AID to deal directly with universities, foundations, research institutions, and other professional organizations. The report suggested that this new unit be a mechanism for providing technical-professional program reinforcement to the regional bureaus. The unit would assist in feasibility studies, in locating contractors, in periodic reviews of the contract program, and would promote and conduct research on educational and human resources development throughout the world. It would also be responsible for assisting universities in strengthening their work in the development field. The report emphasized the importance of the new unit having adequate personnel strength (both numbers and quality) and of enjoying the full support of the administrator.

Finally, the Gardner report ends with a proposal for the creation of a "semiautonomous government institute to handle certain aspects of technical assistance, particularly those aspects dealt with by universities." (71) The report suggested that the new organization might be called the National Institute for Educational and Technical Cooperation (NIETC). It would be a separate corporate entity under its own board of trustees and would have an independent budget, but would ultimately be responsible to the AID administrator.

Agriculture

The report makes the following suggestions as to the institute's functions:

1. It would take over all of the basic and most of the applied research on the development process . . . .

2. It would use both grants and contracts to strengthen the international capabilities of the universities. . . .

3. It would develop intimate knowledge of the institutional and trained manpower resources in the United States for overseas work in educational and human resources development . . . . It would explore new means of mobilizing talent, such as the university consortium.

4. Because of the expert knowledge the Institute would develop in pursuing the above functions, it might often be turned to by the [AID] regional bureaus [and the new staff unit] to help select a contractor or do a feasibility study. In addition to such service, the Institute would have its own funds to write contracts with universities and other organizations for long term projects in educational and human resource development. It would, of course, be in close touch with regional bureaus and the new staff unit in AID, so there would be ample opportunity to coordinate programs, but the Institute would be free to act on its own. (72)

In 1965 EWA conducted another study which resulted in the publication of The University Looks Abroad: Approaches to World Affairs at Six American Universities. (73) Six EWA teams conducted in-depth case studies of universities with substantial, long-standing involvement in world affairs. The principal information-gathering technique for these case studies was intensive campus visits of several days by three or four members of the EWA staff. The universities studied were: Stanford, Michigan State, Tulane, Wisconsin, Cornell, and Indiana. In selecting these universities, an attempt was made to find a reasonable diversity in the type of institution, size, geographical location, and character and stage of development in the international field. The major issues and problems identified through the six case studies were then analyzed and discussed by this EWA report.

The first point raised in the concluding chapter was that the international dimension of an educational institution, if it is to be meaningful, requires long-range planning and assessment and reassessment of the institution's goals and objectives. It is stressed that an effective program, above all, requires commitment and leadership from its highest levels, "at least from president, trustees, deans and key faculty. . . ." (74) This leadership role is to make it clear both to the university community and to the public that the international dimension is a permanent, integral part of the university's total educational mission.

On the other hand, the study found that some universities were making such large international commitments that they were spreading their available resources too thin in view of U.S. requirements. Evidence of individual overcommitment was also found. A new breed of faculty was discussed - the "jet set professor," described as "one who is on so many panels, has so many consultancies and administers so many contracts that a student can only talk to him on the way to the airport." (75) It was concluded that in universities where existing staff was not supplemented for university international activities, university teaching was diluted. The study also saw a need for national overall feedback in which:

> . . . all university technical assistance programs abroad - the unsuccessful as well as the successful - are systematically studied and evaluated. The foreign aid files in Washington and at key universities are bulging with field reports, end-of-project surveys, and other materials brought back from projects abroad. The time is long past for a thorough analysis and assessment of what the university contract system has and has not accomplished, both toward overseas development goals and toward the academic development of American institutions. (76)

The investigators found a multitude of problems regarding foreign students who received educational training in U.S. institutions. The principal issue raised dealt with the increasingly large number of foreign students in the U.S.; in 1965 there were about 85,000. Many universities were considering quotas. It was not clear what impact this large number of students had on U.S. institutions. It was clear, though, that the students who came to the U.S. were experiencing many problems, including: 1) the majority of the students were chronically hard-pressed financially; 2) few universities had very coherent plans for relating their interests to foreign student needs or competencies; 3) in some cases, foreign students were admitted in a haphazard manner; 4) many foreign students arrived with academic deficiencies sufficient to hinder successful progress in their education (an insufficient command of English also hindered educational development); 5) a high rate of "nonreturnees" was found; and 6) perhaps most seriously, the study raised the question of whether the academic programs prescribed for foreign students would in fact prepare them for their roles back in their home culture. A related question was whether the educational and manpower needs of the students' home countries were given careful consideration.

In technical assistance projects, the study found that faculty members working overseas, although competent in their respective academic fields, were not effective when engaged in direct managerial operations that were only vaguely and remotely related to teaching and research. The study also found a tendency, on the part of the universities, to assume that American principles and operating experiences could be transferred overseas wholesale. It was found that universities brought a "'made-in-America' solution to problems that simply do not respond to this appraoch." (77)

In general the overseas contract operations of the AID-supported university were found to be good but some problems also existed. The study's main criticism was that AID awarded contracts to the same universities over and over again. It was concluded that this placed an impossible burden on the few institutions involved and also overlooked the possible contributions of many other institutions. The study cites an example of the difficulties encountered in entering overseas service by a highly qualified and motivated faculty member of a small liberal arts college. Another problem was that very little was being done by universities to adequately prepare personnel for overseas assignments. Not one of the universities visited operated a systematic program of orientation for educational service abroad. Faculty resistance was found to be a major obstacle to collaboration among U.S. universities. Even among established consortia, very little real collaboration was found to emerge from their efforts.

Finally, the EWA teams judged one of the most serious of the unresolved problems to be a communication gap "within universities, among universities, and between the universities and the world outside." (78) It was found that American educational institutions generally lacked intimate knowledge of their counterparts abroad.

Two additional EWA studies, <u>The Professional School and World Affairs: Report of the Task Force on Agriculture and Engineering</u>, (79) and <u>U.S. Universities: Their Role in AID-Financed Assistance Overseas</u>, (80) were also reviewed. We found these studies to be relevant to U.S. universities in international agriculture as well. Both of these studies made similar observations and came to many of the same conclusions as the earlier EWA studies regarding the role of U.S. universities in international agricultural development. Some of the common areas of concern were that (1) there was a lack of coordinated planning between all parties involved, (2) the universities moved too quickly without adequate preparation, and (3) the U.S. universities should develop their own capacity to work in development.

CIC-AID Rural Development Research Project

In 1968, the International Rural Development Subcommittee of the National Association of State Universities and Land-Grant Colleges joined with AID in sponsoring an "Analytical Study of AID University Contract Projects in Agricultural Education and Research." The Committee on Institutional Cooperation (CIC) agreed to undertake the study, which lasted three years. Thirty-five senior staff members of nine land-grant universities were engaged in various aspects of a broad study of factors affecting the success of the AID-supported university projects. These nine land-grant institutions had extensive experience in providing overseas technical assistance of the type to be studied. The general purpose of the project was to analyze the accomplishments under contracts between AID and U.S. land-grant universities for technical assistance in institution building in agriculture and related fields in developing countries. Each of the cooperating universities completed a portion of the study; a final summary report which presented the principal findings was also prepared.

The final report, <u>Building Institutions to Serve Agriculture</u>, concluded that although there were wide variations in the effectiveness of various projects, the university contract program in agriculture made important contributions abroad to building institutions, at comparatively small cost to the United States in money, manpower, and in interruptions of domestic programs. The report suggests that the overall past record demonstrated that the use of U.S. university teams to assist a less developed nation build an institution to serve agriculture could be productive abroad and well managed at home. (81)

Benefits to the host country were assessed largely in terms of institutional change, i.e., the extent to which institutions in the developing countries became problem-solving, service-oriented institutions for agricultural education and research. In addition, the study team assessed the changes in physical facilities, staff, study body, programs in teaching, research and extension, attitudes, and the integration of these institutions with society.

The study found that in almost every technical assistance project there was a marked improvement in physical facilities during the life of the contract. New buildings were built; classroom, laboratory, library, and office space was increased; library holdings were increased; laboratory equipment was modernized; and experimental farms were improved and expanded. These improvements were, in large part, funded by the host country; the study suggests that the technical assistance activities stimulated the host institution to commit its own resources to improved physical facilities, while the U.S. made commodity inputs of those items requiring scarce foreign exchange. These items included: books, laboratory equipment, and

Agriculture   113

field equipment made in the U.S. It was further noted that the professional and technical capability of the host institution's staff and the orientation and understanding of LDC agricultural leaders were vastly improved in practically every project. In general, the assisted universities experienced large increases in student enrollments and in the numbers of graduates at the bachelor's level during the period of the contracts.

Finally, the study found that eight associated universities had permanently changed their basic attitudes and philosophy regarding their purposes, responsibilities, and programs. Substantial progress in this direction was also evident in other universities. Thus, the study teams conclude that the overall progress in building institutions had been encouraging, even though slow.

It was reported that most of the host institutions made changes in teaching methods, examination procedures, and faculty-student relations which indicated changes in institutional attitudes on educational matters. These changes were usually accompanied by more practical laboratory work and better library use. In addition, few of the host institutions established effective linkages with the secondary schools from which the students came, but university relationships with the government agencies, which employed most of the graduates, improved slightly. The investigation concluded that technical assistance was valuable in increasing the research capability of the host institutions and in focusing their research efforts on practical problems of importance to LDC agriculture. They report finding examples where the results of research projects were being put to use by the farmers and where the ministry of agriculture showed interest in supporting such activities at the university.

Using U.S. land-grant colleges as models, the U.S. teams often recommended that the LDC universities carry out extension and public service activities. In many developing countries, the government itself, usually through its ministry of agriculture, had traditionally carried out these activities. The study found that in most of the projects, U.S. university personnel experienced great difficulty in reaching an agreement with host-country government ministries of agriculture as to the appropriate role of the host university in this regard. Correspondingly, progress in extension was found to be the least evident result of U.S. university efforts.

The report's overall assessment of the effectiveness of U.S. universities in technical assistance projects indicated that when the activities "are in operation long enough a very worthwhile contribution [can be] made to the building of indigenous agricultural institutions, but not all have achieved such success, and few progressed as rapidly as they might have." (82) In evaluating the effectiveness of these programs, the report points out that the host-government commitment and host-institution leadership were of critical importance. "Many poor records are due to local conditions and many of the successes owe much to strong host country commitment and leadership." (83)

Another important objective of this CIC-AID investigation was to analyze the working relationship between AID and the universities. John M. Richardson, Jr., of the University of Minnesota, divided the time span, 1949-1965, into six different periods and made the following capsule descriptions.

1. Genesis (January 1949-Summer 1953). During this period, the University contract program was conceived and the first projects were initiated. A number of government agencies were associated with project administration: The Technical Cooperation Administration (TCA), Economic Cooperation Administration (ECA), Mutual Security Agency (MSA), Institute of Inter-American Affairs (IIAA), and the Office of Foreign Agricultural Relations (OFAR) of the Department of Agriculture (USDA). Projects were exclusively oriented toward agriculture and rural development. The program was experimental in nature and extremely limited in scope.

2. Proliferation. (Summer 1953-July 1955). The period of proliferation coincides with the tour of Harold E. Stassen as administrator of foreign assistance. During this period, responsibility for the administration of foreign assistance was consolidated under a single agency, the Foreign Operations Administration (FAO). The university contract program expanded rapidly, due largely to Stassen's policies.

3. Retrenchment. (July 1955-September 1957). The period of retrenchment gets its name from the philosophy and policies of John B. Hollister, who served as director of the reorganized agency now called International Cooperation Administration. Between July 1955 and September 1957 the expansion of university participation was halted as Director Hollister attempted to cut back the foreign assistance program and establish more uniform financial, legal, and administrative procedures. There was a major deterioration in agency-university relations.

4. Inertia (September 1957-Summer 1961). During this period, there were four administrators, but no major organizational or policy changes in the agency. Numerous conferences were held between high ranking agency and university officials concerning the university contract program. However, there

was little discernible improvement in agency-university relations at the level of the participating universities.

5. <u>Interregnum</u> (Summer 1961-December 1962). We also considered calling this the period of confusion. The arrival of the "New Frontier" in Washington resulted in major organizational and personnel changes in the agency. The short-term consequences of these changes for agency-university relations were, to quote one observer, "chaotic."

6. <u>Harmony</u> (December 1962-?) The appointment of David Bell as administrator of AID marked the beginning of a period of harmony in agency-university relations. During Bell's tenure, university participation was expanded and diversified. A major study and conference focusing on the university contract program increased mutual understanding between agency and university representatives. The most salient characteristic of this period, however, was the administrative style of Mr. Bell himself. Although we conclude our history in the spring of 1965, the period of harmony does not, as yet, appear to have come to an end. (84)

The CIC-AID summary report comes to the following overall conclusions:

AID should continue to move in the direction of cooperative partnership with the universities in a flexible, long-term program which is guided by carefully conceived goals, but adaptive to differing human and environmental conditions. The changes in the program which were recommended by the Gardner Report and the Conference on International Rural Development and endorsed by AID officials and the university community have not yet been fully realized. Moreover, evidence gathered by this project indicates that in some areas these important documents did not go far enough in their proposals for remedial action. Our findings also indicate that the universities should realistically examine their capabilities to participate in long-term programs of international technical assistance. It is clear that at the departmental level especially, the costs of such participation have often exceeded the benefits.

Finally, and perhaps most important, it is essential that the level of public understanding about technical assistance programs be improved. The educational role which the universities can perform in this area is of crucial importance. Bridging the gap between affluence and starvation is both consistent with the national interest of the United States and an essential and necessary component of national policy. Certainly the U.S. will have technical assistance programs for many years to come, although the form and character may change.

We must strive continually to make these programs more effective, but at the same time they must not be jeopardized by being aimed at unrealistic expectations. We must profit by experience, not only by making our programs more effective but by making our goals and expectations more realistic. (85)

The CIC-AID report goes on to make ten specific recommendations for action. Many of these recommendations have in some form played an important role in policy decisions over the last ten years and therefore warrant citing:

1. There should be a stronger commitment on the part of all participating agencies to an expanded and long-term program of building institutions to serve agriculture.

2. More flexible project agreements and improved liaison between AID and the university community would effect needed improvements in AID-university relations.

3. Research on the institution-building process should be significantly increased and existing knowledge should be utilized more effectively.

4. The basic ideas that underlie the land grant type institution are highly relevant in technical assistance projects if properly understood and employed.

5. Agreement on goals and commitment to an overall strategy by host and U.S. personnel should be strengthened by wider participation in project planning and review.

6. Those aspects of technical assistance programs which have contributed to the highly negative attitudes of many university staff members and department heads should be changed.

7. There should be fundamental changes in orientation programs in order to prepare team members adequately for their overseas assignments.

Agriculture

8. Programs of participant training should be more carefully planned and more adequately supported so that they conform to the development needs of host institutions.

9. The university community should exert its leadership in developing a fuller public understanding of international technical assistance.

10. AID and the universities should cooperate in strengthening the international capabilities of U.S. universities. (86)

Analysis

This section has discussed evaluations, carried out primarily in the 1960s, of U.S. university involvements in international development. While not all of these evaluations deal specifically with agricultural development, all of the observations, criticisms, conclusions, and recommendations are relevant to U.S. universities involved in international agricultural development.

These evaluations at times suffered from being too general, but they are valuable in that they provide a broad perspective from which to formulate opinions and policy with regard to U.S. involvement in international agriculture. Although all the evaluations conclude that some good was accomplished, it often is not made clear (A) what specifically was done, and (B) why it was good. The CIC-AID study provided more data than the others regarding accomplishments but fell short in clearly evaluating the impact of the institution-building projects on the agricultural development of a particular developing country. Furthermore, the study is probably the least independent of those we have summarized and the most subject to the "evaluators evaluating themselves" syndrome.

What is dramatically clear from reviewing these studies is that the same basic problems persisted throughout the time period covered by these evaluations. Difficulties in the AID-university relationship; lack of social, cultural, and political sensitivity by the university staff; lack of adequate planning and coordination by the funding source (most often AID), U.S. universities (most often land-grant colleges), and LDC institutions (most often host universities); lack of appropriate education for LDC participants in the U.S.; lack of long-term programs; and other problems - all disturbingly appeared over and over again. Little progress in correcting these deficiencies in the university activities seemed to have been made in the 15-20 years that these evaluations covered.

The same basic recommendations appear rehashed with each new study team, task force, or single investigator, apparently with a very small amount of, or at best slow, implementation of their respective suggestions. For example, the National Institute for Educational and Technical Cooperation, recommended by John Gardner in 1964, bears a striking similarity to the "newly" proposed Foundation for International Technological Cooperation. On the more positive side, the CIC-AID report had a significant influence on the creation of the Title XII legislation, and many of the recommendations of that report are now being incorporated into the various Title XII programs.

The seventies have seen very little of these types of evaluations that deal specifically with the U.S. universities in international agricultural development, (87) but there are indications that many of the same problems still exist and that there has still been relatively little progress made since the first evaluations were reported over 15 years ago. U.S. universities, AID, and the developing countries involved in international agricultural development activities appear to be slow learners; or else, the problems are extremely difficult - or both.

CURRENT ISSUES AND QUESTIONS

Several issues and questions emerge as being of importance in considering the future role of U.S. universities in science and technology for development in general, and agricultural development in particular. A discussion of these issues is based on: (A) the analysis of involvements and legislation presented in earlier sections of this report; (B) current thinking by individuals who, for the most part, are professionally recognized in international agricultural development and, in some cases, are people in policy-making and/or influential positions; (C) interviews with agricultural university administrators and faculty members currently involved in agricultural development; (D) views advanced by participants in a workshop on the role of U.S. universities in science and technology for development held on July 13-14, 1978 (see Appendix A); and statements presented to the annual meeting of the Association of U.S. University Directors of International Agricultural Programs (AUSUDIAP) held on June 21, 1978, in Logan, Utah. The issues are not necessarily analyzed in order of importance.

Agriculture 119

"New Directions" and U.S. University Involvement

The thrust of the "New Directions" policy is toward directly meeting the basic needs of the rural "poorest of the poor." The 1973 amendment to the Foreign Assistance Act, in part, states:

> United States bilateral development assistance should give the highest priority to undertakings submitted by host governments which directly improve the lives of the poorest of their people and their capacity to participate in the development of their countries.
>
> ... greatest emphasis shall be placed on countries and activities which effectively involve the poor in development, by expanding their access to the economy through services and institutions at the local level, increasing labor-intensive production, spreading productive investment and services out from major cities to small towns and outlying rural areas, and otherwise providing opportunities for the poor to better their lives through their own effort. (88)

Furthermore, one of the basic congressional mandates regarding implementation of Title XII specifically commits Title XII development assistance programs in food and nutrition to "meeting the basic needs of the poor majority. . . . " (89)

U.S. land-grant agricultural universities that draw upon their experiences in the U.S. may find that those experiences are of limited value in dealing with landless and small farmers in developing countries. Agricultural commercialization and increasing food production efforts often focus on the productive capability of larger agricultural enterprises with their potential for profits and increased output; emphasis on rural development and social equity focuses on smaller farmers and landless rural workers, and is concerned with their earnings and equity in the emergent, commercial agricultural system of the developing countries. Although increased commercialization in agriculture may also lead to increased benefits for small farmers and rural workers, the record thus far indicates the need for special efforts to reach these people.

Douglas Ensminger, president of the Mid-Missouri Associated Colleges and Universities and professor of rural sociology, University of Missouri, who has himself been extensively involved in agricultural assistance activities for many years, states:

> ... with few exceptions, transfers of science and technology the past thirty years have benefited the larger farmers simply because they had the resources to apply the advanced technology.

The small food crop farmers continue to follow traditional practices and function outside the institutional structures which serve the large farmers. (90)

Furthermore, Buford L. Nichols, professor in the department of pediatrics and physiology at Baylor College, questions how capable U.S. professionals are in dealing with LDC problems.

> Trained individuals from developed countries often are not capable of dealing with local problems in developing countries for the same reasons that most well trained foreigners could not design solutions for our problems - our training just does not optimally prepare us for dealing with problems in environmental conditions culturally foreign to our experience. (91)

In addition, recent literature has suggested that introduction of modern cash-crop production techniques may not have helped landless and small farmers as intended. (92) Indeed, now being questioned is whether or not the U.S.-sponsored agricultural programs have adversely affected the poorest of the poor who are the primary target group for U.S. assistance.

> Issues: Are the two objectives of encouraging growth of a commercial agriculture and of obtaining greater equity from smaller farmers and landless farmworkers compatible? What trade-off should be struck, if necessary, between ameliorating current deprivation among the rural poor on one hand, and stimulating more production and earnings for the general economy on the other? To what extent will U.S. university intervention adversely affect small, poor farmers at the expense of large farmers more receptive to innovation? Should the U.S. university be concerned or even held responsible for these issues or is this solely the concern of and controlled by the host government? To what extent will U.S. university personnel be able to contribute to the solution of local problems faced by the poorest of the poor?

Impact on Women

In Africa, in particular, women have traditionally done the farming. New farming techniques introduced by change agents such as U.S. university representatives have been primarily introduced to the men. [For a further discussion on the role of women in development, see Tinker. (93)] This change process, which tends to eliminate many of the roles that women have traditionally held without developing new roles for them to fill, seems to have effectively worsened the status of women in these countries.

The major finding of a report on the roles of women in rural development in seven countries done in 1974, reveals that,

> In six of the seven countries studies, women take part equally with men in basic agricultural production but, despite this fact, external development projects designed to transfer technology to rural people seldom incorporate women as participants. Furthermore women are rarely members of project planning groups. (94)

The summary report of the Conference on Women and Food, held in Tuscon, Arizona, in January 1978, described the shared perception of the conferees which provided a common ground: (1) each individual person has the right to adequate food and water, and each woman, man, and child has a stake in whether this right materializes; and (2) women, as well as men and children, are individuals in their own right, and must be provided the conditions and safeguards necessary to develop their own individuality. However, of equal importance is the fact that women are integral members of families and of societies, and they must participate in development activities within the context of their families. Women must become equal, active partners with men in the process of development. (95)

> Issues: What unique impact does agricultural development have on rural women? To what extent does U.S. university involvement influence this impact? What steps can U.S. universities take to insure a positive and equitable development process that will benefit all members of a rural, agricultural society?

Relevance of the Land-Grant University Approach

The basic premise under which U.S. universities have been involved in international agriculture programs since their earliest activities is that the U.S. land-grant model which worked successfully in helping the United States can also be of help in the LDCs. Yet, from the very beginning of this involvement, many have seriously questioned this premise and today questions continue. Dale Hathaway, assistant secretary of agriculture and former head of the International Food and Agriculture Policy Research Institute, stated in 1977:

> ... one of the major activities of American universities abroad has been institution building that attempts to transplant the Amer-

ican system of land-grant universities, agriculture experiment stations, and extension services to developing countries. In the American system, the functions of research, teaching, and adult education of farmers and related persons are combined into a single institution. This institutional structure has worked successfully for American agriculture. There is substantial question, however, whether or not it is transferable. <u>There is evidence that it is not only not transferable but also perhaps has little relevance in many developing countries with major food problems</u> [emphasis added]. I am not asserting that all American university efforts at institution building abroad in the field of agriculture have been a failure. Indeed one can point proudly to many successes.

I do suggest that the American universities' efforts at institution building have been substantially less than successful because relatively few societies have a social and economic structure where an American type of university system can survive, thrive, and produce technology for the agricultural sector as it has in the United States. (96)

In essence, this American land-grant model entails a three-pronged integrated approach: teaching, research, and extension. Transferring and/or adapting this approach to agricultural institutions has been at the heart of international agricultural programs in which U.S. universities have been active, and it appears it will continue to be in future programs.

> Issues: How applicable and directly transferable is the uniquely American land-grant approach? How much do U.S. universities investigate economic, social, political, cultural and other specific host-country characteristics before making program recommendations? To what extent are host countries allowed to define their own needs and requirements? Does this type of approach lead to a strong indigenous capability.

AID-Graduate Countries

Cooperative scientific and technical relationships between United States institutions and those in developing countries can be financed by AID, if the country falls within AID guidelines, or can be financed by the developing country if it is wealthy enough and wishes to do so. So called "AID-graduate countries" (former AID recipients with annual income/capita more than $520) cannot receive U.S. aid but in many cases are still in need of technical assistance. Many countries

Agriculture

may be somewhat developed industrially and economically, but are still very much agriculturally backward and can ill afford the full cost of collaborative relationships with U.S. universities.

Ralph Smuckler, dean of International Studies and Programs, Michigan State University, and now head of the Planning Office for the Foundation for International Technological Cooperation, maintains that a "dangerous gap is developing between our institutions and those with which we should now be moving in active and cooperative ways - the centers which are emerging essentially in the middle-income countries . . . . " (97) Smuckler illustrates his point with an example:

> There are few people who question the future importance of Brazil as a force in world affairs and as an important trading partner to the United States and other nations. As its economy has grown, it has become a "graduate" country, no longer eligible for U.S. development assistance. Over the years AID or its predecessors have supported strong relationships between Brazilian agricultural universities and those in the U.S. to the point where there are now five or six institutions in Brazil with a reasonable number of faculty members holding advanced degrees from Western Europe and the U.S., with ongoing research interests, and program and personal ties to the U.S. On June 30, 1978, the final program of the U.S. Government in support of sustained relationships between these agricultural institutions and parallel institutions in the U.S. will come to an end. In this most recent effort a cooperating group of universities in the U.S. led by Michigan State University and working with six advanced graduate centers in Brazil has provided about 200 man-months of U.S. agricultural science personnel over the past 12 months. After June 30 the number will drop to a trickle. Of course there will always be some coming and going. About 150 Brazilians are still being trained at American agricultural institutions under the program. The Brazilian Ministry of Education and Culture has a program which will continue to support Brazilian exchange with the U.S. at a comparatively modest level. But in spite of all our efforts to make use of Title XII or other reasonably progressive pieces of supporting legislation, there is at the present time no sign of a means of sustaining an active interchange, a cooperative relationship with these emerging centers of agricultural science in Brazil at a level that would be in their and our national interest. (98)

> Issues: Should mechanisms and/or legislation be developed that would allow U.S. universities to work with agriculturally developing nations who do not meet current AID guide-

lines? To what extent will AID-graduate countries be able and want to assume the financial responsibilities of scientific collaboration? To what extent will U.S. universities contribute their own funds to support effective exchange relationships with "emerging" centers?

Program Evaluation

Throughout this investigation, a recurring experience was the difficulty of obtaining independent evaluations of many university involvements. Occasionally, final progress reports or evaluations by the sponsoring agency were found, but it was very difficult to find evaluations that could be considered truly independent. This appears especially true of the last 15 years or so. Project directors and agency sponsors have biases such that their evaluations cannot be totally conclusive, although they are, to some extent, helpful and can shed considerable light on what happened. Some have suggested, though, that in-house evaluations will rarely reflect any negative aspects.

Much controversy exists as to how to evaluate programs. Universities insist that programs are complex and long range making it difficult to show short-term progress. AID, on the other hand, must justify their expenditures to Congress and the American public. Lowell H. Watts, director of Extension and Community Services, Colorado State University, suggests what he feels is the bottom line on evaluations: "To meet Congressional objectives of Title XII our future programs must stand evaluation on whether or not <u>food production is increased and whether or not human nutrition is improved</u>" [emphasis added].... (99)

Issues: Should independent evaluations be required as part of contractual agreements with U.S. universities? What is a reasonable time period for a program to exist before being evaluated? Can some evaluations be carried out during program execution? When evaluations are done, should they be made more publicly accessible?

Implementation of Title XII

Title XII provides long-term authorization for greater university involvement in all phases of technical work on agricultural development. Title XII activities have not been underway for over two and

Agriculture

one-half years; early experience with the implementation of Title XII to date suggests some problems. According to a current AID progress report (April 1978), AID-university understandings regarding the scope and implementation of Title XII have greatly matured in the past year; (100) yet informal conversations with agricultural university leaders suggest that cumbersome and time-consuming procedures still hinder program implementation.

H. F. Massey, chairman of the International Collaboration Committee of the American Society of Agronomy, in January 1978 expressed some concerns regarding implementation of Title XII.

> Implementation efforts have been slowed by jockeying for position between AID and the universities, and to some degree between the universities themselves or groups of universities. If these matters cannot be resolved rather soon, the legislation should be amended to more clearly define operating procedures which will allow the full participation of universities as was originally intended. (101)

AID, in the 1978 report on Title XII to Congress, claims that "the past year (1977-1978) was indeed one in which the spirit of partnership was given an even greater emphasis," but goes on to admit:

> Bringing the BIFAD (the Board, JRC, JCAD and staff) into a fully participatory relationship with AID in the program and policy planning, budgeting and implementation process has been one of the most complex issues confronted in the implementation of Title XII. (102)

One AID official recently remarked:

> There is a lot of opposition to [Title XII] still within AID. I think that some of the program technical people feel that this is a threat to their authority and power . . . . There are a lot of serious problems yet to be overcome with [AID] lawyers and contract officers who just can't understand contracting in a collaborative research mode rather than AID's traditional system. . . .

> The question of eligibility is still one that they're wrangling with; the question of competition, as well. How do you respond to a White House directive to insure competitive contracts when you're in bed with the universities to begin with and you've decided not to have competition but to declare certain universities eligible? The last point is that I still think we have got some serious hurdles

to overcome with our regional bureaus and field missions who feel that they haven't had enough of a say.

To date, according to the 1978 AID report, progress has been largely of an organizational nature, referred to as "tooling up" with very few actual assistance-related programs underway. Two problem-oriented collaborative research programs are finally scheduled to begin in October 1978. The next two research programs will most likely not begin until 1979.

There has also been skepticism by Congress which is anxious to see some results. A strong congressional criticism is directed at the lack of LDC input to the program. Neither of the two Title XII projects scheduled to begin in October, 1978 had much LDC input in the design stages. Congress wants to see more U.S. interaction with agricultural scientists and producers in LDCs.

Comments of the Board for International Food and Agricultural Development (BIFAD) in the AID 1978 report indicate overall satisfaction with AID's efforts to make Title XII work, although in their judgment, one major problem is that "the Administration's [AID] budget request for the Collaborative Research Support Program is inadequate." Specifically, BIFAD calculates that the administration's budget requests for $9.5 million are of "only one-half the amount needed." (103)

Other unresolved problem areas which continue to hinder full implementation of Title XII include a conflict-of-interest issue since the universities will tend to benefit from their own legislatively mandated input into the AID decision-making process and the fact that there has been difficulty achieving "an improved understanding of the objectives, rationale and scope of Title XII among the participating parties – the university community and AID." (104)

> Issues: Will full consensus be reached between AID and the universities as to the intent of Title XII? To what extent will independent evaluation of activity be carried out rather than in-house evaluation to ensure that the purposes of the act are being met? Why is it necessary to exclude a large number of colleges and universities from participation by requiring them to have _extension_ and research and teaching capability? Is it not possible to make contributions to building an indigenous S&T base in agriculture by contributing to one or two of these areas and not all three? What will be the distribution between funds spent in the U.S. and LDCs? Is Title XII a good model for similar activities by U.S. engineering and science academic institutions?

Agriculture 127

State Level Support

By and large, the American public, in particular in agricultural states, has not fully endorsed the activities of agricultural colleges involved in international programs. It is likely that a majority of U.S. citizens feel that state colleges owe primary allegiance to their states. This is most evident in state universities and land-grant colleges where some farmers feel that the international activities of state college agriculture departments are helping foreign farmers to compete with domestic farmers. A good example of this exists in Illinois where concern has arisen regarding the competition being felt from foreign soybean growers in Brazil. This public attitude is reflected, in part, in state governments where there appears to be a dearth of legislative mandates legitimizing the role of state universities in international development activity. In most cases, internal budget support from the overall university or its respective departments of agriculture for international agriculture programs is still minimal. Rewards for faculty to participate in international activities are often outweighed by disincentives. This lack of support by the state results in having to depend on federal funds for international activity which in turn creates many problems. Whitaker and Boyd Wennergren state that:

> Inadequate and uncertain funding of international technical assistance programs at U.S. agricultural universities is manifest in three principal internal conditions. These are (i) failure to recognize technical assistance to international agriculture as a legitimate university mission, (ii) inefficient internal organizations for international programs, and (iii) lack of incentives at administrative and staff member levels to support and accept technical assistance assignments in developing countries. These conditions are highly interdependent. (105)

Much of the problem lies in the original mandate of universities. T. Kelly White, director of International Education and Research, Purdue University, writes:

> The vast majority of major universities with significant research programs are state chartered and state supported. Their mission, as determined by charter and funding, is to train domestic students and conduct research to solve domestic problems. Not only is the focus on domestic students and problems, but even more frequently, it is on students and problems of the individual state. (106)

But many universities are now investigating the possibility of changing their original charters. To date, few have been successful, but the University of Minnesota, for one, has responded. The university's Mission and Policy Statement, adopted by the Board of Regents in July, 1975, states that:

> . . . the mission of the University of Minnesota is to serve the people of the state, through teaching, research and public service. <u>Beyond this is the commitment to contribute as fully as resources permit to needs both national and international.</u>
>
> To the people of Minnesota it is their University. Yet, in the broader sense, <u>it is an institution of worldwide responsibility, scope and impact, one that Minnesotans share unselfishly with others.</u> (emphasis added) (107)

Few states have similar mandates.

> Issues: Do U.S. land-grant colleges or other universities with agriculture programs have a legitimate role in international activities? What efforts should U.S. universities undertake to make citizens of their states aware of the importance of international agriculture programs? Is it reasonable to expect state legislation to allocate significant funds to the development of institutions in the poor countries of the world?

Relevance of Graduate Education and Participant Training at U.S. Universities

One of the major activities of U.S. universities in international agricultural development is the teaching of foreign students in the United States. <u>Open Doors</u> reported 3,240 foreign graduate students studying agriculture in U.S. universities during 1975. (108) These U.S.-trained men and women constitute a significant amount of potentially valuable manpower to LDCs as well as a significant portion (22 percent) of all graduate agricultural students in the U.S. Although LDCs are building their own institutions (in many cases with U.S. help) in which they can train their own students, there will very likely continue to be a need to train foreign students in the U.S. for some time to come. However, there is serious concern regarding how relevant U.S. education is to foreign student needs. As has been pointed out, U.S. education is oriented toward U.S. farming which is large

Agriculture

scale and capital intensive, whereas in most LDCs the agricultural sector consists of predominantly small-scale, labor-intensive subsistence farming. (109) In most LDCs, production is under tropical conditions, whereas in the U.S., temperate conditions prevail. In the U.S., most crops or livestock are raised under conditions of affluence and sophistication in contrast to the crude facilities and scarce resources found in LDCs. Dale Hathaway expresses some concern when he states:

> . . . one shortcoming is the nature of the applications that foreign students are asked to make. If American universities are to be serious in the training of foreign scientists for problem solving in the developing countries, the students should have in their training, classroom, and field experience more opportunities than now afforded to apply the techniques, scientific knowledge, and tools they have acquired to the kinds of problems faced in developing countries. (110)

Further:

> American universities are especially ill-equipped to deal with certain problems which arise in my own field – economics applied to policy planning – and related fields such as political science. The United States has had little experience in centralized or even decentralized public planning and management. Foreign scholars came to the United States assuming that they will learn government planning techniques and other tools related to the problems of centrally managed economic systems in which the government plays a major role. (111)

It has also been observed that few U.S. faculty can relate to an individual foreign student's specific agricultural problems in his home country. Many LDC graduate students end up doing research on problems totally unrelated to their country's needs.

In addition to actual teaching, there are extracurricular support activities that should be provided by the university to insure a positive and successful experience for foreign students. These activities include administrative activities such as providing campus and community orientation and assistance to meet the personal needs of new students, providing assistance to participants in the interpretation of AID regulations, advising the U.S. faculty and department chairmen on program objectives, conducting program evaluations and providing feedback to those involved in the training program. One source has calculated expenses incurred in operational programming, including staff input and related costs, at over $400 per student per year in almost all U.S. universities (1978). (112) Universities are now claiming that due to university budgetary restrictions, they are largely unable to continue subsidizing these functions and are looking for additional support from other financial sources.

> Issues: Should U.S. agricultural universities make their curricula more relevant to the needs of LDC students studying in the U.S.? How can this be achieved? What impact would such activities have on the domestic programs? Should a ceiling be placed on the number of foreign students admitted to a program? What can be done about "brain drain"? Who should subsidize the programming costs - AID, the university, the participant?

Rationale for U.S. University Involvement

There are some who feel that helping to build an indigenous S&T base in LDCs is injurious to U.S. interests and causes us to lose the big advantage in technological capability that we have. There are others who feel that such a mandate is self-defeating because heavy U.S. involvement precludes truly indigenous development. Still others believe that humanitarian reasons alone justify participation by the U.S. universities - especially in helping to solve world food and nutrition problems.

Other justifications for U.S. university involvement include the fact that: (A) American activities, and (B) international involvement strengthens our universities. For many, there are a variety of rationales and justifications for U.S. university involvement. W. D. Buddemeier, associate dean and director of International Agricultural Programs, University of Illinois-Champaign/Urbana, cites examples of potential benefits to our economy and standard of living resulting from international cooperative research:

> It was by work abroad that our entomologists discovered an insect predator of a serious insect pest in the United States. Continuing research abroad may pave the way to introducing the predator into this country.

> One faculty member's short period of service overseas gave impetus to the efforts of our food scientists to develop various low-cost, nutritious foods from soybeans. This project may be of incalculable importance to the market for soybeans as well as to the problem of improving human nutrition at home and abroad.

Agriculture

> Another potential benefit lies in the work of agricultural economists. Through studying problems of economic development abroad, they should gain new insights into problems of the United States.

Buddemeier goes on to cite benefits to the university.

> A benefit which cannot be easily measured but which is immensely important is the enrichment that a faculty member brings to his work on campus after he has returned from overseas service. He is better able to see problems in a world framework instead of in terms of the state or the nation only. Also, his wealth of material and experiences can be used to vitalize his teaching. (113)

Case studies were made by AID on benefits to the United States from American technical assistance activities abroad. (114) For example, one study looked at a "Food Grain Drying, Storage, Handling and Transportation" project carried out by Kansas State University. The purpose of the project was to solve problems involving the losses of food grain and food grain quality during postharvest handling, drying, and storage. These are major areas of concern for many LDCs. Throughout the project, which lasted from 1968 to 1975, Kansas State identified grain storage and marketing problems, recommended solutions, evaluated remedial programs, and designed and conducted training programs to improve grain storage and marketing practices in Korea, Ethiopia, and Panama.

According to the report, the benefits to United States interests were multiple. Initial assessment of storage and marketing conditions existing in the areas studied have provided U.S. firms an opportunity to participate in large-scale feasibility studies which in turn have identified foreign opportunities for U.S. investors, and the improved storage, marketing, and processing recommendations have resulted in a demand for U.S. equipment and supplies. In addition, research results such as development of storage structures and drying facilities for adverse tropical conditions and a projection model of operational and managerial problems for long-range projects, are being applied to U.S. situations.

Finally, it has also been suggested that international trade is improved by university efforts to develop a country's agricultural sector. Developed and emerging middle-income countries provide the best markets for agricultural commodities and manufactured goods from the United States. In addition, exports to many countries contribute heavily to our national economy and make possible positive inputs to our balance of payments.

Issues: Should U.S. universities be involved in activities that may lead to the development of potential competitors for U.S. farmers? Does this activity benefit the U.S. more than the developing countries? Will the benefits derived to both U.S. universities and for the U.S. citizenry at large from U.S. university involvement in international programs be sufficiently convincing to justify expanded involvement? Will mutual benefit become an important rationale for future U.S. university involvement?

Contribution of Science and Technology

Of central importance to this study is the issue of whether or not science and technology contribute to development, especially in meeting the basic needs of the poorest of the poor. Melvin Blase, director of International Programs at the University of Missouri, made the following comments at the Project Workshop:

> Prior to the mid-1960s there was real question about whether yield increasing technology was the appropriate avenue to pursue in trying to increase the amount of food produced in developing countries. Most developing countries traditionally had emphasized increasing the land base rather than increasing the yield per unit of land as the appropriate strategy for increasing their agricultural production. Major investments were not being made by LDCs in science and technology generation.

> But as a consequence of the breakthroughs at the international centers, particularly the short, stiff-strawed wheats, and to a lesser extent with rice, as a consequence of the work at CIMMYT and IRRI and the resulting Green Revolution, there was then a demonstration - a very clear, highly visible demonstration - that the yield-increasing technology strategy would work not only in the U.S. and other industrial countries but would work in LDCs as well. That was a very strategic thing in terms of selling the possibilities for technology generation. (115)

But after initial euphoria with this strategy, many serious questions were asked. The Green Revolution, in particular, which at first was hailed as the answer to the world's food problem has come under extensive criticism. Douglas Ensminger, who worked in India for almost 20 years, agrees with a statement in a recent U.S. News and World Report:

Agriculture

> In India and many other countries with a history of hunger, it's still the <u>weather</u>, not ... miracle ... seeds or fertilizer, <u>that determine harvest results</u>. And <u>distribution</u>, not crop size, is the key to how much people <u>get to eat</u>. (116)

Ensminger clarifies the point:

> Science and technology cannot contribute to 'overcoming poverty' until, and unless, there is a political commitment to legislate and implement land and institutional reforms [in LDCs] that will make the nations' resources more equitably available to all the people. (117)

Robert McNamara noted that for some 40 percent of the developing countries' populations, development is

> not reaching them in any decisive degree .... Their countries are growing in gross economic terms but their individual lives are stagnating in human terms. ... <u>The miracle of the Green Revolution may have arrived, but for the most part, the poor farmer has not been able to participate in it.</u> [emphasis added] (118)

A recent National Academy of Sciences study highlights the role of research in development:

> A strong research base is essential to all activities needed to increase the food supply, reduce poverty, and moderate the instability of supplies and prices. The role of research is to broaden the range of choices available to all those who affect world food supply and nutrition such as farmers, consumers, and government officials. In stressing this we do not assume that research and development will solve world food problems, only that it is an essential part of the solution. (119)

In light of specific U.S. congressional mandates to direct the benefit of U.S. aid to the poorest of the poor, this is a highly important concern. The Title XII program is expected by the U.S. Congress to contribute substantially to preventing malnutrition and starvation.

> Issues: What contribution will science and technology, research and development make directly to achieving freedom from hunger? Are science and technology neutral? Who will benefit most by the development of "better" science and

technology? How much choice of science and technology do the poor have? How much involvement and participation in development projects is there by LDC agricultural scientists and by LDC poor? Should scientists and technologists be held responsible for how their science is applied?

1890 Land-Grant Institutions

One relatively new issue concerns the 1890 land-grant institutions. In 1862, the first Morrill Act established one land-grant college in each state; by and large these were predominantly white institutions. In 1890 Congress passed the second Morrill Act establishing land-grant colleges in the southern states primarily for the predominantly black population.

It is now suggested that the 1890 land-grant institutions, as a consequence of working predominantly with disadvantaged populations here in the U.S. (primarily poor, rural blacks), have developed a unique strategy and a proven expertise for meeting the social and economic problems of limited resource people in less developed countries. T. T. Williams, director of the Unemployment-Underemployment Institute, summarizes what he feels is the uniqueness of the historically black land-grant colleges:

1. Historical commitment to, and an involvement interrelationship with, "the people left behind."

2. Over 90 years of experience in reaching and serving the needs of the disadvantaged population.

3. A comprehensive cadre of professional and scientific expertise possessed by people who emerged from the target or similar populations and thus are capable of communications in a manner that cannot be matched by persons having more sophisticated backgrounds.

4. Ethnic similarity - especially in Africa, South America, and the Caribbean Islands - facilitates social acceptance which in turn reduces skepticism and enhances communication with limited resource people.

5. Though the historically black land grant college possesses unusual commitment and ability to serve the needs of the

disadvantaged, they are professionally and scientifically equipped to provide technical assistance in the most sophisticated disciplines essential for community and human resource development. (120)

In addition, Williams summarizes the strategies of the 1890 land-grant colleges as follows:

1. Involvement of the people to be served in planning for community and human resource development.

2. Identification and utilization of indigenous leadership in program implementation.

3. Helping people to help themselves ("do with" as opposed to "do for").

4. Demonstration teaching as opposed to theoretical teaching.

5. Community Organization:

    a. Extension methods b. Family involvement
    c. Cooperatives. (121)

Mechanisms for bringing the 1890 institutions into the mainstream of international development activities are now being created, including a consortium of some 31 universities. The consortium, which is based at the North Carolina Research Triangle, is the South-East Consortium for International Development (SECID). The Title XII program is also developing ways to incorporate the 1890 institutions into international activities. For example, according to an AID report, minority institutions (the 1890 land-grant colleges and Tuskegee Institute) are not required to provide matching funds to participate in the university-strengthening program. (122) Furthermore, plans call for these institutions to participate at some time late in 1978 or early in 1979 in a workshop to discuss their role in science and technology for development as a possible input to the 1979 UNCSTED.

> Issues: To what extent is the 1890 land-grant institution relevant to African agriculture? Can the large U.S. Hispanic-American population develop special ties with Latin American and Caribbean agriculture or the Oriental-American population with Asia?

CONCLUDING REMARKS

This chapter has reviewed U.S. university involvement in science and technology relevant to agricultural development. It was found that over the past 25 years, U.S. university involvement in international agricultural development programs has been extensive. The state universities and land-grant colleges have been and continue to be today the major thrust of this involvement. Such activity in international agriculture is justified by the potential contributions that U.S. universities can make to the present world food and nutrition situation, but there are many benefits that accrue to the U.S. economy, U.S. agriculture, and U.S. universities as well.

The role of U.S. universities in international agricultural development has been significantly influenced by U.S. legislative mandates and policy trends. The U.S. government, primarily through AID, is the major source of funding for these activities. Although U.S. university assistance is an important element, it is of secondary importance to the developing country's own efforts to build and sustain an agricultural base that will meet the food and nutrition needs of its people. This relationship is becoming truer with the passage of time as the cadre of developing country agricultural scientists becomes increasingly more capable of carrying out many S&T activities.

There are a significant number of cases in which U.S. universities and U.S. university personnel appear to have contributed to building an indigenous developing country S&T base that could be utilized in agricultural development; yet the success or failure of the technology transfer process in aiding the poorest of the poor appears to depend heavily on the social, political, and economic infrastructure within the developing country. Other critical factors in achieving success or failure of a particular involvement are: (1) the selection of U.S. university personnel; (2) adequate planning by all actors involved in the activity; (3) the commitment of the U.S. university institutions to support their international agricultural programs; and (4) the interdisciplinary nature of the projects.

Finally, it is concluded that U.S. universities are a valuable resource that should be increasingly utilized and can make significant contributions to the S&T agricultural base of developing countries. The Title XII programs appear to have a significant potential for facilitating the effective involvement of U.S. universities in international agricultural development.

4 Science

As defined earlier, science includes both <u>basic</u> and <u>applied</u> aspects within the confines of <u>natural</u> science. A further differentiation can be made between applied <u>scientific</u> research and <u>technological</u> research. The former produces knowledge that is hoped to be applicable to technological and other developments, whereas technological research directly produces gadgets, processes, and prototypes. The example Moravcsik uses to illustrate this distinction is the development by the Asian Institute of Technology of a solar energy driven water pump. (1) The applied scientific research measured the annual average solar energy falling on different areas in Thailand, while the technological research took this information and produced the actual pump most suitable for the purpose. In this section, we will consider only scientific, not technological, research.

By including both basic and applied science in the science rubric, the range of activities becomes very broad indeed. The development of a physics or chemistry or biology department in an LDC is clearly a science involvement. But science also underpins a great deal of the activity in engineering and agriculture. Thus, building of agricultural and engineering institutions in LDCs of necessity involves science. Similarly, 211(d) grants to strengthen capability in the U.S. in such fields as watershed management (University of Arizona) or soil fertility (North Carolina State University) also fall within the limits of applied science. The areas chosen for emphasis here (due primarily to limited time but also due to their relatively high importance), are most of the basic natural sciences, i.e., biology, chemistry, geology, hydrology, physics, as well as some of the basic science underpinnings of agriculture and engineering. A few examples of applied science will also be mentioned.

Perhaps even more so than in agriculture and engineering, getting data and evaluations on science involvements has been extremely difficult. At this point, we have a relatively good overview (with a bit more detailed information on some projects), but we have virtually no evaluations at all. In light of this, the following five sections will attempt to convey as much of an overall picture as possible.

TYPES OF INVOLVEMENT

U.S. university science involvements fall into four categories: institution building, cooperative research and development, resource base development, and graduate and undergraduate education.

Institution Building

Institution building involves many types of projects. Three rough divisions appear to be useful: (a) projects that were concerned with overall development of an institution; (b) projects designed to improve science education both in secondary and upper level schools; and (c) projects concerned with strengthened capabilities in one department or specific field.

Overall Institutional Development

Several types of overall institution-building projects have been developed by U.S. universities. U.S. university scientists have been instrumental in planning, establishing, and operating a number of international centers of research. Generally, the centers are regional and based on the cooperation of an international group of established scientists working together, some on a part-time basis, on selected research problems. The areas covered include primarily agriculturally related research (see chapter 3), but also insect physiology, aquaculture, and nutrition (see fig. 4.1). For example, Carl Djerassi, a chemist at Stanford University, was a prime moving force behind the formation of the International Center for Insect Physiology and Ecology in Nairobi, Kenya. (2) An international consortium of national science academies partially funded by the Ford Foundation and the National Science Foundation helped in its establishment. Dr. Meinwald, professor of chemistry at Cornell University, and Dr. Nakanishi, professor of chemistry at Columbia University, both served as research directors for the ICIPE Chemistry Research Unit from 1970 to 1977. (3)

1) International Center for Agricultural Research in Dry Areas (ICARDA), Cairo, Egypt, est. 1976.

2) Asian Vegetable Research and Development Center (AVRDC), Taiwan.

3) International Center for Insect Physiology and Ecology (ICIPE), Kenya.

4) International Center for Living Aquatic Resources Management (ICLARM), Philippines.

5) Central Food Technological Research Institute (CFTRI), Mysore, India.

6) Institution of Nutrition of Central America and Panama (INCAP), Guatemala City.

7) Nutrition Center of the Philippines, Manila.

Fig. 4.1. International centers of research excellence, (excluding agricultural centers est. before 1974).

Another example of an overall institution-building project was the University of Pennsylvania-Pahlavi University (Iran) project. (4) In this project, science fields were only part of the areas that were strengthened. In 1960, the Iranian government invited the University of Pennsylvania to undertake a survey of higher education institutions in Iran in order to determine if any could be transformed into a modern, U.S.-style university. Funded by AID, a University of Pennsylvania team undertook this project and recommended the transformation of the University of Shiraz. Based on their recommendations, the Iranian government commenced the change from faculty and chairs to colleges and departments, and the new university was endowed with the status of a private institution governed by a board of trustees.

In 1962, the University of Pennsylvania received an AID contract to provide advisors and faculty to the Pahlavi University in medicine. The first team worked to solve across-the-board problems. This generalist spirit diminished as time wore on and more specialists were added in the areas of engineering and arts and sciences. Some team members felt they should only be advisory, that their advice should be automatically heeded, and that they should have little teaching responsibility. According to Copeland, "This feeling of detachment and lack of response to immediate need was perhaps in part engendered by the fact that salaries were paid to Pennsylvania personnel through AID funds, not from money provided by Pahlavi University." (5)

But as Iran soon became an AID-graduate country, funding was ended in 1967. Since both universities were interested in continuing the relationship, in January 1967 a five-year direct contract was signed under which the Pahlavi University provided $300,000 annually. The University of Pennsylvania also contributed an additional $125,000 annually, although not required by the contract, primarily to support Iranian graduate students studying for their doctorates in Philadelphia and for Iranian interns and residents in the U.S. This contract made possible a continuing exchange of professors and students, the development of joint research proposals, and reciprocal recognition of degrees. Also, the University of Pennsylvania initiated a project for the recruitment of Iranians in the U.S. for faculty positions at Pahlavi University. This last item was quite effectively accomplished. Extensive recruitment over the period 1966-1970 generated applications from over 400 Iranians in the U.S., judged to be 25 percent of the Iranian graduates in this country as of 1970. Slightly over 100 of these actually accepted offers of the Pahlavi University and returned to Iran.

As of 1971, 14 University of Pennsylvania faculty were teaching in Iran, and a total of over 54 Iranian students had come to the University of Pennsylvania. Two joint research projects had also been initiated. The success of these programs seems to be somewhat connected with: (1) attractive salaries and fringe benefits offered by the Pahlavi University, (2) adequate money to maintain the University of Pennsylvania cooperative arrangement, and (3) support and encouragement of top administrators of both universities. One unusual note is that as of 1971 only three U.S. students had taken advantage of the exchange to Iran even though their total costs would be less than tuition and room and board here, and all the courses there would be taught in English.

Still another project was the University of California-Universidad de Chile program. The objective of this program as stated in a survey of "Inter-American University Cooperation" of the Organization of American States was:

> To demonstrate that a multi-campus university such as the University of California and a multi-faculty university such as the Universidad de Chile can develop efficient methodologies to provide support to each other in required areas. (6)

By mutual agreement, counterpart subcommittees were established to examine problems in the following areas: (1) agriculture and veterinary medicine, (2) arts and literature, (3) natural sciences, (4) social sciences, and (5) administration. All activities within each field were jointly agreed upon and personnel from both California and Chile were involved in each. During the years 1965-1975, the Ford Foundation provided a total of $10 million. According to Ford Foundation annual reports, the primary activities accomplished were exchange of students and faculty and a strengthening of the research and teaching capacity of the Universidad de Chile.

An interview on June 6, 1978, with Dr. William Weir, department of nutrition, University of California-Davis, indicated that although Ford Foundation funding has ended, cooperation is continuing. Dr. Weir was involved in the Chile project for some years and has made numerous visits to the Universidad de Chile. When asked for an overall evaluation of the project he stated that he thought the project was successful, with one of his criteria for success being that it is still operating after outside funding ceased. According to him, some of the contributing factors were:

1. The "convenio" was strongly supported by top administrators in all of the University of California campuses.

2. The ten-year grant allowed for long-term ties to be developed.

3. Over a period of years, communications improved as students and professors came to know each other better. Occasionally, Chilean students would come back for a second degree, further improving contacts.

4. The multiplicity of campuses and departments involved generated more broad support than possible in a single, one-to-one arrangement.

This project merits further study. Lack of time and access to files limited its coverage in this report.

The University of Houston-Universidad de Guayaquil, Ecuador, project was supported by USAID. Its objectives as stated by the OAS "Inter-American University Cooperation Survey," were: "Development of a business college, a basic sciences institute, a chemical engineering college, and a central library, plus financial and administrative modernization of the university." (7) The general techniques used were: participants from Ecuador received training in the U.S., on-the-job training in Ecuador followed, and U.S. consultants later visited Ecuador briefly to check the progress and help solve specific problems. This procedure was followed for all three areas: business, chemical engineering, and basic sciences. The project budget was $1.14 million from 1960-1970. Six staff of the University of Houston were assigned, four in Ecuador and two in the U.S. As of 1967, ten Ecuadorian students had studied in the U.S. Three Fulbright scholarships and one Latin American Scholarship Program of American Universities also contributed to the program. Although the University of Houston was requested to send more recent information on this project, no new data were received.

Science Education Improvement

This area has received a great deal of attention, and numerous projects have been funded by AID, NSF, NAS, the Ford Foundation, the Rockefeller Foundation, and others. (8) There have been programs in undergraduate, secondary, and graduate level education since the early 1960s. The amount of emphasis on one level or the other, however, has varied over time. A discussion of some projects in each level follows. (9)

Undergraduate level

A project of science programs in India was jointly implemented through the National Science Foundation and the Indian National Council for Science Education. Its three main tasks aimed at undergraduate education development were stated in the 1970 NSF annual report as:

1) the training of personnel through summer institutes, workshops, seminars, and short courses;

2) the development of new teaching materials, including syllabi, textbooks, examinations, laboratory equipment, handbooks, and journals; and

3) the development of institutions which can sustain the improvement effort. (10)

From 1966 to 1970, the major U.S. contribution was providing consultants to summer institutes. Funding was provided jointly by AID and the India Ministry of Education. The number of consultants ranged from around 60 to 180 per year. (11) Although it is not known exactly how many of these were U.S. university associated, it is highly likely that a fair percentage of the consultants were connected to U.S. universities because their functions were teaching, lecturing, and supervising.

NSF science education programs in Central America, from the early 1960s until around 1968, aimed at improving science and math teaching at the undergraduate level in a number of universities. (12) Activities were coordinated through the Superior Council of Central America (CSUCA) headquartered in San Jose, Costa Rica. The activities involved: curriculum improvement; selection of laboratory equipment; assistance in design of buildings; improved lab instruction; design and construction of simple lab equipment; selection of textbooks; and training in the use of A.V. equipment, electronic lab equipment, and modern chemical analysis techniques. As in the case of the India project, the U.S. university contribution was in the form of consultants and visitors.

One observer of the program, who has been involved in other science education programs, commented that this program unfortunately collapsed primarily because of international jealousies among the various countries of Central America about whether and how the aid would be fairly distributed.

Funded by the Ford Foundation, the Midwest Universities Consortium for International Activities (MUCIA)-Universidad Agraria del Peru ran from 1964 to 1969 to strengthen the faculties of basic sciences of the Universidad Agraria del Peru in both teaching and research. Through the $618,000 contribution, laboratory equipment and books were purchased; U.S. professors in the fields of chemistry, physics, taxonomy, botany of grasses, meteorology, statistics, and ecology visited Peru to cooperate in the teaching and development of research projects; six study grants were awarded to Peruvian graduate students to study for two years in the U.S.; and three U.S. staff members worked on the project full-time in Peru. (13)

A conversation with K. N. Rao, who was with the Ford Foundation at the time of this involvement, revealed that there had been some problems with the consortia approach. Generally Rao expressed a belief that the consortium mechanism was too big, had too diffuse an authority structure, and was difficult to coordinate. He also indicated that the eruption of political dissension at the Peruvian university contributed to the difficulties. (14)

Secondary level

In its overall patterns of spending on science education, the Ford Foundation initially spent greater amounts on undergraduate and graduate levels than on secondary levels. In the period 1959-1965, $7,401,000 was spent on university programs in basic science, $5,154,000 on university programs in engineering, and only $1,050,000 on secondary education science and mathematics programs. The Ford Foundation's early projects in secondary science education grew in scope, however, during the mid-1960s. (15)

Although the Ford Foundation has been involved extensively in science education around the world, we chose to look at programs in five specific countries because of the availability of information on them through a study by Robert H. Maybury of the UNESCO Field Science Office for Africa. His book Technical Assistance and Innovation in Science Education, published in 1975, is a comparative appraisal of projects in Argentina, Brazil, Lebanon and other countries in the Middle East, the Philippines, and Turkey. (16) Each country's program was unique in its own way; however, the common thread through all of them was the attempt to transfer U.S. science curriculum projects in one form or the other to upgrade science education in secondary schools. These curriculum projects were integrally involved with U.S. universities. The chemical education materials study was based at the University of California as was the science curriculum improvement study. The physical science curriculum study started at MIT, the school mathematics curriculum project had Stanford University as its base, and the biological sciences curriculum study (BSCS) started at the University of Colorado. These were programs initially begun as a national effort to upgrade U.S. secondary science education during the 1960s. Some or all of these were utilized in the foundation's five programs.

Beyond the transfer of the texts themselves, various representatives from the projects did visit some of the countries involved. In some cases, educators from the developing countries visited the projects in the U.S. One other common factor was that in all five projects the special science educational consultants to the Ford Foundation were U.S. university professors. In one case, through the suggestions of the U.S. consultant, the University of Wisconsin became involved as a resource base for Lebanon, giving workshops and curriculum development advice, and performing other functions.

A discussion with Dr. Richard Tolman, one of the staff at the BSCS project in Colorado, revealed that even today, no evaluation has been made to examine the effectiveness or the impact on the students of the massive diffusion of U.S. science curriculum materials. He did, however, identify certain criteria for success in transferring the BSCS to LDC secondary schools:

1. Economies of the LDC must be strong enough and committed to secondary science education.

2. A person from the national Ministry of Education must be connected with the program, because national control over secondary schools is generally very tight in LDCs.

3. The university and national exam structure must be changed to emphasize the higher cognitive questions found in the BSCS, rather than rote memorization traditionally stressed in LDCs. Otherwise, there will be no encouragement to utilize the BSCS or its approach.

4. International financing has significantly helped in some projects - the Ford Foundation, NSF, and USAID have all contributed.

5. BSCS does not promote its materials. LDCs have requested their use; the bulk of the royalties stay in the LDCs. (17)

A recent article, "Factors Influencing International Curricular Diffusion," by Arnold Grobman, who served as Director of BSCS from 1959 to 1965, also indicates a number of criteria that affect the adaptation process.

It appears that the diffusion of curricular innovations proceeds most rapidly and easily among countries of large population size that invest substantial amounts of money in public education.

The rate of diffusion is less for countries with educational goals
that are perceived to be different from those of the model materials, for those where there are barriers to the free exchange of
information and persons with the innovating nation, for those
where the publishing technology is underdeveloped, for those
where the program of educational innovation falls into the hands of
entrepreneurs, and for those where government officials assume
a dominant, rather than a supportive, role in the curricular diffusion process. On the other hand, the rate of diffusion is greater
for countries where certain favorable historical accidents occur,
for those where a strong leadership role is taken by one or more
educators, for those where a diversity of models is available and
a climate favorable to innovation exists, for those where the continued support and interest of a variety of foundations is to be
found, and for those smaller nations where educators approach
opportunities for innovation on a regional basis. Language differences among nations and the activities of UNESCO seem to
play relatively unimportant roles in the process of international
curricular diffusion. (18)

A conversation with K. N. Rao, who was with the Ford Foundation during much of the time of the secondary school programs, added some additional perspectives. He felt that in some cases, such as the Turkey Science Lycee program, there was controversy over whether one of the major problems was due to the approach of establishing a new institution outside of the national educational structure, as opposed to changing an existing school. He also believed that of all the different science curriculum studies, the biology one (BSCS) was most successful primarily because its different versions (some stressing ecology, some highly theoretical, etc.) facilitated adaptation to the local conditions. Rao felt that the secondary level science education project in Brazil worked out best due to the fact that it had (1) good adaptation of the U.S. materials, (2) strong leadership, (3) adequate funding because of the entrepreneurship of its administrators, and (4) good connection with industry through its equipment and textbook production. (19)

One final interesting analysis of the secondary science curriculum projects came from the conversation with Dr. Richard Tolman about what the specific role of U.S. universities has been, and can be, in this area. Summarizing his remarks:

Of the Past: Transfer of U.S. science curriculum materials has
had a significant impact on the developing world. The BSCS is
used in more than 60 countries and is in 20 languages. The

departments that evolved these studies and helped in their adaptations were relatively autonomous from the university structure, although within it officially. Most LDCs are in their second or third generational use of the science materials and are largely independent of both BSCS and U.S. funding. Many of the original study groups have since left their respective universities and established separate institutes. There is some argument that they are <u>more</u> effective outside the university. The nitty-gritty help in adaptation of the materials to local conditions was actually not done by the BSCS as an organization, but by the individual U.S. people who traveled and worked abroad. Of course, many of these were U.S. university professors.

Of the Present and Future: (1) Faculty exchange in the science teaching area is desirable; (2) joint research in teaching methods is still needed; (3) LDC graduate students, in many cases, <u>can</u> get a science teaching education in the U.S. that is relevant to their needs back home; (4) LDC science educators need help with gaining competence in the use of computers in evaluations, both of the curriculum materials, and of the student's progress. Tolman suggests they could come study in the U.S. either through sabbatical leaves, summer institutions, or other mechanisms. (20)

A description of two of the Ford Foundation involvements in science education can be found in Appendix C.

Graduate level

The NAS-COLCIENCIAS project in Colombia was primarily in the form of a study rather than actual implementation of graduate programs. From November 1970 to December 1972, $40,600 was allocated to the Board on Science and Technology for International Development (BOSTID) to direct the convening of study groups on the potential and problems of graduate education in chemistry, mathematics, engineering and applied sciences, and biological sciences, in Colombia. Over 15 U.S. university professors were involved in one or more of these four study groups. (21) (See Appendix C for further discussion of the involvement of U.S. university personnel in BOSTID activities.)

By the late 1960s, the Ford Foundation was increasing its interest in graduate education programs. (22) According to K. N. Rao, the two most successful projects have been the development of the Graduate Education Centers at UNAM in Mexico City and at the Monterrey Institute of Technology. Both of these have graduate schools in basic sciences and engineering and both received direct and substantial support from the Ford Foundation during the 1960s. There were no formal institutional connections with U.S. universities in these projects; however, U.S. university professors served as consultants at various times. Rao believed that the success of these projects was in part attributable to not having an entire U.S. university linked to the LDC institution. (23) This position is supported by Kidd who indicates that too close identification with U.S. organizations can lead to charges of external domination, (24) and by Hazeltine who feels that it would be very difficult politically for some African universities to be too closely associated with large institution-building efforts centered on one U.S. university. (25)

Projects Emphasizing One Specific Field or Department

Of projects emphasizing one department or specific field of study, the best example we have come across is the NAS/CNPq chemistry project in Brazil. (26) In October 1969, the U.S. National Academy of Sciences (NAS) and National Research Council joined with the Brazilian National Research Council (CNPq) to initiate an experiment in transplanting to Brazil an advanced research capability in selected fields of chemistry regarded as important to Brazilian development. The project was conceived of and the rationale further laid out in an article by Carl Djerassi, a Stanford University chemistry professor. (27) This program was implemented through NAS's Board on Science and Technology for International Development (BOSTID).

American postdoctoral fellows in chemistry, with the guidance of their U.S. professors, pursued their own research in leading Brazilian universities in collaboration with Brazilian faculty and postgraduate students. Specifically, a senior Brazilian chemist would propose cooperation with a U.S. colleague based on compatibility of interests and availability of graduate students, instrumentation, and laboratory space. With approval from a joint U.S.-Brazilian committee, young Ph.D. chemists from the U.S. were selected to conduct research and teach in Brazilian universities for two to three years. The U.S. professors participated through continued communication and semiannual visits.

As of 1976, 13 eminent U.S. chemists from Stanford University, the University of Michigan, California Institute of Technology, Indiana and Northwestern Universities had committed themselves to this program. The Brazilian universities consisted of Federal University of Rio de Janeiro, University of Sao Paulo, and the Centro Brasileiro de Pesquisas Fisicas. Funding of NAS costs was provided by an AID host-country contract with CNPq from NSF, and from a number of private sources.

Positive aspects of this program, according to the NAS BOSTID report, are:

1. The degree of high commitment from U.S. professors looks promising for the prospects of this program as a prototype for future cooperative ventures.

2. There is a strong emphasis on training Brazilian doctoral candidates in Brazil, rather than abroad.

3. The quality of the research has been consistently high: 2 books published, 64 papers published in international scientific journals, 32 more papers being written.

4. Brazilian industry has begun to benefit from the program; particularly Petrobras, the national oil company. (28)

This program definitely looks attractive for helping to build an indigenous S&T base. One physics professor with strong interests in science development feels that there is incentive for U.S. faculty to work cooperatively with faculty in LDCs in areas such as chemistry, botany, and biology because there is much of interest to be learned by both sides in a truly cooperative program, due to the strong variation of phenomena in these fields with location. Conversely, LDCs stand to gain by focusing on local products which give them a comparative advantage. A discussion of the NAS/CNPq chemistry project at the July 13-14, 1978, workshop in connection with our project indicated that there was some concern about high costs. A NAS in-depth evaluation of this project is scheduled to be completed in September 1978.

Cooperative Research and Development

Again, there are many types of involvement in the area of cooperative research and development. One example is ongoing collaborative research at, or with, the International Centers of Excellence. The Universities of California, Oregon State, and Washington State have worked with a staff member at CIMMYT on creating superior varieties of wheat with semiwinter characteristics. Cooperative work between CIMMYT and Michigan State University has been examining the protein and lysine content of triticale to find high nutritive strains for livestock. (29) ICIPE continues collaborative work with faculty at Columbia University.

Next, there are examples of universities that have specific projects in a developing country. The University of Arizona and the Universidad de Sonora, Mexico, have cooperated in the construction of a pilot desalination plant at Puerto Panasco, Mexico, to be used by the two institutions for research purposes. The University of Arizona also uses the marine biological station owned by the Universidad. This project began in 1965 with an indefinite terminating date. (30) Another specific example, funded by NSF's Special Foreign Currency Program, provided in 1976 for Colorado State University engineers to do research to test classical theories of flow and sedimentation in alluvial channels by working with colleagues of the Water and Power Authority in Pakistan. Here, binational teams had access to the Link Canals of the Indus Valley for collecting data. (31)

University of Wisconsin-Universidad de Chile

A more complete view of a cooperative research project is provided by the University of Wisconsin-Universidad de Chile project. (32) From 1962 to 1969 the Laboratory of Neurophysiology, Medical School, University of Wisconsin and the Instituto de Fisiologia, Universidad de Chile participated in a joint research and research training project. The total budget, a series of paired grants from the National Institute of Neurological Diseases and Blindness (NINDB) and the National Institute of Health (NIH), came to approximately $200,000. Each of the two universities provided funding for their project personnel, physical facilities, and materials.

The University of Wisconsin assigned part-time work to four staff members in research training, to four in research, and to one in medical electronics service and training. Two performed their duties in Chile, five in the U.S., and two in both places. The University of Chile assigned four part-time teachers, four part-time researchers, and one director. The project also involved two to four years study in the U.S. by three Chilean students, one graduate and two postdoctoral. Some of the activities included: comparative studies on the mammalian nervous system using staff and facilities at both places; assemblage at Wisconsin of equipment for a Chilean neurophysiology laboratory and shipment to Chile; joint discovery of improved research methods at Wisconsin; and a special study of cerebral localization in sensory systems in Chile, partially led by a Wisconsin professor.

NSF's Cooperative Science Program in Latin America

A more extensive cooperative research and development involvement is NSF's Cooperative Science Program in Latin America.

Established in 1972, it involves Argentina, Brazil, and Mexico in programs designed to "foster and support mutually beneficial scientific and technological cooperation." The scope of the programs encompasses most fields of science and engineering and the forms of activities include cooperative research projects, joint seminars, and scientific visits. This program was a formalization of part of earlier NSF joint research projects in 13 Latin American countries. (33) Individual country projects include programs in materials science between U.S. and Argentinian researchers and in Guayule rubber research between Mexican scientists and chemists at the University of Akron's Institute of Polymer Science. (34)

NAS Activities

The National Academy of Sciences has also had some joint research projects, although more in the line of studies than laboratory research. These have been part of BOSTID's Special Studies/Advisory Panels program. Ten of their studies over the period 1970-1976 had significant international participation. Eighty-six U.S. university professors served on the panels which ranged in subject matter from "Solar Energy for Developing Countries" to "African Agricultural Research Capability" to "Remote Sensing and Development." The bilateral workshops of NAS might also be considered a type of cooperative science activity. These have been conducted primarily from 1970 to 1977 in many countries of Latin America, Africa, and Asia. Usually they are oriented around one subject such as, "Research Priorities and Problems in Execution of Research in Ghana." These workshops involved over 150 U.S. university professors, with a budget over the seven-year period of $5 million. (35)

International Research Projects

A further, broad type of involvement has been with international science research projects such as the International Biological Project, the Global Atmospheric Research Project, the International Hydrological Decade, and World Geophysics Research. Within these programs, scientists from developed countries have had the primary roles, but much of the research done has been relevant geographically to some LDCs, i.e., "Structure and Function of Desert Ecosystems." (36)

OAS PRDCYT

A last, important involvement is the Regional Scientific and Technological Program (PRDCYT) of the Organization of American States

(OAS). (37) The PRDCYT has been supporting training activities, courses, and projects for member states at regional centers in fields such as physics, mathematics, chemistry, genetics, and biochemistry. More recently, OAS has carried out projects concerned with transfer of technology, applied technological research, and science and technology policy. (38) The total budget for the PRDCYT grew from $8.3 million in 1961 to $56.5 million in 1975. The U.S. contributes about two-thirds of this amount.

A typical sequence of events for any project would be as follows: Interested Latin American countries approach OAS with a request to initiate a project, through their national science agencies. Each country then sets up its own budget (to cover primarily administrative expenses) and designates a center to be the active participant. At OAS's expense, the participating countries are brought together to discuss the project, set individual goals, and coordinate their efforts. OAS money then goes to buy equipment, provide travel money for researchers to visit various participating centers, cover research expense, allow student fellowships, facilitate meetings and seminars, among other things. Preference is given to projects that are multinational. (39) M. Naraine, in the article "Science for Progress," has summarized the PRDCYT situation as follows:

> Ironically, PRDCYT's main problem has been its association with the OAS. The programme was developed through political actions and it has never lost that association in the eyes of its critics. The United States' dominance of the OAS and its manipulation of the region to secure hemispheric solidarity has raised many doubts about the worth of any OAS programme.
>
> Undoubtedly, the benefits have not been as large as they might have been. The more developed countries have benefitted most from the programme, but this should be expected since they have a better science and technology infrastructure. The Mar del Plata account has helped to redress the balance, but the main issue here is that it limits the amount of money countries can obtain. The newly created Simon Bolivar fund, financed mainly by Venezuelan petrodollars, will concentrate its efforts on rural development - PRDCYT's main problem is that it has not been closely linked to direct economic development. Criticisms of it on technology transfer conditions, patent legislation and types of development, however, are spurious, since these are issues that have to be settled internally by national policies. Awareness of these problems on the part of some national agencies has actually stemmed partly from PRDCYT-organized courses, but these agencies find it diffi-

cult to convince their superiors of the need for integrated science policies.

PRDCYT has been important in the development of the scientific and technological infrastructure of the Latin American nations. Through its efforts many centres of excellence now exist that are the prerequisites for the generation of indigenous technology. Many institutions rely on the programme for financial support for research, since they receive very little money directly from their own governments. According to Article 45 of the OAS Charter, governments should encourage science that is oriented toward the overall improvement of the individual - that is a foundation for democracy, social justice and progress. In too many Latin American countries money is used for the very opposite purposes. (40)

The present involvement of U.S. universities in OAS-PRDCYT is that quite a few visiting consultants, joint researchers, guest teachers, etc., are professors from individual U.S. universities. Some universities represented are the University of Florida, the University of North Carolina, Northwestern University, and MIT. Dr. Michael Greene, deputy director of Scientific Affairs at OAS, would like U.S. universities to play a greater role in OAS-PRDCYT projects. He suggests that one U.S. university could link up with each project and participate in the same manner as all the rest, except that outside funding would come from some separate U.S. agency. This would guarantee that OAS funds would continue to support Latin American interests and not be diverted to U.S. universities. The advantage to this, however, would accrue to both sides; U.S. universities have much expertise to contribute, but using OAS as a linkage allows U.S. universities access to and participation in research areas that might otherwise be closed. (41)

Resource Base Development

Since most of the resource base projects came out of AID's 211(d) funding, and because AID invested most heavily in applied science as opposed to basic science activities, the relevant examples here are of the science underpinnings of agriculture and engineering. The 211(d) projects have included Water Resources Management, Tropical Soils Management, and Aquaculture and Fish Farming. (42)

North Carolina State 211(d) Tropical Soils Program

The North Carolina State-Physical and Chemical Properties of Tropical Soils Program started in 1971 and is still functioning. (43) The program places major emphasis on soil fertility and management, relating plant nutrition to the physical and chemical properties of tropical soils. The geographical focus has been on Latin America, although much of the information generated is applicable to similar ecological areas of other continents. Of the soil science department's 40 professors, an equivalent of 12 full-time senior professors are involved in international work, including several stationed in Latin America. In accordance with the objectives of the program, 6 senior faculty members conducted on-site studies of soil properties in tropical zones, courses were modified to add emphasis to tropical soils, a research project in three major ecological regions of tropical Latin America was initiated, and soil scientists and agriculturalists in official positions of tropical zones were consulted.

One of the aspects of this program which stands out is that the relevant departments already had a long history of involvements in international activities. Members of the soil science faculty had had more than 15 years of direct research and operational expertise in the tropics, including the areas of tropical soils classification in relation to soil fertility, tropical savannah management, and tropical corn fertilizer.

The original 211(d) grant in 1971 was for $500,000. The grant has a current active status until June 30, 1978, although according to the September 30, 1977, report on 211(d) grant status, it is likely to be extended. Other AID money had supported and continues to support contract research at North Carolina State in related areas. An example of this is "Agronomic Economic Research on Tropical Soils," from 1970 to 1975. As indicated in <u>Benefits to the United States from American Technical Assistance Activities Abroad: Some Case Studies</u>, AID, 1972, this project has particular benefits to the U.S. in the formulation and sales promotion of U.S. fertilizer industries, and also has applications to soils in the southeastern United States.

MIT's International Nutrition Policy and Planning Program (INP)

A more recent example of resource base development is the international Nutrition Policy and Planning Program (INP), MIT's interdisciplinary approach to generate more knowledge required for improving the nutritional status of people in the poor countries of the world. (44) Established in 1972 with Rockefeller Foundation support, AID soon provided funds for further collaboration between the MIT department of nutrition and food science and the MIT Center for International Studies. Other participants are the departments of economics, political science, and urban studies. At the beginning of 1978, a joint MIT/Harvard International Food and Nutrition Policy Program was established through the Harvard School of Public Health. The resultant synthesis of all these components provides strength in a wide variety of disciplines including nutrition, food science, anthropology, international policy analysis, maternal and child health, epidemiology, tropical public health, and much more. The core activities of the program are research, graduate education and advanced training. Of particular interest is the Advanced Study Program designed primarily for officials of governments, of bilateral agencies, and of foundations whose careers are already committed toward strengthening governmental nutrition policy and programming. Fellows from Morocco, Colombia, Brazil, Zambia, Ecuador, and USAID have recently been or are currently involved with the program. Under the sponsorship of the UN University, foreign officials will also study in this program (for further information on the UN University, see Appendix C).

The INP also provides advisory and research services to governments and international assistance agencies. For example, INP has worked, through AID funding, with the government of Pakistan's Nutrition Cell in a multidisciplinary nutrition planning project. Also, training workshops have been carried out for national and international agency officials, and research activities have centered on problems in low-income countries.

Other Involvements

Studies administered by BOSTID's Advisory Committee on Technology Innovation (ACTI) of the National Academy of Sciences may be regarded as helping to build a resource base for international activity in the U.S. During the period 1971-75, ACTI studies have involved 76 professors in 12 projects. Examples are "The Winged Bean as a Potential New Plant Crop for the Humid Tropics," October 1974-June 1975, budget, $17,745, and "Underexploited Tropical Plants with Promising Economic Value," June 1975-August 1977, budget $35,185. Also, some of NAS's special studies such as the world food and nutrition study and the UNCSTED study have involved U.S. professors. (45) These projects serve to develop and sustain interest and competence for U.S. science faculty in international activities.

Our final miscellaneous project, performed in the U.S. but with potential for use in LDCs, was an NSF-funded study at the University of Maryland on the "Study of Inexpensive Science Teaching Equipment Worldwide." This project received $117,000 over a two-year period from 1968 to 1969. (46)

Role of U.S. Universities in Science and Technology

Graduate and Undergraduate Education and Exchange

As an introduction to this type of involvement, graduate and undergraduate education and exchange, it is useful to examine some definitions set by the NAS/NRC Board on International Scientific Exchange (BISE). The following is an inventory of possible objectives for international scientific exchange:

1. Building world science

    1.1 Training graduate students and post-doctoral fellows

    1.2 Communicating existing knowledge through lectures, meetings, symposia, summer schools, etc.

    1.3 Generating new knowledge through both short-term and long-term collaborative research.

    1.4 Developing a global strategy for scientific research in which each country would be able to optimize its scientific effort.

2. Working toward the solution of global problems: Food, disarmament, energy, environment, etc.

3. Building an international community: Maintenance and strengthening of a scientific community which reaches beyond political and economic boundaries.

    3.1 Development of close personal ties.

    3.2 Encouraging innovative scholars working under repressive or otherwise depressed conditions.

    3.3 Training of scholars in devoloping countries.

    3.4 Gaining new cultural perspectives, both through an increased knowledge of other cultures and the added insight such knowledge provides concerning one's own [culture].

4. Foreign policy objectives: using scientific and technological interchange as a way of reducing political tensions and [of] increasing trade. (47)

Science

BISE has also identified four forms of international scientific exchange:

1. Survey visits; short term ($\lesssim$ 1 month) visits by individuals and groups under an organized program.

2. Short-term collaborative research ($\sim$ 1 day to 1 year)

   2.1 Visits by individuals who take part in an ongoing research program at a host institution (graduate students, post-doctoral fellows, visiting professor)

   2.2 Group interaction; participation in national and international meetings, summer schools, "Garden" conferences, summer workshops or joint research groups, of 3-12 weeks' duration, informal interaction among individuals working during the summer (or the academic year) at regional, national, or international research centers.

3. Long term ($\gtrsim$ 1 year) collaborative research, with substantial contributions from both (or all) partners.

   3.1 International "years": International Geophysical Year, etc.

   3.2 Continuing joint research programs involving institutions and individuals in both (or all) countries.

4. Joint ownership and operation of physical facilities. (48)

Foreign Students and Faculty to U.S.

According to the Open Doors 1975/76-1976/77 Report on International Education Exchange, approximately 22,000 foreign students (excluding Europeans and approximating the LDC total) were studying the natural and physical sciences during the 1975-76 academic year. This represents a 6 percent increase over the previous year; however, total enrollment also increased by 5.5 percent, so the overall percentage of students studying these sciences remained the same. (49)

In the 1976 fiscal year, 505 foreign scholars came to the U.S. in connection with the Senior Fulbright-Hays Program. Of these, 21 percent were in the physical sciences (chemistry, earth sciences, engineering, mathematics, and physics) and 23 percent were in the life sciences (animal and plant sciences, bio-sciences, and medical sciences). The large majority of visitors in the sciences came principally as research scholars with only about 20 percent coming primarily as lecturers. Of the 505 foreign students, 24 percent were from developing countries. (50) Through its India exchange program, NSF helped bring over 117 Indian scientists to the U.S. during 1970-1976. (51)

Several multilateral agencies sponsor fellowship programs that bring foreign scientific personnel to the U.S. The U.S. contributes significantly to these agencies; for example, U.S. contributions in FY 1977 totaled $28.8 million to OAS and $100 million to UNDP. (52) In 1976 the UN Development Program financed a total of 5,198 fellowships for personnel from developing countries, at a cost of $32.5 million. The U.S. hosted 851 fellows, the largest number in one country. (53) It is unclear how many of these fellowships were in the sciences.

An unusual form of exchange occurs through the East-West Center in Honolulu, Hawaii. It was established by the U.S. Congress in 1960 to promote better relations and understanding between the U.S. and the nations of Asia and the Pacific through cooperative study, training, and research. The center is on land provided by the University of Hawaii. Graduate degree students from the university participate in programs along with visiting scholars, researchers, and leaders and professionals from the academic, government, and business communities. Over 1,500 people participate in programs each year; for each center participant from the U.S., two participants are sought from the Asian and Pacific areas. Program subjects include problems of communication, culture learning, environment and policy, and resource systems. Mechanisms of establishing contacts are through a variety of institutions in most Asian nations, as well as in Australia, New Zealand, the islands of the South Pacific, and other areas of the U.S. (54)

Of particular interest is the East-West Resource Systems Institute - one arm of the center. As of 1978, its program consisted of broad studies of East-West problems of food, energy, and raw materials, and their relationship with each other. Three projects, one in each of the areas, are being pursued, with emphasis on interinstitute cooperation. Again, we see an overlap of types of involvements here; the center conducts both exchanges and cooperative research. (55)

U.S. Students and Faculty to LDCs

One program for U.S. students is the International Association for the Exchange of Students for Technical Experience (IAESTE). IAESTE is a nonpolitical, independent, nongovernmental organization having consultative relationship with the UN Economic and Social Council, UNESCO, ILO, and OAS. Of the IAESTE member countries, 17 are developing countries. In the U.S., 106 universities and 72 employers participate in this program to exchange students during their long vacations to provide practical experience in technical fields. In 1977, the U.S. sent 88 students abroad, 14 in sciences, 11 in agriculture, 51 in engineering, and the rest in other assorted fields. Of the total, only 9 went to developing countries. (56) While this program appears to have potential as an industry-university international linkage and exchange mechanism, it certainly has not yet emphasized or exploited the U.S.-LDC exchange possibilities.

The Fulbright-Hays program provides awards for U.S. faculty to lecture and do research and for U.S. graduate students to pursue research studies in other countries. The program is intended to support the international exchange of scholars; the foreign scholar counterpart was described previously. The history and development of the Senior Fulbright-Hays Program with emphasis on awards for faculty, is described in an April 30, 1976, report, which also documents the complex relationships between government and private organizations in administering the program. (57)

Table 4.1 indicates the number of applications and awards made to U.S. participants in the Senior Fulbright-Hays Program over the years. There has been some recovery from budget cuts in 1969-1971, but according to Wilburn, both financial support and the quality of the terms of the grants have decreased since the late 1960s. About 20 percent of the awards are in science, agriculture, and engineering. (58) In 1975-1976, about 40 percent of all grantees went to LDCs. (59)

The Fulbright-Hays program is also described in "U.S. Scientists Abroad: An Examination of Major Programs for Nongovernmental Scientific Exchange," prepared by the Science Policy Research Division, Congressional Research Service, Library of Congress. (60) A summary of the number of U.S. scholars recommended for awards in the natural and applied sciences in the regions of Africa, Latin America, East Asia, and Near East Asia is presented in table 4.2. As can be seen, there was a sharp decrease in recommended candidates interested in science in LDCs over the five-year period indicated. Furthermore, the bulk of cuts for research awards in the sciences has been in the developing countries. (61) Table 4.2 may indicate that there has been some recovery.

NSF financing in the form of international travel grants has taken more than 10,000 scientists abroad over the period 1952-1973. (62) This number is all inclusive, and it is unclear how many scientists came from U.S. universities as opposed to other U.S. institutions

Table 4.1. Applications and Awards to U.S. Participants In the Senior Fulbright-Hays Program, 1948-75

Year	Applications	Awards	Year	Applications	Awards
1948-49	108	33	1962-63	1,995	602
1949-50	771	166	1963-64	2,045	607
1950-51	1,580	206	1964-65	2,451	632
1951-52	2,267	226	1965-66	2,253	690
1952-53	2,304	328	1966-67	2,109	650
1953-54	2,225	391	1967-68	2,098	611
1954-55	2,009	409	1968-69	2,397	590
1955-56	1,839	411	1969-70	2,261	297
1956-57	1,510	380	1970-71	1,346	381
1957-58	1,482	419	1971-72	1,780	536
1958-59	1,665	435	1972-73	2,400	547
1959-60	1,740	445	1973-74	2,563	494
1960-61	1,900	493	1974-75	2,774	522
1961-62	1,851	572	1975-76	2,629 (prelim.)	455 (prelim.)

SOURCE: "Stewards for International Exchange: The Role of the National Research Council in the Senior Fulbright-Hays Program, 1947-1975," National Research Council.

Table 4.2. U.S. Scholars Recommended for Senior Fulbright-Hays Awards to Study in LDCs in the Natural and Applied Sciences

Year	Lecturers	Research Scholars	Total
1966-67	175	47	222
1967-68	174	45	219
1968-69	151	25	176
1969-70	102	27	129
1970-71	68	5	73

SOURCE: Science, Technology and American Diplomacy, U.S. House of Representatives, Committee on International Relations, p. 901.

or whether they visited developing countries or other developed countries. Some clue to the latter question may be provided by data on NSF fellowship awardees who took foreign tenure. Between 1960 and 1970, 19 fellowships were awarded for study in foreign institutions in Latin America, 12 went to Africa, 12 went to the Middle East and South Asia (excluding Israel), for a total of 43 U.S. scientists tenured in developing countries over a ten-year period. (63) This compares with a total of 2,790 such awards for study in all foreign countries. Also somewhat relevant are the international travel grants for science education. From 1966 to 1970, the foundation awarded approximately 50 individual awards; however, most of the recipients traveled to attend conferences in Western Europe. (64)

The Scientists and Engineers for Economic Development (SEED) program, started in 1972, provides funds for U.S. scientists and engineers from academic institutions to conduct research or teach, or both, in a participating developing country. (65) Grants may also support discrete development projects designed by U.S. scientists and engineers in collaboration with foreign counterparts. International travel grants are also available. Applicants must be U.S. scientists and engineers from U.S. academic institutions who have at least five years of postdoctoral or equivalent experience in teaching or research. About 150 people participated between its start in 1972 and 1977, working in over 13 countries in Africa, Asia, and Latin America. There is some indication that it has been difficult to attract enough people to the program. The operating budget as of 1977 was approximately $400,00 per year. (66) Originally the program had separate funding by AID and was only administered by NSF. Now it appears as though the entire responsibility will be shifted to NSF after a transition period where both agencies provide funds. (67)

Recently, NSF supported an internal evaluation by Kidd to help determine the future shape of the program. The report concluded that the SEED program has been a success "primarily in terms of the stimulation of research and graduate education and less in terms of promoting continuing institutional relationships." (68) Apparently NSF is still determining its future role in international development activities; the SEED program will be affected by the outcome of this determination. Kidd recommended that a number of elements in the SEED program should remain unchanged while a number should be modified. Excerpts from these recommendations include:

1. Retain existing procedures and criteria for selection of grantees.

2. Revise program goals and criteria for evaluation.

3. Place greater emphasis upon research and less on technique.

4. Involve LDCs more effectively.

5. Review the administrative guidelines. (69)

Specifics within these recommendations that might affect U.S. universities are suggestions to 1) possibly finance visits to the U.S. by faculty and students of host institutions, thus making the program a truly two-way exchange, and 2) not make institutional grants to U.S. or foreign universities. One recommendation encouraged the practice of sending grantees to universities as well as to non-university institutions such as the Korean Institute of Science and Technology (KIST) and the International Rice Research Institute (IRRI), which combine advanced training and research. (70)

One additional NSF involvement was a $330,200 grant in 1969 to enable a consortium of 25 U.S. institutions to send faculty and students to study tropical biology in Costa Rica. (71)

Other Types of Involvements

The area of science also seems conducive to some interestingly different types of involvements, not quite fitting into the categories listed above. One such involvement is U.S. exhibits in international science conferences or fairs, partially constructed by U.S. university-connected people, notably the NSF-funded Brazil Exhibit in 1967. (72)

Another is the Physics Interviewing Project, which is designed to bring about a better match between prospective foreign graduate physics students and U.S. physics departments. The project started in 1969 with UCLA, Michigan, Oregon, and Pittsburgh as participating universities. (73) Some 20 universities now support two U.S. physics professors who travel periodically through developing countries that have no Ph.D. physics programs doing interviews and evaluating prospective students for assistantships in the U.S. A recent article in Science suggests that such a model should also be useful for sciences other than physics. (74)

MECHANISMS

The various mechanisms used by the science involvements we have summarized fall primarily in the bilateral as opposed to the multilateral mode, although there are several exceptions. By examining all of the involvements mentioned earlier in this section, we have generated figure 4.2, a breakdown of science involvements according to the mechanisms used. As with types of involvements, the distinctions between categories are somewhat fuzzy. This is true particularly in a number of cases for bilateral research projects which do tend to be primarily technical assistance at the start but which move more toward truly cooperative research and development as the LDC institution is strengthened.

We do not know if the science involvements discussed above are entirely representative of the broader set of involvements which undoubtedly exist. Additional work to collect such information would be useful.

CURRENT THINKING ON SCIENCE INVOLVEMENTS

Michael Moravcsik

Dr. Michael Moravcsik, a theoretical physicist by training, has certainly been one of the strongest forces in writing on and demanding attention for the area of international science involvements. His two most comprehensive publications on this subject have been <u>Science Development: The Building of Science in Less Developed Countries</u>, and "Science and the Developing Countries," a contribution to the U.S. Country Paper for the UNCSTED. (75) Moravcsik has covered so much ground that it is difficult to condense his work; however, with an effort not to leave too much out, a summary of some of his conclusions follows:

1. Generally, international science programs are overbureaucratized. There is too much red tape, particularly in light of Moravcsik's view that scientist-to-scientist contact is the most conducive to positive results. Emphasis should be on decentralization of projects.

2. Usually institution-to-institution links have been too large. AID style in particular favors involving the whole university, or even the multiple-university consortia approach. Moravcsik feels links need to be made in small units. "The organizational task is to find ways for large organizations to assist many small bilateral links, all generated by the scientists themselves." (76)

1) Bilateral

   a) Individual

      SEED Program
      NSF Cooperative Science Projects in Latin America
      NSF U.S.-India Exchange
      NSF Science Education in India and Latin America
      NAS Studies and Workshops
      Fulbright-Hays Fellowships
      Ford Foundation (secondary level science curriculum projects)
      Physics Interviewing Project
      IAESTE

   b) Institutional

      NAS/CNPq Chemistry in Brazil
      UCLA/Universidad de Chile
      University of Houston/Universidad de Guayaquil, Ecuador
      University of Pennsylvania/Pahlavi University of Iran
      University of Arizona/Universidad de Sonora, Mexico

   c) Consortia

      MUCIA/Universidad Agraria del Peru
      NSF Tropical Biology Consortium

2) Multilateral

   a) International Research Centers of Excellence
   b) Colorado State and Binational Teams in Pakistan's Indus Valley
   c) International Research Projects
   d) OAS Fellowships and PRDCYT Project
   e) UNESCO Projects
   f) UN University
   g) International Scientific Conferences

3) Cooperative Research and Development

   a) University of Pennsylvania/Pahlavi University
   b) University of Wisconsin/Universidad de Chile
   c) East-West Center

4) Resource Base Development

   a) 211(d) Projects in Water Resources, Tropical Soils, and Aquaculture and Fish Farming
   b) University of Maryland (science equipment grant)
   c) NAS Studies
   d) MIT's International Nutrition Policy Planning Program
   e) NSF/NAS/Ford Foundation Funding (U.S. science curriculum materials development centers)

Fig. 4.2. Science involvements arranged by mechanism used.

3. Exchange of scientists is good, but the contact of developed country scientists in developing countries is too brief, and not enough. Scientists in developing countries need opportunities to visit the developed countries.

4. A crucial new focus should be the encouragement and mobilization of more scientists in developed countries to get interested and involved in international science. Correspondingly, there needs to be stimulation of LDC scientists, as well as improved communications among LDC scientists and between scientists of developed and developing countries.

5. The most significant involvement of U.S. universities in science and technology for development has been through the education of many foreign students who have come to study in the United States. Some problems in this regard are: (A) U.S. curricula are not relevant to needs of LDC students; auxiliary education is needed. (B) Selection of students continues to be haphazard. Lack of information exists in LDCs about developed country institutions. Applying students are not evaluated uniformly or fairly. Personal interviews in LDCs are needed. (C) Developed country education is too expensive. (D) Once here, U.S. advisors are insensitive. Advisors are needed with experience in LDCs who will follow up after the student receives a degree. (E) Students need ways to keep in touch with the developed country institution after returning home. (F) The "brain drain" is a continuing problem.

K. N. Rao

K. N. Rao, senior research associate, Center for Policy Alternatives, MIT, in his article, "University Based Science and Technology for Development: New Patterns of International Aid," makes some comments specifically relating to science. In analyzing past developments, he states:

> stimulated by fellowship and exchange programmes of international agencies and domestic scholarship programmes, there has resulted a rapid build-up of scientific cadres in countries such as India, Korea, Brazil, Ghana, Colombia, and Mexico. Many African and Middle Eastern countries need continued help in building their science communities. Political emigres from certain countries are helping others in temporarily filling positions in universities,

government and industry. Unfortunately, while scientific and technical competence is in place in many countries, innovative policies to put them to <u>effective use in development tasks</u> [emphasis added] are lacking. (77)

Later on, in talking about agricultural, industrial, and scientific research he says:

Agricultural research by its very nature tends to be directed at local problems, but industrial research and development is still in its infancy in most LDC universities. Scientific research, on the other hand, tends to follow international currents. Basic scientific research at the universities unfortunately gives the impression of being isolated from national needs. Basic research in such areas as the taxonomy of local flora and fauna, geology, hydrology and climatology of the region, vectors of endemic diseases, entomology of agricultural pests, biochemistry of national products or ecosystems in the different geographical regions of a country -- all these provide the base of knowledge necessary for economic development in the near and long term. Support for such "small science" projects is lacking in most regions. Fully operational communication networks among LDC scientists working on common problems do not yet exist. (78)

The article then proceeds to classify developing countries into three categories based on the level of development of their science and technology infrastructure and postgraduate courses. This analysis along with some general conclusions will be discussed in more detail in chapter 5 on the future role of U.S. universities. One other area of concern, though, does touch specifically on science involvement. Rao asks the question "Mission-Oriented Universities?" and then deals with it thus:

Educators have long fought over the distinctions between education and training. That universities should be concerned with the universe of knowledge and that artificial barriers separating disciplines must be torn down has become the accepted dogma. Mission-orientation and the unfettered pursuit of knowledge may thus appear contradictory in meaning and practice. But development even in its broadest meaning must deal with objectives, the process itself and with end results affecting the lives of people now and in the future. When viewed in this light, pursuit of knowledge and its application are both parts of a single mission.

This suggests some revolutionary paths for institutional development, and examples abound in the developed countries. (79)

Rao then goes on to list various connections between industry, the private sector, and other organizations already in existence, such as degree-granting programs of U.S. hospitals.

Michael P. Greene

Through a three-month Fulbright fellowship and a six-month OAS fellowship, Michael Greene, a physicist from the University of Maryland, taught in Peru at the National Engineering University (UNI) and Catholic University (PUC) from June 1970 to February 1971. He also taught at the University of La Plata, Argentina, and the Faculty of Sciences of the University of Chile, the first in October 1971 and the latter in July 1971. His relevant conclusions, found in <u>Physics in Latin America: Peru and Chile</u>, will be summarized here. (80)

1. There was much dissatisfaction with the University of California/University of Chile convenio. The Chileans complained that most of the money was controlled and spent in California. Several Chilean Ph.D.s from Berkeley obviously benefitted from the program, but some of the people sent to Chile did not work out well and Chile could not get all the people they wanted in many cases. The Davis cyclotron brought to Chile was claimed by the Chileans to be unsuitable for research; although its operation drained time, money, and energy from faculty and staff, it had not resulted in a publication in seven years. According to Dr. Greene:

The Ford Foundation evidently is aware of this dissatisfaction. I was told by an official that they have no intention of expanding this Compact nor of initiating similar arrangements in other countries. (They had an unhappy experience with a similar Compact in Peru also.) They will in the future tend to provide support directly to the Latin American institutions leaving it to the beneficiaries to decide to send their students and invite their visitors all from one U.S. university or several. (81)

2. The presence of foreign visitors can be a great stimulus, helping to prevent isolation of newly graduated Ph.D.s.

3. The Fulbright fellowship program seems to have chosen fellows without regard to local needs, and generally the visits are inadequately planned and too short to be useful. However, the Fulbright policy of sending visitors to conduct week-long short courses (cursillos) is an excellent one.

4. U.S. universities could contribute by providing LDC educators with experience in lecture demonstration techniques. This is important as class size increases in the primary basic science courses.

5. The Peruvian and Chilean governments are committed to basic research, and the Ford Foundation has continued to play a valuable role in both countries; therefore the question of material support does not seem to be the major problem.

H. Harry Szmant

H. Harry Szmant, associate dean for science, College of Engineering and Science, University of Detroit, is concerned with the relationship between scientific and economic growth in Latin America, particularly in the area of chemistry. Through an extensive look at Latin American students in the U.S., at the numbers of published research projects worldwide (including Latin America); at growth per income per capita, and at various other areas, Szmant arrives at several conclusions and recommendations. In summary form, these include:

1. Many administrators of the educational aspects of foreign aid have been made to feel their activities were getting minimal attention from top administrators and that their projects were "window dressing" to sell larger economic packages.

2. When foreign aid was channeled into science education, little attention was paid to which science area would be most closely related to or most directly connected to a significant impact on economic growth in the LDCs.

3. Specifically in chemistry, it is clear that hardly any effort was expended to build into the programs the potential practical technological and economic ramifications of chemistry.

4. The result of this practice is that an LDC student's learning and research experience in the U.S. produces little economic impact on his country when he returns.

5. This is not surprising, as the U.S., until recently, has also had a separation between studies and practical, economic applications.

6. Therefore, our infusion of scientific research and training into developing countries must no longer be done in an economic and industrial vacuum.

Carl Djerassi

Carl Djerassi, in "A Modest Proposal for Increased North-South Interaction Among Scientists," comments on the viability of the International Center for Insect Physiology and Ecology (ICIPE) type of institution building in the future, and considers the needs of the types of countries that he feels, in a scientific and technological context, are worse off now than in 1965. He questions whether the ICIPE model is still valid in the context of present economic conditions and changing attitudes of both LDCs and developed countries. Briefly, his principal conclusions are:

> The ICIPE model . . . the creation of a new center of research excellence - is probably not realistic in 1976, nor even desirable for the type of country that I am discussing. It is not realistic because the present economic constraints among donor countries are such that only very special institutions with narrow mission-oriented purposes have any chance of being funded . . . .

> The undesirability is associated with the fact that building a new institution with better working conditions and salaries in a country which does have existing scientific institutions that are run down or lack in funds can be disruptive. Frequently, it simply results in cannibalization of existing institutions and creates professional jealousies in a comparatively small scientific community, which more than anything else needs cooperation and mutual commitment to a common goal . . . .

> The attractive and unique aspect of ICIPE was and is the cooperative effort of a large number . . . of national academies . . . or similar bodies . . . to assure the creation as well as continuing

operation . . . of a first-class advanced research institute in a country lacking such institution. (83)

Djerassi goes on to ask "how can this cooperative feature of ICIPE be expanded to the current cases under consideration?" His answer is an elaborate proposal that professional societies become the catalysts for a broad-based, person-to-person, cooperative effort which he feels will have significant impact and be of a long-lasting nature. The details of the proposal will not be considered here; however, there is an indirect connection to this study in that, generally, many members of professional societies are from universities.

Other Views

We have not been able in the time available to summarize LDC thinking on science development to any extent. Some LDC views on universities with some relevance to science are presented in studies by Thompson and Fogel (84) and Harrington. (85) In addition, the American Association of Advancement of Science has sponsored sessions at their annual meetings which have led to the formation of an Association for the Advancement of Appropriate Technology for Developing Countries, consisting of graduate students and faculty from LDCs studying in the U.S. This latter group has been addressing a number of issues, including ways of improving the orientation of LDC students at U.S. universities and focusing their programs more on development needs. (86) Finally, there is a large body of foreign literature and a number of foreign scientific associations that need to be examined.

At the workshop held in connection with this project on July 13 and 14, 1978 (see Appendix A), R. Walker emphasized that pure science is tremendously important to LDCs and quoted from an article by H. Bhabba of India to reinforce his views. He stated that LDC scientists need to do first-class science and be a part of the international scientific community; to support this activity, initiatives are needed such as programs of frequent visits by LDC scientists to the U.S., efforts such as the physics interviewing project, and increased contacts between developing and developed country scientists through lecture programs, overseas visits, etc. (87) F. Long, while generally sympathetic to this view of science, questioned whether smaller LDCs could afford much basic science. He also stressed the importance of international science and technology policy research and the role of a new Council on Science and Technology for Development in facilitating such research. (88) K. N. Rao pointed out the importance

of individual, nongovernmental initiatives in international science such
as the FORGE program and the International Foundation for Science.
(89) V. Walbot discussed the role of the Washington University Plant
Biology Program in international science and development, commenting upon the ability of the program to sustain strong international ties
with little if any U.S. government support. (90)

ANALYSIS

Again, because of lack of evaluation of the projects, it is difficult
to judge the overall success of science involvements. It may be hazarded, however, that the successful completion of a project's objectives, e.g., strengthening undergraduate science education, did contribute to building an indigenous science and technology base. Particular examples of projects that have appeared to accomplish some or
all their objectives are the NAS/CNPq chemistry project, the international centers of research excellence, and some of the secondary
science education projects. These are visible because they involve
a whole department or institution. The impact, and therefore "success," of the exchanges, small cooperative R&D programs, consulting visits, etc. is generally less visible but may be equally significant, if not more so. This may also be true for LDC students trained
in the U.S. who return to their own countries.

Therefore, the factors that can be most clearly identified affecting
the success or failure of a project are those relating to the successful
completion of a project according to its internal objectives. Given this
limited definition of success, we have summarized these factors in the
science area in figure 4.3.

Glyde has done one of the very few analytical studies known to us
of institutional links in science and technology. (91) Although focused
on cooperation between the United Kingdom and Thailand, and based
upon a small sample, the results presented in table 4.3 may provide
useful information on cooperative international S&T activity.

ISSUES

Some issues in science involvements such as funding levels,
length of projects, and lack of independent evaluation, center on areas
of general concern to all three fields we have chosen to examine,
namely science, engineering, and agriculture. Other issues are

Criteria and Conditions for Success
- Good leadership; often a personal responsibility of one individual, as was the case in science education in Brazil.
- Commitment of LDC government to basic and applied science - shown necessary in many cases.
- Organized authority, not too diffuse, as was the case in MUCIA/Universidad Agraria del Peru.
- Good planning of exchanges.
- Connection with industry beneficial in several cases.
- Overcoming feeling of isolation in LDC projects through continued contact with personnel from developed countries - relevant in a few cases.
- Bureaucracy and governmental restrictions kept to a minimum.
- Orientation toward innovation and flexibility in both LDCs and U.S. - occasionally facilitated progress.
- National political atmosphere needed to be suitable; however, some contacts continued throughout political strife, i.e., Brazil, Iran.
- An LDC institution must want to develop a particular area for a project to be successful, e.g., physics in Peru.

Limitations to University Involvement
- Consortium mechanism not very successful.
- Involvement of an entire institution created an unresponsive arrangement, e.g., University of California/Universidad de Chile.
- Lack of encouragement within universities for interdisciplinary work that might best provide the framework for international projects.
- Publish or perish syndrome discourages faculty from leaving for a substantial period of time at the risk of being "out of touch."
- Universities do not always encourage striking out in innovative ways.
- A strong department within a university framework may be restricted and function more effectively outside the university, e.g., BSCS and other U.S. science curriculum development centers separating from their respective universities.

Fig. 4.3. Factors relating to science involvements.

Table 4.3. Factors Distinguishing Successful from Unsuccessful Links

FACTOR	Low	Medium	High
1. Mode of Initiation (a)			
DC → AC (b)		6	4
AC → DC		1	
Via third body (c)		1	
By third body (d)	3		
2. Method of Establishing Objectives (e)			
Use of DC centre recommendations		3	3
Long-term collaboration by DC-AC personnel working in DC centre		4	2
Survey by the AC	2		
3. Type of Visit by AC Personnel to DC Institution			
Repeated short visits (small group)		2	2
Repeated short and long visits (small group)		2	2
No visits		4	1
Single long-term visit	3		
4. Type of Funding (f)			
Core funding or internal support	1	5	5
No core funding	3	3	
5. Management Strength of DC Institution			
High		2	5
Medium		6	
Low	3		
6. Magnitude of Funding (in £ 000's)			
Less than 10		3	3
10-100		5	2
Greater than 100	3		

(a) In one case, the method of initiation is unknown.
(b) DC → AC: Initiation by direct approach from the DC institution to the AC institution.
(c) Via third body: Initiation by one institution through a third body, which located the other institution.
(d) By third body: Initiation by a third independent group, which brings the two institutions together.
(d) In two cases, the sources of objectives were unknown.
(f) One link falls into both categories.

SOURCE: H.R. Glyde, "Institutional Links in Science and Technology: The United Kingdom and Thailand," *International Development Review/Focus*, 1973, Vol. XV, No. 1, p. 8, with permission.

specific to science such as the relevance of scientific research to
local and/or national needs, the size and administrative complexity
of science projects, the feasibility of carrying out true collaborative
scientific research, and the question of the relative emphasis on basic,
applied, and technological research.

General Issues

Funding of science projects seems to have a different dynamic
than in engineering and agriculture. Although there have been complaints from U.S. and LDC universities alike that funds have been too
few, it has seemed possible to accomplish a great deal with relatively
small amounts, judiciously allocated. Particularly, the bilateral
mode of interaction seems conducive to maintaining relatively high
performance at low levels of funding. Moravcsik (92) and others have
argued that even with the current level of U.S. funding, a shift in priorities toward more emphasis on bilateral, one-to-one type science
development projects would have a significant impact on LDC's S&T
infrastructures. This does not necessarily invalidate the case for
more spending on science projects, but it raises the issue of which
mechanism, bilateral or multilateral, can utilize funding in a more
effective way.

Length of projects has been of particular concern to science involvements for a number of reasons. First, the process of building
research and/or teaching capacity in the basic and applied sciences
is definitely a slow one. It is a process that must occur at many levels - secondary, undergraduate, and graduate - therefore it is extremely complex. Second, visits by U.S. faculty or students need to
be long enough so that they can become accustomed to local conditions
and have time to make a significant impact. Also it seems that U.S.
institutional involvements with a long-range commitment encourage
faculty and students to become involved. This is because it appears
to generate greater acceptance at the U.S. university and occasionally
will also provide opportunities of work and/or research for the U.S.
personnel when they return home.

Although the above analysis has some general validity, it seems
to apply primarily to institution-building efforts. Table 4.3 indicates
that other approaches - short-term visits, a mix of visits, even no
visits - can yield some success in cooperative R&D activity.

Lack of evaluation, as in engineering and agriculture, is overwhelmingly evident in most science involvements. The diffusion of
U.S. science curriculum materials around the world has received
much attention, but we know of no specific evaluation of the role of

U. S. universities in their development. More effort is needed to gain access to science project evaluations that may exist in agency or foundation files and to build evaluations into new projects.

Issues Specific to Science

There is much argument that LDC universities are still ivory towers, irrelevant to practical needs, and that basic research must be more effectively used in industrial and public works applications. This relates directly to the debate over the relative emphasis on basis vs. applied vs. technological research. Moravcsik maintains that this argument has become somewhat of a red herring and stands in the way of accomplishing anything. To him, any science of high quality as judged by the internal criteria of science itself is worthwhile in an LDC, especially in the initial stages of development, and the order and the emphasis on short-term applications is not so important. (93) Others feel just as strongly that scarce scientific resources need to be targeted toward development needs. The international centers of research excellence seem to straddle this argument and manage to accomplish basic and applied research which is put to practical use. Regardless, the debate still continues with respect to the importance of basic science background for assisting in the choice of technologies transferred to LDCs, and that debate does not appear to be subsiding nor destined for early resolution. However, these are choices that developing country governments, and not the U.S., must make. The area in which the U.S. does have control is the relevance of U.S. university curricula and training to LDC students and faculty. Here, too, there is disagreement about the relative importance of the "basic" versus "practical" emphasis.

The size and complexity of administration of science projects is a major concern as well. There appears to be a distinct preference for smaller, more personal, less bureaucratized, often bilateral projects in the science area. This helps encourage individual initiatives and is apparently most conducive to effective research/laboratory arrangements. Perhaps the issue then becomes how can multilateral involvements manage to be kept in small enough packages to lend themselves to this mold? Or is there another form that multilateral projects can take?

One final issue is the question of whether the "mutual-benefit" theory can actually work and collaborative research among equals be accomplished. It would appear desirable for numerous economic, scientific, and diplomatic reasons for this to occur. As LDC institutions grow and expand, they become capable of providing facilities, personnel, and funding for joint projects. The East-West Center, the international research centers, the OAS PRDCYT program, and a few of the bilateral joint research projects seem to indicate that this potential is developing in LDCs and should be carefully gauged for possible future U.S. university science involvements.

5 Future Roles for U.S. Universities

In this chapter, future roles for U.S. universities and university personnel in science and technology for development are discussed under four types of involvements: institution building, cooperative R&D, recourse base involvement, and graduate and undergraduate education. Mechanisms for bringing about these involvements are described and three alternative scenarios are sketched. The analysis of this chapter builds upon the previous analysis of past involvements and the legislative mandate. General policy issues and options are considered in chapter 6, along with specific legislative changes that could affect the role of U.S. universities.

TYPES OF INVOLVEMENT

Institution Building

Future involvement in institution building on the part of U.S. universities is likely to reflect past efforts but with some significant differences. In the 1950s, U.S. land-grant universities were involved in a major effort to build agricultural institutions patterned to some extent after their own image. In this connection, the term "institution building" is usually defined as a model or as an approach, for example, "The objective of the institution-building <u>approach</u> [emphasis added] is to develop an indigenous, long-run technical assistance facility that can provide, or create, the techniques for solving problems relevent to its environment." (1) The literature on institution building seems to focus heavily on the theory of the institution-building process or

Future Roles for U.S. Universities 179

model as opposed to the actual results or impact of institution-building efforts.

Major institution-building programs extending from the late 1950s to the early 1970s included the work of five U.S. agricultural colleges to help build agricultural colleges in India, and the efforts of two university consortia to build engineering colleges in India and Afghanistan (described in chapters 2 and 3). Large-scale institution-building involvements are continuing, particularly in the OPEC countries which are replacing the U.S. government as the major source of support for these bilateral involvements. The 1977 annual report of the Education Development Center describes three such efforts, involving consortia of U.S. engineering and engineering technology schools, to build the National Institute of Electricity and Electronics (INELEC) in Algeria, the Institutes for Polymer Science and Technology in Algeria, and the Port Harcourt College of Science and Technology in Nigeria. (2)

At the same time, there appears to be a shift in emphasis of U.S. government funded projects away from large-scale U.S. university involvement in building LDC institutions, particularly those of higher education. The new emphasis in the Title XII program (see chapter 3) is on collaborative research projects as opposed to institution building. This shift is a result of several factors: (A) science and technology institutions and infrastructure in LDCs are more developed than they were in the 1950s and 1960s; (B) large-scale U.S. involvement in institution building in some LDCs is less politically acceptable than it was previously; (3) and (C) U.S. government policy and funding are not presently focused on large, higher education, institution-building efforts. The collaborative R&D mode and consulting visits by individual faculty are more in keeping with these three conditions than are large, U.S. institution-building involvements.

Rao has provided a perceptive analysis of the role of universities in developing countries that is useful in considering what types of U.S. university involvement might be needed in the future. (4) Figure 5.1, taken from Rao, lists assistance LDC universities could use, divided into three categories according to various stages of development. An important generalization is that all LDCs are not alike; U.S. university responses must take into account the specific S&T infrastructure needs within individual LDCs or within regions of LDCs.

Rao also outlines several areas in which opportunities for action are possible. He feels that regional centers of excellence in the developing countries are desirable links in building S&T infrastructure in LDCs. (5) However, although regional centers have many advantages, they are sometimes difficult to implement and administer politically. Therefore, Rao states that what is required are national centers (see. fig. 5.1, category I) with graduate level research and

Category I. Countries in which the infrastucture of science and technology is reasonably well developed and universities have moved forward to establish post-graduate courses

Assistance universities could use and functions they could perform include:

a) continued development of post-graduate programs of teaching and research.
b) selectively targeted loans and grants to complete the development of specific departments.
c) contracts for the training of professors from other institutions in the country and in the region.
d) contracts for curriculum development in collaboration with other universities and secondary schools.
e) support for contract research and consultancies for industry, government and international agencies.

Illustrative examples of countries in which such conditions obtain are: India, Pakistan, Korea, Philippines, Brazil, Argentina, Chile, Nigeria, Ghana and Singapore.

Category II. Countries in which the infrastructure of science and technology is only partly in place and university development has not passed beyond the stage of undergraduate education.

Assistance program for this category of university would include:

a) university systems planning and institutional planning assistance.
b) loans and grants to complete undergraduate programs.
c) assistance for the training of local professors at home and abroad.
d) completion of buildings and acquisition of equipment.
e) technical assistance including the provision of expatriate professors and university managers.

A partial list of countries in which such needs exist are: Indonesia, Tanzania, Malaysia, Thailand, Egypt, Kenya, Ghana, Colombia, Iran, and Turkey.

Category III Countries in which both the science and technology infrastructure and the system of higher education are in rudimentary stages of development.

Programs of assistance in this class of universities would typically include:

a) planning assistance.
b) investment in physical facilities, including equipment.
c) advanced training of prospective professors during the period when physical facilities are being constructed.
d) substantial technical assistance including expatriate professors, university planners and managers.

Examples are: several African and Middle Eastern universities, Bolivia, Ecuador, Burma, Vietnam, Cambodia, Laos, and Central American universities.

Fig. 5.1. Role and Needs of LDC Universities

SOURCE: K.N. Rao. "University Based Science and Technology for Development," *Impact of Science on Society* 28, no. 2, © UNESCO, 1978, reproduced by permission of UNESCO.)

Future Roles for U.S. Universities

training programs that can serve a regional role. He cites Nigerian and Ghanian institutions in the West Africa region, Mexican universities in the Central American region, and Egyptian and Lebanese universities in the Middle East as already serving this function to some extent. A recent article by Szmant indicates that strong national centers for chemistry are emerging in Latin America with focus on regional problems. (6)

It may be desirable to shift emphasis, where possible and where desired by the developing country institution, to programs that seek to couple the local LDC universities to the productive sector. The RITA program and Georgia Tech's cooperation with the Technology Consultancy Center of the University of Science and Technology, Kumasi, Ghana, are two examples of such involvement. A key element here seems to be the presence of an indigenous LDC institution which can interact with the productive sector in a significant way.

In the science area, we have come across very little information on large-scale institution-building efforts of the type found in agriculture and engineering. There has been some recent discussion of the desirability of supporting the building of "MIT or Cal Tech" type institutions in selected LDCs centered on research areas likely to be of importance to these countries. Such centers could contribute to both national and regional development. An interesting prototype of such efforts is the NAS-Brazil graduate chemistry project, which led to the strengthening of chemistry departments in Brazilian universities, staffed in part by U.S. postdoctoral researchers with part-time senior U.S. faculty supervision. On the other hand, Carl Djerassi, who spearheaded the Brazilian effort, warns against the disruptive effect that the creation of new centers of excellence may have on existing institutions in countries with significant technological and scientific shortcomings. (7)

Although institution building often comnotes the building of an LDC university or of specific capability within the university, e.g., a department, other types of institutions can be envisioned. Hathaway seems very positive about the impact of the international agricultural research centers (e.g., IRRI, CIMMYT) which were built with U.S. foundation support, and less positive about the contribution of U.S. universities to institution building. (8) However, U.S. universities were a primary source of initial staffing for these centers. Furthermore, under the Title XII program, increased cooperation among U.S. universities and the international agricultural research centers is expected to develop.

If LDC institutions are to be relevant to short-term, immediate basic needs and problems, they need to develop outreach programs and learn to work directly with the people most affected. (9) This effort could involve other institutions as well as universities. Required skills can be obtained at the level of technical schools, or through nontraditional and nonformal education methods. It may be that U.S. community colleges or technical institutes and their personnel could play a role here. In addition, there is no reason why selected four-year colleges and universities with graduate programs in both the U.S. and LDCs could not develop a focus that would be responsive to these concerns. (10)

According to C. Barker of AID, the largest role of U.S. universities has been in providing advisors to LDC country governments (ministries, planning agencies, etc.). (11) Consultation, visits, and visiting appointments by U.S. faculty and other U.S. university personnel to these organizations as well as to LDC universities are likely to continue to be requested by LDCs for some time to come, with the demand exceeding the supply. Supporting commodities such as scientific equipment and books will also be needed. These aspects of institution building are less visible than the large "bricks and mortar" projects but are of considerable importance.

It may be time to retire the phrase "institution building" and substitute for it something like "institution strengthening." It also may be that the cooperative R&D mode will be more in tune with the 1980s than the institution-building approach. However, the large, multimillion dollar contracts involving consortia of U.S. universities in institution building in OPEC countries indicates that the institution-building era is by no means over.

Cooperative R&D

Cooperative R&D undertaken jointly by LDC and U.S. universities, departments, and/or faculty has been written about and discussed for some time. (12) Our examination of available information revealed little good documentation of such activity in the past, which may indicate that little has actually been done. However, discussion at the Project Workshop and our examination of current thinking indicate that cooperative R&D activity may be a major focus for future U.S. university involvement, in spite of some concern and skepticism related to the difficulties associated with performing truly cooperative R&D.

The term "cooperative" or "collaborative" R&D can encompass a broad spectrum of activity. Cooperation between partners who fully share in the conception, funding, and execution of R&D projects is one possibility. Alternatively, one partner may be more advanced than the other and yet collaboration still serves to transfer skills and knowledge. The nature of the collaboration would depend upon the skills, resources, and needs of the cooperating institutions. Another distinction that might be made concerns the objectives of the collaboration. A cooperative international science project might generate basic knowledge which may or may not prove useful some day. Another cooperative project might focus very specifically on a development problem requiring immediate attention.

Most international science projects have involved collaboration among developed countries. Many U.S. scientists and engineers prefer to collaborate with developed country research centers. These preferences are reflected in requests for travel grants and research support to government agencies like the National Science Foundation. NSF has an international cooperative science program but, with the exception of a small number of Latin American countries, there appears to be little focus on science and technology for international development. This NSF program might prove to be an effective vehicle for future activity, particularly in the science area.

A major new thrust of U.S. university involvement could be an expanded program of collaborative research on development problems between U.S. and LDC institutions. Such a program would seek wherever possible to match institutions with common interests so that both participants gain from the experience. In agriculture, the Title XII program is a source of support for collaborative R&D on food and agriculture problems among U.S. universities that will be linked to LDC institutions. Such activity is just getting underway and needs to give more emphasis to involving LDC institutions. Some arrangements should also stress enhancing the capability in the LDC for carrying out development-oriented R&D.

Several new cooperative R&D thrusts involving U.S. universities have been recommended in the recent National Academy of Sciences UNCSTED study report. For example, included are recommendations for joint research on soils and water management, on the industrialization process, on small-scale technologies based on renewable energy resources, and on the marine environment. (13) Such initiatives could very well be an important part of an expanded international S&T effort.

There are certain limitations and obstacles to cooperative R&D. Some were discussed at a May 10, 1978, talk at Washington University by Dr. M. S. Gore, director of the Tata Institute of Social Sciences, Bombay, India. Although they need not apply to all LDCs and although they are based upon experience with social science research, they do have some relevance to cooperative science and technology research. Often, the cooperation is unbalanced. There is a one-way flow of funds, operating with a "patron-client" syndrome. Gore feels that collaboration must be at the intergovernmental level to be politically acceptable. Moravcsik states that there are many examples that could be cited of collaboration not involving governments. (14) In India, U.S. money is channeled through an Indian body, the U.S. Educational Foundation in India, which has both U.S. and Indian representatives. The Ford Foundation has insisted that all proposals received be screened by the Indian government before they consider funding them. Mechanisms for joint research are lacking. Those that are in place are cumbersome and slow. Gore feels that ingenuity is needed to find new institutional patterns. He favors a strategy of low-profile collaboration (not talking too much about it) and is generally opposed to large projects.

A potentially important future role for U.S. universities that relates both to cooperative R&D and to resource base development involves expanding the amount of research carried out in U.S. universities that is relevant to LDC development problems. With such expansion is likely to come more opportunities for U.S. researchers to visit LDC centers and for LDC researchers to have more frequent contacts with the broader international research community. Such an effort might make substantial contributions to building LDC S&T infrastructure and to solving development problems, provided that it can be done in a way that does not drain scarce LDC human resources away from the LDCs when and where they are most needed.

U.S. Resource Base Development

The concept of building capability in U.S. universities to contribute to the international development effort has had a rocky path. AID's 211(d) program appears not to have been adequately funded and, according to a GAO report, not fully accepted and utilized by AID. (15) Although AID's authority to make 211(d) grants is still in effect, the program has rapidly wasted away since 1975, with very few new grants being made. As of the end of March 1978, it seemed likely that the number of new grants would stay very small and be focused on selected U.S. minority institutions.

This somewhat passive role of AID with regard to 211(d)'s future is at odds with recommendations of two major reports. The National Academy of Engineering's study on The Role of U.S. Engineering Schools in Development Assistance recommended a major expansion of 211(d) activity in engineering. (16) The industrialization panel of the National Academy of Science's UNCSTED study proposed the revival and extension of the 211(d) program. (17) It is possible that AID's loss of interest in 211(d) came about due to a variety of factors, including GAO criticism and the creation of the new Title XII program. The latter may provide substantial future resource base support but it will be focused solely on food and agriculture. Furthermore, only land-grant universities or universities with teaching, research, and extension capabilities in food and agriculture may receive Title XII institution strengthening and collaborative research support grants. The net effect of the switch from 211(d) to Title XII, once the latter is fully implemented, will be to strengthen the potential U.S. university capability for involvement in international agriculture. However, in the fields of engineering and science, the change has resulted in reduced possibilities for resource base support from AID.

An important reason for continuing to provide resource base or institution strengthening support to U.S. universities is to make the education of the thousands of LDC students who come to study engineering, science, and agriculture in the U.S. more relevant to their home-country needs. Furthermore, such support provides a base for collaborative R&D involvements with which to involve those students in close, continuing relationships with institutions in their home countries.

The resource base idea is most effective if the base is integrated into an overall set of development-related activities, including the provision of relevant education and/or training for foreign nationals in the U.S., cooperative R&D, and institution building or strengthening. If funds are restricted to only one of these activities, then the base is more likely to be isolated and unused. Our evaluation of the Georgia Tech experience (see chapter 2) indicates that such integrated programs can be successful.

Washington University's School of Engineering has been interested in the resource base idea since the mid-1960s and has attempted to develop a multifaceted program involving courses, curricula, and cooperative R&D. (18) A framework for international cooperation now exists at this university, but it was necessary to move the program away from its initial, exclusive international development focus due to lack of financial support. A small level of international effort is now maintained within a broader technology and public policy framework.

A major problem with the resource base idea as implemented in the 211(d) program, particularly with regard to engineering programs, is that relatively large grants were made to a very small number of institutions out of a low total budget. This concentrated, limited support served to limit the involvement on the part of U.S. faculty in international development work. Institution-strengthening grants will be a future part of the Title XII program. We do not know what form this will take or how widespread the participation will be. Mechanisms are needed to provide for broader participation. For example, many U.S. universities have strength in areas of science and technology which might be directed toward LDC concerns by an adequately funded, competitive program of relatively small grants.

It can be argued that providing support to U.S. universities to build capability for development work diverts support that could be used directly in LDCs, and that the main task of S&T infrastructure must be done by and within the LDCs themselves. Although we are sympathetic to this argument, it seems to us that the resource base idea is worthy of future support, particularly because of the contributions it can make to the relevant training of LDC students in the U.S. and to cooperative R&D.

Education and Training

LDC Students to U.S.

We are likely to see continued involvement of LDC students in U.S. educational institutions. Moravcsik has listed a number of steps that could be taken to make the U.S. experiences of such students relevant to development back home, such as more pertinent curricula, more stress on exchange visits, and shorter visits. (19) Other useful activities discussed at the Project Workshop include summer courses on development-related topics for LDC graduate students already in the U.S., better mechanisms for matching LDC students to departments and programs, such as the physics interviewing project, and opportunities for thesis research in home countries.

In 1976-1977, foreign student enrollments in U.S. institutes of higher education represented only 1.8 percent of total enrollments. (20) However, the total number of foreign students studying in the U.S. topped 200,000 for the first time, rising rapidly over the past three years. In 1976-1977, engineering, agriculture, and the natural and life sciences accounted for 24.1 percent, 3.2 percent, and 11.3 percent, respectively, of all foreign student enrollments. There has been an increasing number of students enrolling in two-year as opposed to four-year curricula.

The trend in graduate engineering enrollments is particularly interesting. We estimate, based upon <u>Open Doors</u> statistics, that in 1975-1976, the percentage of graduate engineering students who were foreign students fell somewhere between 29 percent and 50 percent, depending upon whether just full-time or full-time plus part-time students are counted. According to the <u>1976 Digest of Education Statistics</u>, 41.3 percent of doctoral degrees in engineering in 1974-

1975 were awarded to individuals of foreign citizenship. (21) We have no detailed breakdown of funding sources for foreign engineering students. For foreign students in all fields, funding is now coming primarily from home-country governments and from private sources available to these students rather than from the U.S. government (see table 2.4).

Since the buildup of engineering research and graduate capability following the launch of Sputnik, engineering schools have competed for graduate students. Foreign students have played and continue to play a significant role. The current high level of foreign graduate engineering enrollments may reflect a period in which the fall-off in total engineering enrollments in the early 1970s was partially compensated for by an increase in foreign student enrollment, particularly at the graduate level. However, one might now expect the percentage of foreign graduate students to decrease as the recent new surge in undergraduate engineering enrollments reaches graduate school. Countering this trend is the start of the reduction in the total college-age population which the U.S. is beginning to experience, as well as large starting salaries for engineers with bachelor's degrees.

The future role of U.S. universities will also be affected by the policies that both LDCs and the U.S. adopt with regard to visas and immigration. Canada, (22) the United Kingdom, and certain Western European countries are moving to restrict foreign student enrollments by charging higher tuition for foreign students than for native students. However, the overall percentages of foreign students enrolled in universities in those countries are considerably higher than in the U.S. Kidd has pointed out that immigration of scientists and engineers to the U.S. from LDCs has fallen off sharply since the early 1970s when U.S. immigration laws were changed to remove preferences for those professions. The LDCs may also take steps to help prevent "brain drain." (23) At the 1979 United Nations Conference on Trade and Development (UNCTAD V), there will be discussion of a resolution calling for the developed countries to pay compensation for negative effects of the "brain drain" on LDCs. (24)

We have not undertaken a detailed analysis of the impact of the "brain drain" issue on the future role of U.S. universities. Nor do we favor increased immigration restrictions on foreign scientists and engineers wishing to study in the U.S. because overall it seems to be in the U.S. national interest not to limit such participation. Foreign students coming to the U.S. have contributed greatly to our graduate and undergraduate programs and, in the case of those who stayed in the U.S., to our society. In the case of engineering graduate programs, foreign students have kept many programs afloat by providing research assistants and students to teach. Many foreign students who have gone back to their countries to significant positions in government, industry, and the universities have maintained friendly contact with counterparts in the U.S.

On the other hand, there are three steps that U.S. universities can take, aided by outside financial support, that could be useful in helping the LDCs build an indigenous S&T base. First, because of the very large enrollment of foreign graduate engineering students in the U.S., and because we can expect continued involvement of this kind in many fields, it seems important to develop curricula, courses, and/or summer workshop experiences, preferably with external financial support, relevant to home-country development needs. There has been reluctance on the part of engineering institutions to do so in the past but the large foreign student enrollments argue for greater attention to such activities.

Second, although many LDC students will pay their own way or have it paid for them by home governments, financial aid awards will continue to be of some importance in determining how many foreign students will come to the U.S. for graduate study in science and technology. Often departments make such awards to foreign students because they lack suitable U.S. candidates. U.S. universities need to support efforts to develop and attract more U.S. graduate students in science and technology, particularly those from traditionally under-represented populations, namely women and minority groups (i.e., Blacks, Spanish surname, American Indian). Wilburn has pointed out that foreign students coming to the U.S. are politically sensitive to the fact that we are not doing all that we could for our minorites here at home. (25) We are not arguing for a total cutoff of U.S. financial support for foreign students; there are good reasons why some support should continue. However, we do think that at some institutions, the proportion of foreign students in some fields may have grown larger than is desirable for all concerned. Such a situation can lead to difficulties. (26)

Third, U.S. universities can aid in building an indigenous LDC S&T base by encouraging prospective students to explore opportunities in their own countries. Sometimes, correspondence can reveal institutions in LDCs that are reasonable alternatives to study in the U.S. An exchange of views on the student's objectives may enable the U.S. institution to provide a form of international career guidance. Contacts with counterpart faculty in LDCs may also be helpful in this regard. By so doing it may be possible for the U.S. university to aid the LDC to stop thinking automatically about sending students to study abroad and to gradually shift the emphasis to strengthening the educational infrastructure within the home country.

U.S. Students and Faculty to LDCs

As LDC institutions develop, there could be continuing demands for involvement of U.S. faculty, and in some cases, students to serve in various capacities. In 1978 a major recruiting effort was underway for INELEC. Ads often appear in Science for individuals to staff new OPEC institutions. Generally, these assignments pay well but are somewhat detached from the professional involvements of U.S. faculty at their home institutions.

NSF's SEED program has been undersubscribed; reasons for this need study. It may be that there are enough potential U.S. participants but that the nature and terms of the involvements are limiting interest. The Senior Fulbright-Hays Program has suffered in recent years from a slippage in the attractiveness of the support and duration of awards. Improved and varied opportunities are needed to attract U.S. faculty to the LDCs.

U.S. students in the sciences, engineering, and agriculture at many levels as well as postdoctoral research associates can play a variety of roles. In the University of Wisconsin-Monterrey Tec exchange program, undergraduate engineering students in their junior year are exposed to the education and culture of a counterpart institution. Peace Corps volunteers with engineering, science, and agriculture training can sometimes earn academic credit toward advanced degrees for their service experiences. The involvement of U.S. postdoctoral research associates in the Brazilian chemistry project might be a useful model for other sciences. The industrialization panel of the NRC-UNCSTED study has recommended that a "Technology Corps," for experienced engineers and mid-career consultants, be developed, as proposed by the U.S. at UNCTAD IV, (27) and the Anti-Nuclear Proliferation Act of 1978 calls for the Department of Energy to carry out a feasibility study of a "Scientific Peace Corps" to work on small-scale energy sources.

Training Programs

We use the phrase "training programs" to denote an experience different from a formal degree program. Such programs usually are of limited duration and have specific objectives. As such, they are less likely to lead to "brain drain." Training programs both in the U.S. and LDCs have been a major part of AID-supported activity. They are likely to be needed for some time.

MECHANISMS FOR FUTURE U.S. UNIVERSITY INVOLVEMENTS

In order to sustain the types of involvement described above, a variety of mechanisms are required. As always, we use the term "mechanisms" very broadly to include programs, organizational forms, and sources of support created by legislation as well as arrangements between or among LDC and U.S. institutions and/or individuals. Mechanisms can be either bilateral or multilateral. We also explore possible relationships between U.S. universities and other institutions in the U.S. that might prove useful in helping to build an indigenous science and technology base.

Bilateral Programs

Programs for Individuals

There are a variety of mechanisms either proposed or established for involving U.S. students and faculty in S&T for development. Diversity is desirable. However, it is our impression that existing opportunities leave much to be desired, both in terms of the type of experience and the amount of financial support offered which in turn limits participation.

There are some programs that are specifically geared toward involving U.S. faculty and other researchers in development projects. One is the SEED program, Scientists and Engineers in Economic Development, which provides support for short- or longer-term visits by U.S. faculty to developing countries. This program, which up until now was funded by AID and administered by NSF, appears to have contributed in modest ways to institution-building efforts. (28) It is a mechanism that appears to be somewhat unique, useful, and worthy of improvement efforts.

In his evaluation of the SEED program, Kidd points out that only one-half of one percent of the $69 million spent by the government in FY 1977 on educational exchange activities was spent on the SEED program. (29) It can probably be safely assumed that much of the remaining 99.5 percent was not focused on building an indigenous S&T base in LDCs. Kidd advocates an expanded SEED program with a two-way flow of scientists and engineers rather than the one-way flow which now prevails. It might also be helpful if the prestige and financial arrangements of the SEED program were improved so that more university personnel would be attracted to it.

Another program is FORGE, Funds for Overseas Research Grants in Education, which Szmant feels has been more successful than massive, multimillion dollar, international U.S. programs for chemical R&D in Latin America. (30) FORGE has focused its attention on support for the individual investigator and supports "relatively modest grants awarded on the basis of highly scrutinized, technically sound proposals." (31) This program gives money directly to Latin American investigators. Such small, competitive grant programs would appear to be very desirable and worthy of continued support. At the Project Workshop, there was some discussion of the FORGE program. Since it has been the effort primarily of one individual, Alfred Kelleher, who recently died, there was some question as to whether the program would continue.

A similar mechanism is provided by the International Foundation for Science, IFS, which is based in Sweden and awards small grants on a competitive basis in several scientific areas relevant to development. (32) Roger Revelle of the U.S. is on the board of trustees but the U.S. makes no financial contribution to this multinational organization. Eilers suggested using existing U.S. organizations with overseas links, such as Georgia Tech and the Denver Research Institute, as channels for providing U.S. government support to LDC researchers and institutions rather than direct grants, as in the case of FORGE or IFS. (33)

Workshop participants generally felt that a small grants mechanism was needed for U.S. researchers. A competitive research grants program could serve to greatly increase the involvement of U.S. scientists, technologists, and their institutions. Currently such a program does not seem to exist. Activities within NSF's International Division might be expanded to provide such a mechanism or it might fall under the aegis of a new Foundation for International Technological Cooperation. Proposals might also be accepted for collaborative research between U.S. and LDC investigators. Some cooperative activity of this kind does take place in the NSF Cooperative Science Program within the NSF's International Division.

Other existing mechanisms for facilitating the flow or exchange of students and faculty between U.S. and LDC institutions include: 1) awards under the Fulbright-Hays program; 2) international travel grants from NSF and other domestic agencies; 3) student exchange and "study abroad" programs between universities; 4) international conferences and large international projects such as the International Geophysical Year; 5) the Physics Interviewing Project. Workshop participants indicated that there is need for a better mechanism or mechanisms to advise LDC personnel where to go to study in the U.S. Existing organizations such as the Institute for International Education, the LASPAU and AFGRAD scholarship programs, and the Council for the International Exchange of Scholars are active in this area; however, more targeted efforts in engineering, science, and agriculture may be required, as in the case of the Physics Interviewing Project. In general, the demand for international travel and research project participation on the part of U.S. scientists is greater in other developed countries than in LDCs. More targeted funding for LDC activity on the part of sponsoring agencies could change the situation.

Links between U.S. universities and other U.S. institutions can provide useful mechanisms for individuals to participate in S&T for development activity (see a later discussion). Other mechanisms include the proposed Technology Corps and Scientific Peace Corps mentioned in the previous section.

Programs for Institutions.

The new Title XII program provides a legislative mechanism for expanded involvement of land-grant and other qualified U.S. universities in international food and nutrition activity. The title itself and its implementation provide several mechanisms of interest for institutional cooperation. Creation of BIFAD gives these universities a mechanism for much stronger involvement in policy making and design of international programs than in the past. The collaborative research support program provides a mechanism for dividing resources among U.S. universities and for relating to the international agricultural research centers. Long-term (five-year) support for universities is provided.

Two of the changes to Title XII that have been discussed at the Project Workshop (and elsewhere) are the extension of the Title XII concept to engineering and science, and the removal of restrictions that currently eliminate many universities from participation in certain aspects of the program. No consensus appears to have emerged. It seems to us that conditions are very different in engineering and in science than in agriculture, making the kind of focused, controlled effort which Title XII represents difficult to implement. Also, it may not be politically feasible for noneligible institutions to expect to gain entry to the program. (For further discussion of Title XII, see chapter 3.)

Another legislative mechanism of interest is the 211(d) program which can provide five-year resource base support for U.S. universities but which has fallen into disuse. That this mechanism is still of considerable interest is reflected by the fact that two National Academy studies have called for its expansion, particularly in relation to engineering programs.

Future Roles for U.S. Universities 193

Although AID has been the predominant source of support for international development activities by universities in the past, legislative mechanisms are coming into place that, with concomitant funding, could expand the international activity of primarily domestic organizations. For example, the Department of Energy in the Nuclear Anti-Proliferation Act of 1978, is charged with carrying out a program of U.S. assistance to developing countries, centered on small-scale, alternative energy sources. Included in this act are provisions for a program for exchange of U.S. scientists, technicians, and energy experts, as well as a charge to carry out a feasibility study of an international cooperative effort to include a Scientific Peace Corps, designed to encourage large numbers of technically trained volunteers to live and work in developing countries. (34)

A National Academy of Sciences study has recommended that "the United States support a modest program to enable universities and other research institutions to cover some of the special overhead costs of international collaboration in research and development and to continue to exchange scholars, subscribe to each others journals, host joint workshops, and the like, in order to cement continuing, productive relationships." (35) A specific suggestion by NAS is that the charter of the Public Health Service be revised to include special authority for engaging in health activities that have global dimensions.

The National Science Board has considered several options for expanded international involvement of the National Science Foundation in international programs, including expanded collaboration with more scientifically advanced developing countries, short courses for LDC students in the U.S., and research in the U.S. and overseas focused on problems relevant to LDC priorities. A modest program of expansion was supported by the Board, but as of this writing, funding prospects were not good. Other domestic agencies such as USDA have international authority but little if any funding.

One mechanism of sorts involves AID paying for the international program involvements of domestic agencies with "pass-through" funds. Although useful, this practice means that programs must be restricted to those that fall within AID's mandate.

The NAS/CNPq Brazilian chemistry program focused on building strength in graduate chemistry study at selected Brazilian universities, employing young postdoctoral researchers from U.S. universities. Some scientists feel that this program represented a very effective mechanism and should be replicated; others feel less positively about it. A parallel idea would be that of strengthening selected LDC regional or national institutions around applied science and technology with potentially large payoffs, like tropical products. An LDC center might cooperate with a "sister" U.S. institution or with individuals from several U.S. institutions, with support from the National Science Foundation. Involvement of U.S. university personnel at several levels of academic development might be considered. Positive impact on LDCs includes the development of applied science and technology capability in fields such as local natural products and resources that will give them a comparative advantage over developed countries. Negative impact of a strategy that focuses on building "centers of excellence" includes possible pulling away of trained personnel from dispersed centers, leading to internal "brain drain." It should also be kept in mind that such centers might very well foster industries competitive with the U.S.

Consortia, Councils, and Networks

Consortia of universities such as MUCIA have been active in overseas institution-building programs in the past. The Education Development Center, a nonprofit organization, played a useful role as the administrative hub for the consortia of engineering schools which helped build engineering schools at Kanpur (India) and at Kabul (Afghanistan). These consortia were useful for recruiting U.S. faculty from a broad range of institutions where it was not possible to get sufficient faculty from one institution. (36) This was probably the main impetus for developing such consortia, which seemed to be favorably viewed in certain parts of AID in the 1960s.

The consortium mechanism is in active use today in projects in Algeria and Saudi Arabia. A new feature is the increasing diversity of consortium members - universities, technical institutions and private industry - which provides a more practical, engineering and technology-oriented, institution-building focus. A new and unique use of the consortium mechanism is the designing and implementing of a project to educate up to 900 students from the Sahel at the B.S. and M.S. level in the U.S. over the next seven years. This effort, funded by the Sahel development office of AID, will involve Rutgers University, Purdue, the South East Consortium for International Development (SECID), and the Consortium for International Development. (37)

Another type of mechanism is the Council for International Cooperation in Higher Education, proposed by Harrington, (38) which would serve as a clearinghouse for faculty and universities, colleges and community colleges desiring international involvement. The council would arrange for linkages, i.e., cooperative relationships with LDC institutions, and be the dispensing organization for a small grants program. In this case, the rationale seems to be that a council of this kind could play a central, facilitating role for its member institutions just as professional associations in Washington serve individual members, e.g., the National Association of State Universities

Future Roles for U.S. Universities   195

and Land-Grant Colleges. Such an organization might be particularly helpful to smaller institutions and "new actors" in the international scene.

The Council on Science and Technology for Development was formed in December 1977 to work with "U.S. government and private organizations in determining how to use science and technology more effectively in advancing international development and in solving global problems of major importance to the United States." (39) The council has been active in advising the U.S. government on preparations for UNCSTED and seeks to serve as a clearinghouse for consideration of policy alternatives, strategies to global problems and to development, and suggestions for new alternatives. It has interest in working with universities on international science and technology policy research.

Several new mechanisms were proposed at the Project Workshop to facilitate linkages between individuals and institutions. Blase proposed that the U.S. take the lead in creating a network of universities, on a worldwide basis, to facilitate the interaction of academic personnel at various levels. (40) Rao suggested that universities be linked worldwide, through an "open university" network, to propose radical new solutions to pressing problems such as energy supply, demand, and conservation. Interaction might be facilitated by satellite communications. (41)

Multilateral Programs

Mechanisms for U.S. university participation in S&T for development via multilateral organizations and arrangements are not as well developed nor as extensive as bilateral involvements. Nevertheless, cooperation through multinational channels such as the United Nations and its specialized agencies, the World Bank, regional organizations, and banks, is likely to be of considerable significance in the discussions at UNCSTED. We will briefly review some mechanisms for U.S. university involvement; more attention needs to be given to this area than we have been able to provide in this project.

The Organization of American States' Regional Science and Technology Program is an interesting example of a mechanism for primarily inter-LDC cooperation. The U.S. is the principal financial supporter of this program. Such regional involvements have the advantage of helping to build self-reliance within the LDC region of interest and respond to the desire for cooperation among developing countries. Too heavy U.S. involvement is likely to be resented but, as has been suggested in chapter 4, individual U.S. universities and researchers can participate in specific projects. An obstacle to this is that apparently OAS would prefer that the U.S. not utilize OAS funds that might otherwise be used by Latin American members; thus, the U.S. might need to supply additional funds for such participation. This limitation may apply in general to many U.S. involvements in multilateral arrangements - the U.S. role is conceived of primarily as providing financial resources.

In a sense, support for regional universities such as the Asian Institute of Technology, is multilateral since these institutions serve a variety of countries. Such activities could benefit from U.S. support and U.S. university involvement, provided that such participation is politically feasible. It may be that the involvement of a consortium of universities from several countries, not just the U.S., in institution building and collaborative research could prevent the LDC university from being labeled as U.S. dominated. The establishment of national universities to serve regional needs is also of interest.

A third mechanism involves U.S. faculty participation as consultants or in other capacities for the World Bank, the United Nations, and its specialized agencies. Many U.S. consultants from universities have been used in connection with such projects and it is likely that such involvement will continue. In addition, these organizations have contracted with U.S. universities with capabilities in science and technology on behalf of UN member states. For example, the facilities of Wisconsin University's Instrumentation Systems Center are being used to help the Singapore Industrial Research Institute develop its metrology capability, supported by UNDP. (42) The U.S. also provides financial support for fellowship programs administered by international agencies such as UNDP, UNESCO, and FAO. A balance between bilateral and multinational support of this kind will probably continue to need to be maintained.

Perhaps the most innovative new mechanism for international cooperation involving universities is the United Nations University. Created in 1975, this university does not have a campus, students, or degree courses, but operates from a central planning and coordinating headquarters in Tokyo via networks of scholars and institutions worldwide. The networks identify critical international problems, undertake internationally coordinated research and advanced training, strengthen research and advanced training resources in LDCs, disseminate research results to decision makers as well as scholars, and encourage mission-oriented, multidisciplinary research and advanced training. (43) Shearer-Izumi has described the natural resources program of the UN University; (44) the other two programs underway deal with alleviation of world hunger and with social and human development.

The MIT International Nutrition Planning Program has established one of the first relationships with the UNU. Working relationships with Ohio State University and the University of Colorado based on specific program areas of expertise also exist. The UNU appears to represent the kind of networking arrangement - international in scope, university focused, centered on urgent development programs - that several of our workshop participants seemed to feel is desirable.

Nevertheless, there are obstacles to U.S. university participation. For one thing, the U.S. has not yet made a financial contribution to the UNU. For another, some university personnel have expressed concern that the arrangement is excessively bureaucratic and inefficient, a concern voiced often about multinational organizations. Finally, the U.S. role in such an organization is likely to be much less dominant than in bilateral arrangements.

In spite of these obstacles, we feel that a strong case can be made for U.S. participation in the UNU. U.S. universities need to be a part of this effort and not become isolated. Furthermore, a less dominant role for the U.S. may be very much in tune with LDC thinking. In preparing for UNCSTED, the question of U.S. financial support for the UNU needs to be carefully considered and resolved.

The Title XII program can eventually provide for greater involvement of U.S. universities in the activities of international centers of excellence (e.g., IRRI, CIMMYT, etc.). This involvement appears to be evolving along the lines of collaborative research with researchers from several universities. Other involvements outside food and agriculture include participation in international cooperative programs (e.g., the Global Atmospheric Research Program), in the International Institute for Applied Systems Analysis (ILASA), and in the programs of the International Council of Scientific Unions. As has been indicated previously, the problem seems to be a dearth of funding for LDC-focused activity as opposed to developed country activity. One useful new mechanism might be a postdoctoral fellowship program for research in LDCs paralleling the NATO fellowships.

Cooperative R&D Programs

Cooperative R&D programs with LDC institutions need major attention. We could like to see such programs supported on a competitive basis, involving peer review and with reviewers from the U.S. and developing countries involved. The new Foundation for International Technological Cooperation could serve as an important mechanism for facilitating such programs. It should be mentioned that this concept of competitive support has not been prevalent in international development activity to date because of the previous emphasis on institution building; yet it seems consistent with the tradition of university-based research and development in science and technology in the U.S.

Gomez has proposed that the U.S. government earmark a specified amount of money for long-term support of several projects involving pairings of one U.S. university and one LDC university or R&D organization. Through an open process of requests for qualifications and requests for proposals, a number of U.S. university projects would be chosen for presentation by the U.S. delegation at UNCSTED. LDCs would then bid for participation and designate one of their own institutions to collaborate in a specific program. A certain amount of matching funds would be required. (45) Although we see some problems with the proposal, it does reflect a desire to provide a specific plan detailing how universities might get involved in UNCSTED.

There also needs to be emphasis on new, interdisciplinary R&D activity that is development focused. Such programs would seem to have potential for contributing to development efforts both here and abroad. As an example, we cite Washington University's Technology and Human Affairs Department and Center for Development Technology, which permit students to focus their educational and research programs on specific development projects in areas such as food, energy, and environment as well as on related technology-policy issues. These programs have potential for working directly with development efforts overseas because their reward structure and their payoff is geared toward how well faculty and students produce on those specific problems rather than how well they function in a traditional discipline. Although these programs are not yet widespread, they should receive increased attention. (46)

There is a real shortage right now in jobs for U.S. young people in development-related activity. Mechanisms for solving this problem need attention. An expanded university effort and an expanded international R&D effort could provide those kinds of jobs, particularly if such efforts can be linked to government or industry.

New Cooperation Among U.S. Institutions

University-Community College/Technical Institute Cooperation

U.S. universities with four-year and graduate programs may not be able to respond adequately by themselves to the heavy emphasis on basic needs, rural development, and practical technical skills in current U.S. foreign assistance programs. Community colleges, technical institutes, and engineering technology programs have shown considerable recent interest in getting involved in international activity. (47) Currently, Nigeria is sending large numbers of students to these institutions in the U.S. (48) It would seem desirable to consider ways of linking U.S. universities with community colleges and technical institutes, perhaps in the same geographic area, to provide the kind of interchange and range of skills that might enhance both institutions' involvement in development efforts. The U.S. consortium in Algeria helping to build INELEC provides for this kind of cooperation. (49)

At the annual meeting of the American Society for Engineering Education in June 1978, discussions were held to explore the establishment of an International Center for Technological Education and Development as a mechanism for facilitating international cooperation in engineering technology (see chapter 2). Such an initiative reflects the growing interest in this aspect of engineering education.

Agriculture-Engineering Cooperation

There is a need in the food area for cooperation between agriculturalists and engineers. Traditionally, this has been provided by agricultural engineering departments in land-grant colleges and universities. However, in the past, these departments have not been strongly oriented toward international development, being more focused on agribusiness in the U.S. Title XII may stimulate such interaction. On the other hand, it may take new, innovative cooperative efforts to provide the kind of interaction that is needed, such as joint research (1) among electrical engineers, computer scientists, and agronomists on applications of remote sensing for monitoring crop conditions, (2) between chemical engineers and agriculturalists on technologies for village level food processing, and (3) between biomedical engineers, agriculturalists, and entymologists on use of time-release capsules for control of tropical diseases and for pest management. Increased cooperation between agricultural colleges and science faculties at non-land-grant institutions might be beneficial as well.

University-PVO Cooperation

A third area involves collaboration between universities and private voluntary organizations (PVOs). For example, Washington University has had interest for some time in collaboration between our Center for Development Technology (CDT) and Volunteers in Technical Assistance (VITA), a leading PVO with an appropriate technology focus. In a suitable master's program, a student might do coursework at a university and then do his or her field work or thesis project in collaboration with a development organization such as VITA which has objectives in science and technology for development that are similar to those of our center. This type of collaboration requires financial support for students so that they can carry on this particular activity. Collaboration between these two organizations is also possible in appropriate technology research. Currently, CDT and VITA are collaborating in a project on renewable resource utilization as part of U.S. UNCSTED preparations.

VITA, an organization comprised of a small staff which matches the skills of a volunteer network of 4,500 scientists and engineers with practical development problems in LDCs, maintains close ties with the academic community. A significant number of VITA's board members and longest-term supporters are affiliated with universities. During 1977, student interns from universities logged nearly 2,500 hours at VITA. Untested designs produced by VITA volunteers and problems posed by people in developing countries are utilized by engineering students and professors in courses. Many VITA volunteers are professors, students, and extension personnel who respond to requests for technical assistance channeled to them by VITA headquarters. For example, professors at Iowa State and Michigan State provide valuable inputs to the solution of many grain-related problems VITA receives. A professor at Pennsylvania State University was able to test a variety of materials found in developing countries - rice hull ash, breadfruit ash, etc. - for their binding properties; his findings are part of a short bulletin now available from VITA. (50)

During the summer of 1978, a student from Washington University, St. Louis, participated in an effort at VITA to review VITA's windpower information for technical accuracy, clarity, etc. As part of this study, the Washington University-VITA intern analyzed VITA's past work in the field of windpower and designed, with staff direction, a project to gain insight into the transfer and diffusion potential of wind technologies as well as the impact of VITA-provided information. The intern will prepare a report which will be compiled and published by VITA. In addition, the student will earn six units of academic credit as part of a larger project effort within the Washington University bachelor's degree program in technology and human affairs.

Still another example of VITA-university collaboration, involving a barbed wire fence-making machine in Botswana, and faculty and students from New Mexico State and Dartmouth Universities, is described in Appendix E.

University-Peace Corps Collaboration

U.S. universities have had a long history of collaboration with the Peace Corps. Early training programs for Peace Corps volunteers (PCVs) made extensive use of university facilities and personnel. It is now possible, through the Peace Corps Strategy Contract Programs, for returned PCVs to be supported financially as Peace Corps recruiters on U.S. campuses while pursuing graduate degrees. Some universities have also created mechanisms whereby PCVs can earn academic credit for their overseas experiences.

It should also be possible for U.S. and/or LDC universities to utilize the overseas experiences of PCVs to focus on specific development projects, and to use what is learned to help develop new educational thrusts both here and abroad. The Scientific Peace Corps being studied in connection with the new, international, small-scale energy program of the Department of Energy should lend itself to this approach. It may also be possible to carry out exchange programs in which PCVs with science, engineering, and agriculture backgrounds work in overseas universities, relieving LDC faculty who come to the U.S. for advanced training. Such an exchange was carried out between a Peace Corps engineer and an engineering professor in Honduras.

It is our impression that the Peace Corps is now emphasizing using generalists as PCVs as opposed to volunteers with more technical backgrounds. This situation may have arisen in part because of difficulties in recruiting technically trained volunteers. New innovative links between the Peace Corps and universities may serve to increase the pool of technically trained personnel.

University-ATI Collaboration

Yet a fifth type of collaboration could involve universities with Appropriate Technology International, the new, private, nonprofit organization recommended by AID and authorized by the Congress to implement appropriate technology activity. As of 1978, ATI did not seem to have much interest in involving U.S. universities in its activity, although individual faculty members were involved in its planning and continue to be consulted. After ATI gets well underway, it would seem desirable for further collaboration to evolve.

University-Research Institute, University-National Laboratory Collaboration

Another form of collaboration might involve interaction between U.S. universities and existing development-oriented research institutes

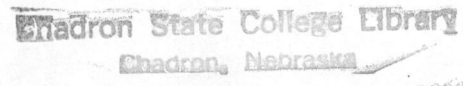

such as the Denver Research Institute (actually a part of the University of Denver), or between universities and U.S. national laboratories, such as Brookhaven, which appears to be developing an international energy program. For example, the lab or research institute might support doctoral candidates to work on development projects that are relevant to their own programs. The national labs in particular have considerable scientific and technological expertise which, if given a mandate for participation in cooperative international science and technology research, could make a significant contribution; they also have a long history of cooperation with U.S. universities.

University-Industry Collaboration

Still another form of collaboration involves the universities and industry. Industry might consider supporting university efforts in the U.S. that are directly related to international development efforts. Industry would be more interested in those projects that are more relevant to their own interests but, in any event, such collaboration needs to be examined.

Witunski has suggested several possibilities for U.S. university-industry collaboration. Multinational corporations (MNCs) might be persuaded to use money they cannot bring back to the U.S. because of tax laws to support research in LDCs. U.S. community and technical colleges and MNCs could collaborate to provide package deals, e.g., the sale of products to LDCs coupled with appropriate training courses at the colleges. (51)

Several workshop participants stressed the desirability of such collaboration, particularly in engineering because of that field's traditional linkages with industrial applications. Such joint ventures would enable U.S. university personnel to act as intermediaries between LDCs and MNCs in connection with technology transfer activity. In addition, Rao suggested that new models for university-industry cooperation in the U.S. such as the Polymer Processing Center at MIT might be good training grounds for LDC students as well as suitable for consideration in LDC settings. (52) Internship/training arrangements with U.S. corporations for students from LDCs and U.S. universities, and new links between U.S. engineering and business schools were also proposed.

Professional Society/Organization Linkages

The heavy involvement of U.S. universities and university personnel in professional societies and organizations has been noted previously. Organizations like the American Society for Engineering

Education, the National Association of State Universities and Land-Grant Colleges (NASULGC), and the American Association for the Advancement of Science are all active in international activities; the latter two have international program offices. NASULGC, in particular, was active in winning congressional support for the Title XII program.

Many of the specialized professional scientific and technological societies (e.g., the American Chemical Society, the Institute for Electrical and Electronic Engineers) have an international membership. These organizations can serve to focus the attention of their members on pertinent issues and to lend prestige and approval to international activity. Annual meetings provide scientists and engineers with opportunities for the useful exchange of information. These organizations bring together representatives from both academia and industry. At the Project Workshop, Miller stated that professional societies, especially in engineering, are badly underutilized in international development activity. He feels they could play a significant role in facilitating technology transfer through their links to the private sector. (53)

Paralleling the U.S. societies and organizations are national and regional groups in other countries as well as international organizations, such as the World Federation of Engineering Organizations and the International Council of Scientific Unions. We have not had an opportunity to examine the role of these organizations in any detail. The U.S. academic community does have contacts and involvements with these organizations.

Although we agree that professional organizations could play an expanded role in S&T for development, they are limited in what they can hope to accomplish. They serve primarily to facilitate activity, and provide mechanisms for sharing information and setting standards, rather than to implement programs.

Other Linkages

Two groups with which U.S. universities have had little contact are U.S. labor unions and appropriate technology organizations. Labor's concern about international matters has grown in recent years, particularly with regard to loss of jobs and technology. Increased communications between U.S. universities and labor might help universities understand those concerns. Appropriate technology groups with a primarily domestic focus, such as the New Alchemists, are devising solutions to problems that may be relevant overseas. Linkages with universities might be beneficial to all concerned.

The Foundation for International Technological Cooperation

Since 1964, several reports have been written recommending the creation of an International Development Institute (IDI). (54) In a letter dated July 1971 from Roger Revelle, chairman of BOSTID, to Joel Bernstein, then assistant administrator of AID's Technical Assistance Bureau, Revelle forwarded an NAS report which

> endorses the concept of such an Institute, and expresses the hope that it will be established as a landmark institution through which the people of the United States will be able to cooperate with the peoples of the developing countries in devising solutions to worldwide critical human problems. It envisions, as a primary role of the Institute, extension of the knowledge and competence necessary for solving, in time, some of the most serious of these problems. This task would require an institution with the capacity to mobilize the resources of the <u>United States universities</u> [emphasis added], research institutions, industrial and service organizations, and the nation's managerial and technical skills to work with and help improve the institutions and human resources of developing countries.

The institute did not win congressional approval.

The IDI idea was carried forward to some extent in a 1977 report from the Brookings Institution to the State Department recommending the establishment of an International Development Foundation to support research, development, and training activities. (55) The proposed functions of this foundation are described in Appendix B. Its objectives include facilitating the application of U.S. and international research competence to the search for solutions to critical scientific and technical problems of developing countries, and encouraging technical cooperation by U.S. institutions with institutions in developing countries on topics of mutual interest such as food production, environmental quality, and population. Such activity is clearly of relevance to U.S. universities.

On May 1, 1978, the President's Office of Science and Technology Policy announced that plans were moving forward for establishing a Foundation for International Technological Cooperation (FITC), as mentioned by President Carter in his trip to Venezuela in April 1978. The objectives of the foundation are summarized in Table 1.2. The foundation's functions are somewhat similar to those proposed by the Brookings Institution and in the IDI. There is emphasis on building LDC S&T infrastructure, on collaborative research, and on orienting activities of U.S. universities toward LDC problems.

In August 1978, the status of the FITC, as summarized at the Project Workshop and gleaned from conversations, was as follows: A planning office had been set up, headed by Ralph Smuckler of Michigan State University. The office reported to John Gilligan, administrator of AID, in his capacity as head of the President's Development Coordinating Council. No decision had been made on final administrative arrangements for the foundation; however, in order for it to move forward rapidly in conjunction with UNCSTED preparations, it was to be part of the AID budget submission for FY 1980. An initial budget of $250 million was being discussed, comprising $200 million from current AID projects which would be diverted to the foundation and $50 million in new funding.

By March of 1979, several developments concerning FITC had occurred. The name of the FITC was changed to the Institute for Scientific and Technological Cooperation (ISTC). Title II of the proposed International Development Assistance Act of 1979 provided for the establishment of the ISTC. A reorganization proposal had also been put forward to establish a new International Development Cooperation Administration (IDCA) in which AID and ISTC would be two separate program elements. The budget for ISTC had been scaled down, with the major portion still coming from transfers of existing AID projects. (56)

The creation of the FITC would fill an important gap for U.S. universities in their efforts to focus on the application of science and technology to international development. Such an organization had been called for in the early 1960s by John Gardner in his study, AID and the Universities. (57) The FITC can provide the missing link in the legislative mandate concerning international scientific and technological cooperation in building an indigenous S&T base. It can create the climate and programs that will encourage widespread U.S. university participation in such efforts.

However, such a foundation/institute has not won congressional approval when put forward previously. Rather, the trend seems to be more toward fragmentation of international R&D/S&T activity among existing agencies, with new programs emerging in high priority areas, e.g., food, energy, appropriate technology. The 1979 UNCSTED may provide the impetus for such an institution/foundation to win approval.

At the Project Workshop, the importance of early and strong U.S. university input to FITC planning was emphasized. Such input seems assured by the involvement of university personnel that has already occurred in planning activities and advisory roles. Discussions also took place concerning models and mechanisms that the FITC might adopt in its operations. Eilers liked the model of having a foundation, in-country, resident representative who has the background to identify scientific and technological opportunities, someone who is a highly innovative, entrepreneurial type. (58) The Ford Foundation experience seems relevant in this regard, as does the work of Bruce Billings in Taiwan. This approach was stated to be particularly effective if the foundation is able to work in AID-graduate, middle-income countries, with people who can relate to both industry and government. Witunski commented that he would like to see the foundation be able to accept funds from a variety of sources, not just the U.S. government. (59)

Another model to consider for the FITC might be the International Development Research Centre (IDRC) of Canada. According to Louis Berlinguet, IDRC has been instrumental in involving certain Canadian universities in programs linked to research institutions in developing countries. (60) Universities are considered a valuable but somewhat underutilized resource; the IDRC and the Canadian International Development Agency maintain close collaboration with the Association of Canadian Universities and Colleges. IDRC operates in a mode in which most of their support goes directly to institutions in LDCs for projects they themselves define. If such a model were adopted for the FITC, the role of U.S. universities would very likely be less significant than a model in which funds are dispensed to or through U.S. organizations. However, the possibility that the LDCs may prefer the IDRC model at UNCSTED should be considered.

At the Project Workshop, participants were asked to rank four administrative options for the FITC. The overwhelming sentiment was for the FITC to be an independent agency, somewhat parallel to the National Science Foundation (see table A.3, Appendix A). Incorporation within AID received the least support, in part because of limitations to the AID mandate which would prevent activity in AID-graduate countries, and in part because of perceptions of its heavily bureaucratic mode of operation. Other options considered included incorporation within NSF activities and incorporation within a reorganized development assistance agency as in Senate Bill 2420 (Humphrey legislation).

CRITERIA FOR SUCCESSFUL U.S. UNIVERSITY INVOLVEMENT

Analysis of past U.S. university involvements suggest that it is important to establish criteria for success or failure, conditions for success, and to be aware of the limitations of such involvement in

Future Roles for U. S. Universities 207

helping to build an indigenous S&T base in LDCs. This section highlights some important concerns in these three categories which might be kept in mind in considering future activity.

Criteria for Success or Failure

In analyzing past U.S. university involvement, there are several criteria upon which success or failure might be based. These include:

1. Did the program meet its specific objectives as they were originally stated?

2. Did the program help build an indigenous S&T base in LDCs? By what measures? (new institutions in place? number of individuals trained? etc.)

3. Did the specific program, if supported by public funds, coincide with U.S. legislative intent? For example, if AID funds were used, were basic needs met? The poorest of the poor reached? By what measures?

4. Were U.S. interests helped, or at least not hurt, by the involvement? Were mutual benefits derived by both parties?

The above list of criteria is by no means all inclusive. Lucas raised some fundamental questions at the Project Workshop concerning the nature and focus of evaluation of U.S. university involvements: (1) Would the effect have come about if the U.S. university had never been involved in the first place? (2) What is the effectiveness of various projects designed to address the same problems? (61) We consider the question of evaluation in more detail in chapter 6.

One immediate difficulty with these four criteria for success or failure is that there is disagreement concerning which are most significant. For example, Liam Finn, dean of engineering at the University of British Columbia recently described a program in which his institution helped develop a master's degree program in irrigation water supply in Cuba. Finn characterized the project as a success because, as stated in the project objectives, a program was developed at the Cuban institution which is now producing the agreed upon number of graduates with the agreed upon skills. According to Dean Finn, what Cuba does with those graduates and what impact they have is their problem. His university did its job successfully. (62) Wilburn supported this line of thinking at the Project Workshop by stating that the U.S. should help strengthen LDC universities as universities first and not jump immediately into the more complex issues of the role of the LDC universities in development. (63) On the other hand, AID's concern with helping to meet basic needs would seem to require very different interpretations of success and failure.

A more fundamental problem we have encountered in this study is that many of the reports and documents describing program involvements do not permit any of the above questions to be answered with certainty. Even as basic a question as "did the program meet its specific objectives as they were originally stated?" is difficult to answer in many cases because objectives may not have been well stated or reports were not available describing outcomes. Dean Finn's brief example as cited above is one of the more informative ones we have come across, in spite of its sharply delineated perspective. Furthermore, even if documents are available, they were generally written by the U.S. project participants themselves. Independent evaluations and first-hand visits overseas are needed for more definitive evaluations. Nevertheless, we will make some comments on conditions for success or failure based in large part upon anecdotal information, with the help of an occasional, well-documented project.

Conditions for Success

Because success may be defined at several levels, conditions for success must vary with objectives. It may be easier to define conditions for success in programs that involve U.S. universities in building institutions modeled after their own image or in undertaking cooperative research at a level with which they are familiar. Some possible conditions for success in these kinds of involvements include:

1. Long-term, assured funding.

2. Strong backing for the activity from a high U.S. university official, e.g., a president or dean active in international activity.

3. Dedicated, capable, culturally sensitive U.S. faculty or other personnel who feel rewarded for their participation.

4. If the project has an international development focus involving extensive travel, some nonconventional base within the university may be required, such as an experiment station, a center, research institute, or an auxiliary enterprise, so

that personnel without heavy teaching commitments can be tapped. This need not be the case in certain programs, for example, a small program involving exchange of chemistry professors or students.

5. The international involvement must be closely tied to an ongoing domestic activity in which the university has strength.

The above are stated at this stage more as hypotheses than as results based upon extensive data, although there is some support for these conditions. The Georgia Tech small industries case (see chapter 2) would appear to satisfy all five.

Some conflict does arise in the literature between those who feel that big, coordinated programs have the best chance of success and those who favor small, individual-to-individual, less bureaucratized programs. The former is reflected in much AID activity. The latter view is well expressed in the writings of Moravcsik (64) and in implementation of the SEED program. It may be that both approaches have their place, depending upon the type of involvement and objectives. A large institution-building agreement might require a consortium of universities to insure the availability of sufficient U.S. faculty to go overseas, whereas small cooperative research projects involving two peers in two different countries might best be fostered by relatively open program arrangements. Individual U.S. universities may be rigid; however, taken as a whole, they show a flexibility and diversity that is needed for successful involvements.

A more difficult situation arises if one defines success as reaching the poorest of the poor and meeting basic needs. The primary missions of our universities are not directed toward immediate solutions to these problems. We have our own poor to whom U.S. universities have difficulty relating. One condition for fulfilling this definition of success then may be that U.S. university scientists and technologists must begin to learn how to involve themselves more in activities to aid the poor in our own country. Such involvement is possible through the programs of the Community Services Agency and the National Center for Appropriate Technology. A new program in appropriate technology which seems relevant to this concern is being designed by the National Science Foundation. Community colleges and technical institutes may be important elements in such activity. The experience of the predominantly black, 1890 land-grant institutions with the rural poor in the South may prove relevant to international development concerns. It is often said that we have more to learn about relating science and technology to the needs of the poorest of the poor than we have to contribute, and that other countries can perform this function much better than we can. This may be an important reason for our getting involved - in order to learn from the experience of others to help with our own pressing problems and concerns.

Another condition for success may be the coupling of a U.S. institution with an institution in an LDC that is integrated into what has been described as a Sabato triangle. Sabato argues that a key to development is the linkages within the LDC of three points of a triangle: one vertex represents scientific and technical researchers; another, government decision makers; a third, producers, i.e., the production sector. If any of these links is missing, the system will not work. (65) If a U.S. university wishes to work with an LDC university or research institute, a condition for success might be that the LDC collaborating institution is a part of that triangle.

We quote here without comment the personal views of a researcher at the University of Wisconsin Land Tenure Center which seem relevant to the discussion of conditions for success:

> . . . go slow. In agriculture, social technology is currently more in vogue than machinery. Many developing countries have been inundated with technology they either can not afford or do not need - the Chinese are beginning to get into the export business on simple or "intermediate" technology - particularly in agriculture - which seems to me to be far better than anything turned out in the U.S. to date. In addition, I think it is useful to remember that the best way to work with technological problems in other countries, other climes, is to be on the spot for a long time and gain intimate, first-hand knowledge of the problems of use and maintenance. Far too much technology, even much of the "appropriate" variety, has not been designed with careful enough reference to the actual needs and constraints. This argues, it seems to me, for an important alliance between organizations such as yours (that is, Washington University's Center for Development Technology) and the foreign area research groups, which have the knowledge and capacity for overseas work. I have witnessed such linkages here on the Madison campus in the two times I have been in residence at the LTC - linkups between the Systems Design people, the Land Tenure Center as the development research agency, and the various area research institutes. Such ties are very profitable, and, I think, necessary in order to maximize university contributions.

Based upon extensive experience in international agriculture programs, Buddemeier has provided some suggestions for successful future programs:

1. For greater economy and effectiveness, move as soon as conditions in LDCs justify it to major emphasis on short-term overseas assignments coupled with heavy emphasis on long-term mutually beneficial, collaborative research.

2. Act as true collaborators with LDCs, not as dispensers of assistance. Joint planning must be an integral part of programs.

3. Outreach must be a positive and planned part of the program, especially in agriculture. Understand the effects of programs on people.

4. Try to work directly, e.g., university to university. Minimize bureaucratic complexity. (66)

Tha analytical study of institutional links in science and technology between the United Kingdom and Thailand by Glyde provide an analysis of steps needed to promote and support successful links. Although Glyde states that his study is limited by the small number of links considered and that some findings disagree with a previous UNESCO study, we briefly summarize his principal conclusions. (67)

To promote successful links, create conditions in which mutual contact between developing country (DC) and advanced country (AC) institutions is encouraged rather than by explicitly making the contact. Those institutions genuinely motivated will make the contact. To improve conditions: (1) provide core funding and a mandate for selected AC institutions to become involved in DC problems, (2) make known to DC institutions that funding for links is available, (3) encourage visits to DCs by AC nationals, (4) support visits by more senior DC institution members to AC institutions, (5) hold conferences, and (6) encourage and support training in R&D management in DCs.

To support successful links: 1) allow the DC institution to set the link objectives, either alone or in collaboration with the AC institution - for long-term links, broad but well-defined objectives are preferred; 2) provide core funding of AC institutions for development problems as part of a program to support links - increase funding for small links, fund at a lower rate for a longer period; (3) emphasize explicit management training and exposure to industrial application, encourage visits in this regard, integrated with the technical assistance activity, preferably at the DC institution, encourage visits by senior DC people for survey of management techniques and operators and to promote contacts; (4) favor small technical assistance efforts involving individuals or pairs - integrate technical assistance with training. (68)

Limitations to U.S. University Involvement

The limitations to U.S. involvements are many. At the individual level, many U.S. university faculty have been reluctant to get involved in these programs which they view as detrimental to their professional careers. For example, professors who have gone abroad for an extensive period of time have found themselves out-of-date in their fields or bypassed for promotion. Little if any effort might have been made to integrate their experiences into their home institution and make use of their expertise. As a result, it is often difficult to recruit sufficient faculty for long-term, institution-building projects overseas.

On a larger scale, bureaucratic and administrative obstacles involving both the funding agencies and host-institution LDCs have been major obstacles to success or effectiveness. Scarcity of money to support a program adequately has also been cited as a major limitation.

Limiting factors faced by U.S. faculty going overseas for purposes of development aid include cultural barriers, insensitivity to local needs, lack of language skills, unawareness of social and political realities, and the sheer distance that separates the U.S. university and host institutions. Some but not all of these limitations could be eased by proper training and orientation programs.

Another growing factor, which may prove a major obstacle to future U.S. university involvement overseas, is a political climate that resists and resents an outside U.S. presence. This seems relevant today in Latin America where, for example, universities are autonomous and where there may be strong anti-U.S. feeling. Such feelings either exist or could develop in other parts of the world as well. Involvement in multilateral rather than bilateral arrangements may ameliorate this situation somewhat, as could small one-on-one projects as opposed to large, more visible institution-building projects. Conversely, there are some areas of international involvement that may prove distasteful for political reasons to either individual U.S. university faculty or institutions - e.g., countries in which there are flagrant human rights violations or countries in which Jews are not permitted.

If it is desired to get involved in activities in which the U.S. entity interacts with the LDC productive sector, it may not be possible to do so in some cases, due to the proprietary nature of the LDC activity. Furthermore, some laws and regulations are beginning to appear in LDCs that regulate the activities of outside researchers. The impact of these laws on cooperative international S&T activity needs further investigation.

If an expanded program of U.S. involvement is contemplated by the U.S., it should be kept in mind that an indigenous science and technology base ultimately means relying primarily on one's own skills and resources. Thus, the more frequent appearance of cases in which U.S. or other developed country help is not wanted may be a healthy sign.

SCENARIOS

The analysis of future types of involvement and of mechanisms for involvement may be used to construct scenarios that illustrate different future roles for U.S. universities. As the prior discussion indicates, there are many options and elements to consider, all of which makes useful scenario writing difficult. With this in mind, we have sketched three scenarios.

1. The "status quo" scenario, which is based upon minimal change from previous years in both activity and budget.

2. The "all-out" science and technology scenario, which incorporates a large increase in funding as well as major involvement of new U.S. scientific and technological resources.

3. The "modest increase" scenario, which involves modest increases in funding coupled with some shift of science and technology activities and priorities and some new programs.

We caution the reader that these scenarios, which were written in 1978, were not meant to be predictions of things to come; (69) they are just a small number of the many scenarios one can imagine. We lack detailed information on past U.S. spending to use in sketching future funding levels. We use the scenarios to illustrate and dramatize possible outcomes of policies and their impact on U.S. universities and developing countries.

The Status Quo Scenario

This scenario is characterized by uncoordinated, unfocused involvement by the U.S. in science and technology for development. Both science and technology as well as the U.S. universities themselves enter as participants in preexisting programs that have objectives other than helping to build an indigenous S&T base in developing countries. As a result, universities are continually reacting to opportunities rather than helping to define them.

In this scenario, primarily domestic, mission-oriented agencies (NSF, USDA, HEW, etc.) commit few if any funds to international development-oriented activity (less than 1 percent of total funding). AID, or its possible successor agency, has little focus on science and technology. The FITC fails to win congressional budget support. Title XII programs evolve very slowly and are marred by excessive bureaucracy, red tape, and the need to constantly justify relevance to a skeptical Congress. Appropriate technology activity continues to stagnate. Narrow interpretation of AID's basic-needs strategy limits opportunities for U.S. universities to participate in development programs.

Large numbers of foreign graduate students continue to study engineering and science in the U.S. Many do not return to their home countries and there are few curriculum experiences in the U.S. relevant to their home-country needs. A small number of U.S. faculty are involved in extensive overseas consulting activity but large numbers of U.S. faculty are not affected. There is still considerable uncertainty in Washington about how effective past U.S. university involvement has been, how to evaluate such involvement, the level of past funding for such activity, and prospects for the future.

Although a variety of specific new initiatives (some of the "technological fix" variety) involving U.S. universities are proposed at UNCSTED in 1979, there is general dissatisfaction by the developing countries about the level of financial commitment and the lack of a long-range plan on the part of the U.S. to help them develop an indigenous science and technology base. Some countries are particularly annoyed at the emphasis the U.S. places at Vienna on appropriate technology. Although skeptical of AT, they would have been inclined to be more favorable toward it if the U.S. had also increased its long-term financial commitment to helping them build an indigenous S&T base. However, this lack of commitment reinforces their belief that appropriate technology is hand-me-down, dependence-continuing technology and an inexpensive way for the U.S. to carry out a development cooperation program.

The All-Out Science and Technology Scenario

In the spring of 1979, prior to the 1979 UNCSTED, the U.S. makes a commitment to increase its contribution of official development assistance to developing countries from the current 0.2 percent of GNP per year to 0.7 percent of GNP per year within five years. Such an increase, to levels that the United Nations has been advocating for years, becomes a major element of U.S. foreign policy toward develop-

Future Roles for U.S. Universities   215

ing nations. This increase in funding level is to be achieved in negotiated response to some of the proposals of the developing countries, which include automatic rather than voluntary resource transfer from rich to poor countries through such mechanisms as a tax on nonrenewable resources, a tax on international pollutants, and royalties from commercial activities arising out of the international commons (such as the ocean beds, outer space, and the Antarctic). In other words, our accounting toward the 0.7 percent is to include any amounts contributed in this manner. This decision was made after it was decided that rather than a flat rejection of the "New International Economic Order" proposals, it might make sense to analyze what the impact of these proposals on the U.S. would be, both in the short and long term. This analysis showed that the cost/benefit ratio to the U.S. of such a policy was surprisingly low compared with other policy alternatives.

The resources devoted by the U.S. and other industrialized countries to activities in the field of science and technology for development are increased by at least a factor of ten as part of the overall increase in development support to 0.7 percent GNP. Direct scientific and technological aid to developing countries is increased to 0.5 percent of the developed countries' GNP and 5 percent of nonmilitary R&D outlays are targeted for R&D directed toward scientific and technological needs of developing nations. This 5 percent is taken from reductions in the military R&D budget. These funds support a total of about 100,000 scientists and engineers per year throughout the world to work on development-oriented problems. The developing countries respond by substantially increasing their own S&T spending. The LDCs are generally pleased with this U.S. response but somewhat puzzled at how such a change came about. They are particularly concerned with how to make use of these new resources without being dominated by the U.S. They are also concerned with lack of prior consultation on program involvements and with the possibility that Congress will not support U.S. presidential initiatives.

The reorganization of USAID in late 1977 involves a decentralization of that agency and a weakening of its focus on science and technology. Partly as a result, a Foundation for International Technological Cooperation is established in 1979 to support expanded research, development, and training activity in the international science and technology area, as had been proposed in a 1977 Brookings Institution report. The foundation is established as an independent agency, somewhat parallel to NSF. In addition, after a shaky start, Congress increases the budget for Appropriate Technology International from $7 million to $20 million per year.

Greatly increased funding for U.S. universities to become involved in international science and technology activities is now available, with the specific objective of building an indigenous capability for science and technology within poor countries. U.S. universities gear up to get involved with relative ease. At least 80 percent of the funds awarded for cooperative R&D, institution building, and other international activity are open, competitive, and subject to a two-tiered review. They are peer reviewed for scientific and technical qualifications. They are also reviewed by government for relevance and conformity with legislative intent such as "basic-needs strategy" and human rights. Programs are also created that emphasize collaborative research with AID-graduate countries through the FITC.

New curricula and courses are developed that deal explicitly with pressing international issues that arise in considering the role of science and technology for development. Cooperative research programs focused on development problems spring up between U.S. and LDC universities. Universities are also involved in efforts to evaluate the effectiveness of past, current, and future institution-building or cooperative R&D programs. In the past, too many efforts had been carried out without serious, independent evaluation.

This shift in U.S. policy did not come about easily. Labor was and continues to be very much concerned with loss of jobs from foreign competition. There is also a continuing concern that U.S. involvement in developing countries is exploitive and undercuts the developing countries in their efforts to become more self-reliant. But increasing pressure from the developing countries coupled with a continuing concern for the plight of people throughout the world bring this policy shift about. It will take at least five years before we begin to know whether we were wise in undertaking an expanded program of international scientific and technical cooperation centered on the problems of development.

The Modest Increase Scenario

Prior to UNCSTED, President Carter, with support from key congressional leaders, announces his intention to develop a long-range program of international science and technology cooperation centered on problems of development. Funding levels rise slowly over a five-year period at about one-fourth the magnitude of the all-out scenario. The main new U.S. proposal at UNCSTED is the creation of the FITC which provides for a significant expansion of U.S. university activity, including small grants programs for U.S. and LDC researchers, and support for limited curriculum development in the U.S. concerned with international science and technology problems. Other primarily domestic agencies acquire new international funding authority as well. The International Education Act is finally funded.

AID renews its interest in helping to build S&T-oriented research institutes and universities in LDCs after negative reactions at UNCSTED to its "New Directions" policy and after intensive lobbying in Congress by U.S. universities. Subsequent AID reorganization and policy changes permit involvement with AID-graduate countries. The FITC becomes a part of the reorganized AID. Appropriate Technology International is disbanded but a new program with that focus emerges from the reorganized AID.

Some U.S. university personnel are generally pleased at developments; some are annoyed that more hasn't been done; many don't notice. However, with time, many more science and technology faculty do get involved in various kinds of international activity - institution-building projects in OPEC countries, small, competitive research grants programs with LDC researchers.

Prior to UNCSTED, the LDCs had been consulted about the FITC and the proposed increases in funding. They responded with suggestions and proposals of their own. Although the FITC remains primarily a U.S. organization, some suggestions are adopted for LDC inputs and participation in planning and implementation. However, there is continuing concern about the reluctance of the U.S. to support multilateral mechanisms for international cooperation. LDCs are particularly annoyed by the failure of the U.S. to make a financial contribution to the UN University.

6 Policy Issues and Options

This chapter contains analyses of several overall policy issues and options which emerge as being of importance in considering the future role of U.S. universities in science and technology for development. The discussion of these issues and options is based wherever possible on the study of involvements and legislation contained in earlier sections of this report. This chapter also considers specific U.S. legislative changes that might be made, along with their possible impact on U.S. universities and LDCs.

FUNDING

Availability of funding will heavily influence the role that U.S. universities will play in helping to build an indigenous S&T base in LDCs. In this section, some issues related to funding are explored.

At the Project Workshop, Segal provided a qualitative analysis of sources of funding for U.S. university involvement in S&T for development. (1) Funding sources include: (A) U.S. federal government, (B) U.S. state and local governments, (C) U.S. private industry, (D) U.S. private foundations, (E) U.S. nonprofit, private development organizations, (F) international organizations, (G) OPEC country and other foreign governments, (H) foreign foundations and other organizations. There has been a recent increase in university involvement supported by OPEC governments and other foreign sources. Support from U.S. private foundations has declined. U.S. state and local government support is nominal. Support from private industry, particularly multinational corporations, will need to be increased if they want to strengthen U.S. university involvements in developing countries.

Policy Issues and Options   219

The principal source of U.S. university international S&T support remains the U.S. government. We will focus primarily on issues involving this source.

Level of Funding

The amount of money spent by the U.S. government and other U.S. sources on science and technology for development is a difficult and important quantity to estimate. It is important because it may be taken as a primary indication of the extent of U.S. involvement in and commitment to assisting the developing countries to build an indigenous science and technology base. The U.S. is highly sensitive to the fact that the UN and developing countries have in the past set guidelines for developed country contributions. (2) The developed countries are likely to be pressed on this issue at the 1979 UNCSTED. The amount of spending for S&T and for R&D for development is a difficult quantity to estimate. There have been few, if any, efforts to do so in the past. Furthermore, it is often difficult to determine definitively if a given S&T and/or R&D activity in the U.S. is of either direct or indirect utility to a developing country. Thus, in many instances, it becomes a judgment call in which the results may be influenced by the biases of the analyst.

The amount of money spent on U.S. university involvement in these areas is some fraction of total spending for this purpose. Specific studies to determine this fraction do not appear to have been carried out.

During the course of this project, at least two studies that we know of have been underway to try to provide a detailed analysis of U.S. activity and expenditures for scientific and technological activities that are performed in the developing countries, or performed in the U.S. but intended for application in developing countries. A study carried out by Schlie under NSF sponsorship was not available to us in time to utilize for our final report; we refer the reader to the Schlie report of the Denver Research Institute entitled "The Quantification of United States Scientific and Technological Activities Oriented Toward the Developing Countries: A Feasibility Study."

We have performed our own crude analysis of United States' spending for R&D and S&T for development. This analysis, which is summarized in Appendix D, indicates that there may be a gap between UN targets and U.S. spending on science and technology for development, especially in the category of direct aid. The gap appears to amount to several hundreds of millions of dollars, if an old AID estimate of U.S. spending quoted by Moravcsik is used. (3) Schlie's study of U.S.

spending leads us to a similar conclusion about the gap. If detailed quantification studies do indicate that such a gap exists and if it is desired to close the gap, new initiatives of the order of magnitude of hundreds of millions of dollars would seem called for. The proportion of these funds channeled through U.S. universities might be bracketed by, at the lower limit, the 12.78 percent of all funding for R&D in the U.S. in which they are now involved, and at the upper limit, their share of AID R&D spending which comes to about 50 percent.

At the time of this writing, uncertainty still exists about the level of U.S. university funding for current programs. For example, Title XII (see chapter 1) appears to be a major expansion of U.S. activity in one area - food and nutrition. Total Title XII money for FY 1978 is estimated at $195 million (see table 1.4). However, whether this amounts to a significant increase in R&D/S&T commitment or merely a redirection, through universities, of previous programs needs further study.

As we said, it is our understanding that in 1978, a first-year budget for the proposed Foundation for International Technological Cooperation was being discussed at a level of $250 million. Although the total amount appeared to be a significant commitment in relation to our crude analysis of UN targets, the fact that four-fifths of the funding was not for new initiatives but funds transferred from ongoing AID programs is of some concern. As of this writing, the proposed first-year FITC budget is less than one-half the amount of earlier discussions. Of equal or greater importance will be the longer term budget trend, if FITC is approved.

Other federal agencies - the Department of Energy, the Department of Agriculture, the National Science Foundation - have either received authorization for new international S&T involvements or are considering requesting such authorization. The amounts involved in the DOE and USDA programs are relatively small - of the order of magnitude of millions rather than tens of millions of dollars. With one exception, the options NSF has considered in recent months have also been of this order of magnitude.

Funding Control and Administration

The major channel for development assistance funds in the United States is the Agency for International Development. Other federal agencies are involved in international programs to varying degrees, for example, the National Science Foundation, NASA's Office of International Affairs, and the Departments of Energy and Agriculture. These latter organizations generally have primary mission objectives

Policy Issues and Options   221

unrelated to international development and in some cases their funding
for involvement in international programs has come from AID.
   In our opinion, AID has never had a strong focus on research and
development or on building an indigenous science and technology base.
AID moved from large, visible capital projects to emphasis on basic
needs; the latter emphasis, narrowly interpreted, may work against
S&T infrastructure building. AID's Title XII program, which has given
land-grant and other qualified universities a much stronger role in the
food and nutrition area, may be an opportunity not only to involve U.S.
universities fully in the decision-making process but to focus on
strengthening the indigenous LDC S&T base. Currently this involvement is restricted only to countries that qualify for AID support. AID's
Office of Science and Technology has a very small budget and staff,
which limits its effectiveness and influence within the agency. The
1978 AID reorganization and proposed legislation introduced by Senators Sparkman and Case, which did not pass, appears to have placed
little emphasis on building an indigenous LDC science and technology
base. However, Title II of the proposed International Development
Assistance Act of 1979 would provide this mandate through the creation
of the FITC.
   A major issue of current concern is the control and administration
of the FITC. Workshop participants strongly supported setting up the
FITC outside of AID as an independent agency, somewhat parallel to
the National Science Foundation (see table A.3) - a viewpoint which
we share. There was concern that current AID policies, procedures,
and attitudes toward S&T for development would not be a setting conducive to the accomplishment of the foundation's objectives.
   This is not to dismiss the long, if at times turbulent, history of
AID-university collaboration and the considerable progress that has
been made in implementing the Title XII program (see chapter 3).
However, the FITC, with its likely involvement in small grants programs and with AID-graduate countries, seems to require an organizational setting that more resembles the activity of a private foundation, like the Ford Foundation, than that of a government agency like
AID.

Distribution of Funds

   One policy issue that does not appear to have gotten much explicit
attention has to do with where and how funds are spent. It is our impression that most funds for international science and technology that
involve U.S. universities are spent in one way or another either at the
U.S. university or on U.S. university personnel working overseas.

A U.S. university might receive AID funds and pass some on to counterpart organizations overseas, as has been the case in Georgia Tech's Small Industries Program and in work of the Denver Research Institute. (4) In some countries, there are counterpart funds that can be used directly in overseas situations.

In the future, there may need to be increased emphasis on support for science and technology within the developing countries themselves. This shift appears consonant with efforts to build an indigenous science and technology base. If this is the case, it may very well be necessary to spend U.S. dollars overseas in increasing amounts, which will create problems for the Treasury and for balance of payments. Spending the money in the U.S. on U.S. universities strengthens that sector of our economy and avoids running into balance of payments difficulties. There clearly are trade-offs and it may be that the U.S. will need to do both, that is, find mechanisms for spending greater amounts directly overseas, while at the same time find ways to strengthen our home resource base for involvement.

Some mechanisms do exist that have emphasized the provision of support directly to overseas institutions and/or individuals. They include: (1) the FORGE program, previously supported by the Ford Foundation; (2) private foundations like Ford and Rockefeller; (3) the Canadian International Development Research Centre; (4) the International Foundation for Science; and (5) the United Nations and its specialized agencies. Planning for the new Foundation for International Technological Cooperation must take into account the possible need for providing support directly to LDC scientists and researchers; further study of alternative mechanisms for accomplishing this objective is necessary.

Criteria for Funding

This is a particularly complex issue. First, what kinds of activities should the U.S. support in the area of science and technology for international development? Presumably, the activities must have something to do with the legislative mandate. For example, under the current AID guidelines, there is heavy emphasis on meeting basic needs and serving the poorest of the poor. It may be difficult to argue that many science and technology activities meet that particular funding intent, at least in the short run.

A second issue has to do with criteria for involving universities and their faculty in S&T for development activity. Ways must be found to encourage as broad a segment of the university science and technology community as possible to participate. This has been done traditionally in the domestic science and technology enterprise either through open competition for submission of proposals or through broadly based block or institutional grants. We have some concern that in the international S&T field not enough use is made of the competitive peer review process; the latter would seem to be particularly suitable for collaborative research activity. A significant portion of past and current university involvements in international S&T seems to be supported by a process in which the government agency makes judgments about the value of a particular institution or institutional program to the agency's mission. Therefore much money is spent based either on past performance or reputation. Although good results can be obtained in this manner, performance has often not been carefully and independently evaluated.

Thus, both responsiveness to legislative mandate and peer review loom as being important. In the case of peer review in international science and technology, methods of involving individuals from other countries in the review process should be considered to encourage genuinely international cooperative efforts rather than efforts dominated by one side. For example, a workshop participant suggested having proposals involving collaboration between the U.S. and an LDC reviewed by a researcher from another LDC.

OBJECTIVES

We will now consider several issues that relate to the objectives of U.S. efforts to help build an indigenous S&T base in LDCs. U.S. universities and faculty participating in such efforts must be aware of these issues in order that they may have a clear picture of what their involvement means.

Building an Indigenous LDC S&T Base

Should the U.S. support building an indigenous LDC S&T base in all countries in all instances? The answer seems clearly to be no. For example, in some countries it may be that the problem is not a dearth of scientists but the orientation of scientists away from development concerns. In other countries, scientists and engineers may leave or be discouraged by government policies that are unsupportive of their efforts. (5) The building of an indigenous S&T base is primarily a problem for the LDCs themselves. External programs of aid or cooperation must be designed with this fact in mind.

There are some who feel that helping LDCs in this way is injurious to U.S. interests and can cause us to lose the advantages in technological capability that we have. There are others who feel that such an effort is self-defeating because heavy U.S. involvement precludes truly indigenous development. Both of these views have some validity and need to be examined carefully.

Should the U.S. be generally supportive of the effort to build an indigenous S&T base under certain circumstances? Here, a more positive reply is forthcoming. If the Third World pushes for this kind of involvement on the part of the U.S. at Vienna, as seems likely, it seems incumbent upon us to respond constructively. We are constantly faced with the poltitical necessity and/or the humanitarian imperative of doing so for individual LDCs. Now there may be a greater pressure, if the LDCs can present a unified front on key issues at UNCSTED as they have been able to do recently in other forums.

An examination of the past legislative mandate reveals a variety of reasons for U.S. involvement in foreign assistance and international education, including helping people to help themselves, helping the poorest of the poor, meeting basic needs, and acquiring a better understanding of the world around us. There appear to be few explicit statements of legislative mandate specifically geared toward building an indigenous S&T base in LDCs. There is some legislation concerned with appropriate technology and "light-capital technology" and there have been unsuccessful efforts in the past to create an International Development Institute with objectives that include building S&T capability in LDCs. It may be that the 1979 UNCSTED will provide the impetus for creating a Foundation for International Technological Cooperation, supported by a focused, "S&T for development" mandate. The rationale for doing so may include the political necessity of responding to the wishes of the majority of nations of the world, as well as the benefits to be derived by the U.S. from international S&T cooperation. However, such a mandate may not be a popular one with the Congress, particularly if it involves new, large budget outlays. Close contact and cooperation with the Congress will be important in gaining approval for the FITC; universities and university personnel who support such an effort need to make their views known.

New Directions and Basic Needs

The focus of much USAID legislation in recent years has been on meeting basic needs, helping the poorest of the poor, and growth with equity. A general issue is whether U.S. science and technology of the kind U.S. universities practice can contribute to meeting those

Policy Issues and Options 225

objectives. One can argue that in the long run, science and technology are important in raising the standard of living. On the other hand, Adelman has argued that certain technologies serve to enhance income inequalities within LDCs. (6) Similarly, new technologies can worsen the position of women in development and of poor farmers. Furthermore, although it may be argued that some support for basic science at the cutting edge of scientific research is desirable for LDCs, there appear to have been instances in which applied science related to development needs was neglected in favor of costly, irrelevant science.

The relationship between AID's basic-needs mandate on the one hand, and LDC S&T infrastructure building on the other is an important one to consider. It can be argued that an emphasis on meeting basic needs at the expense of S&T infrastructure building serves to keep LDCs in a position of dependence on developed countries and makes it impossible for them to escape from poverty in the long run. Such an argument might very well be made at UNCSTED by developing countries. Conversely, it can be argued that building S&T infrastructure without helping to meet basic needs perpetuates LDC poverty and inequality, at least in the short run.

At the Project Workshop, several participants felt that narrow interpretation of the basic-needs mandate by AID was severely limiting support for international science and technology infrastructure-building activity involving U.S. universities. Our own view is that international science and technology cooperation must strike a better balance between these two objectives; if AID is not authorized or does not wish to do so, then there is a compelling case for establishing the FITC, with its S&T mandate, independent of AID. Although we agree with several workshop participants that the principal contributions of U.S. universities in international science and technology cooperation may not be of immediate benefit to the poor majority, we believe that the universities can and should strengthen their efforts to work on development problems which might yield such benefits as energy supplies for rural areas. We agree with several other participants that universities have been and are involved in such development problems and can do more; at the same time, we are concerned that overemphasis on the basic-needs strategy at UNCSTED may draw a very negative reaction from LDCs while preventing U.S. universities from making their optimal contribution to S&T for development.

Appropriate Technology

Probably no single issue elicited more comment or generated more controversy during this project than the issue of what does "appropriate technology" (AT) mean and how is it related to the U.S. university role in S&T for development. We are not able to treat this issue in any detail in this study; one of us (Morgan) has addressed the definitional problem elsewhere. (7) However, we will state our own interpretation of what AT means and then address the university's role in it.

Thomas Fox, executive director of Volunteers in Technical Assistance (VITA), has defined an appropriate technology as one that fits the local needs and the conditions of the user or consumer.

> . . . in VITA's case, this means technologies which assist low-income people to realize their potential as users and consumers. By this definition, an appropriate technology may be of varying size and degree of complexity; it is not defined in terms of any absolute level or function; it is not even necessarily small - although in reality it often is small because the determining factor of appropriateness is the user and not the technology itself. (8)

He goes on to state that the most appropriate technology is one that best meets the following criteria:

> -- It is more labor-intensive than labor-displacing; employment-generating.

> -- It makes the maximum use of local expertise, both technical and managerial, rather than being dependent on expertise from outside the particular local society or community; it uses locally-available resources.

> -- Similarly, it is a technology that can be maintained and sustained locally.

> -- It is a technology that has been requested by and defined by the local consumers' and users' participation.

> -- It is sensitive to the requirements of the environment.

> -- It is sensitive to local mores, local cultures and local values.

> -- Above all, it makes a difference - an impact on the problems facing low-income people. By this I mean it increases production, makes possible a new business, generates employment and income, improves health, and so on.

I would go on to say that appropriate technology is as much a way of thinking and of placing values on technology as it is a specific compilation of things or a specific technology. While this fact is not fully tangible, it is potentially powerful. For it mandates that the choice of technology should result from a _process_ which if carried out appropriately becomes appropriate technology, as we said in our most recent newsletter. And the process, once set in motion, has its own potential for promoting change and development. (9)

Another related definition is that of "light-capital technology." Congressman Clarence Long has offered several amendments that have been incorporated into recent legislation including one in PL 95-105 which requires that the U.S. place "important emphasis" on light-capital technologies in preparing for and participating in UNCSTED. Long's definition of light-capital technology includes the following:

Light Capital Technology should _not_ be regarded as "primitive," "low," "unsophisticated," or "obsolete" technology. Rather it _is_ technology economical of capital. Producing a light capital technology that works, is culturally congenial, and is economic can require ingenious design and careful field testing.

Light Capital Technology should _not_ be regarded as synonymous with _inefficiency_ or _high cost_. On the contrary, if done appropriately, it should represent the _least-cost_ solution by combining factors of production according to their relative scarcities, economizing on capital wherever capital is scarce and expensive and labor abundant and cheap.

Labor intensiveness _is_ a _necessary_ condition by which to define light capital technology, but is _not_ a _sufficient_ condition, since even primitive or labor wasting technologies are labor intensive.

Light Capital Technology is _not_ defined by dividing the total cost of a project by some total of beneficiaries, especially where it is difficult to identify these beneficiaries and to measure their individual benefits. It _is_ defined as a small amount of capital investment _per worker_ using the capital, and preferably by small projects that can be managed by small entrepreneurs.

Light Capital Technology does _not_ necessarily mean the displacement of large scale infrastructure projects. Light capital technologies can be developed in the rural areas or inserted into the interstices of urban sectors of poor countries simultaneously with capital-intensive infrastructural development, especially if the latter are designed to complement the light capital development, e.g., irrigation and rural roads. But so much emphasis has been given in the past to infrastructural projects that it would seem wise, for a while at least, to shift the emphasis to light capital technology. (10)

Fox's definition of appropriate technology and Long's definition of light-capital technology are clearly related, although they differ somewhat. Long's definition stresses economics and smallness; Fox's definition stresses process and local participation. Our own use of the term appropriate technology agrees in large part with Fox's use.

We define AT primarily in relation to low-income people as "users or consumers." The point was made at our Project Workshop that there are many technologies that are or might be appropriate for developing countries that do not fit this definition - petroleum refining facilities for OPEC nations or large fertilizer complexes for the People's Republic of China, for example. We agree that such activity may be "appropriate." However, we restrict our use of the phrase "appropriate technology" to the context of the VITA definition.

Another issue that arises is to what extent the U.S. should emphasize appropriate technology in its foreign assistance programs. Although the legislative intent is clear, funding authorizations are small and program implementation has been slow. Too much emphasis runs the same risk as was pointed out previously in connection with the basic-needs strategy, that is, negative reaction by LDCs.

Appropriate technology represents a new challenge for universities. Fox stated at the Project Workshop that the U.S. university's role in appropriate technology will be much more significant if it is made clear that AT is much broader than small-scale, primitive gadgets. Furthermore, he stressed that U.S. efforts to involve developing countries in activities such as research on small-scale energy sources will have no credibility unless such activity is taken seriously for application in the U.S.

Self-Reliance

The extent to which the U.S. should be involved in helping to build an indigenous S&T base is a difficult issue. Too much involvement can lead to U.S. domination, dependence on the U.S., and transplanting of inappropriate U.S. models, all of which may serve to undercut the very purpose of the assistance, which was to help build an <u>indigenous</u>

Policy Issues and Options   229

S&T base. There would appear to be no hard and fast rule to follow.
It may be helpful to listen to what other countries are saying and then
respond. The needs of individual countries vary greatly.

AID-Graduate Countries

An issue that arose frequently at the Project Workshop was that
of assistance for AID-graduate countries, that is, countries with GNPs
such that they no longer qualify for assistance under AID guidelines.
AID-graduate countries appear to be prominent on President Carter's
agenda, as illustrated by his April 1978 trip to Venezuela, Brazil,
and Nigeria. These resource-rich countries are of growing political
importance to the U.S. It is likely, therefore, that new international
S&T programs will need to be developed with AID-graduate countries
in response to UNCSTED.

Some OPEC countries will be able to pay their full share of international S&T programs. Other countries with gross national products
below a certain level are eligible for AID support. A void exists for
those AID-graduates that may not be able to pay fully. Options for responding to this situation include: (1) broadening AID's mandate and
programs to incorporate some of these countries, perhaps using criteria other than GNP, (2) focusing NSF's cooperative science program
on AID-graduate countries, with some funds to support developing
country participation, and (3) incorporating AID-graduate country
programs within the mandate of the FITC.

Mutuality

We came across many references to the mutually beneficial effects
of international S&T collaboration between the U.S. and developing
countries. In science, cooperative research on tropical products provides unique challenges and information to U.S. scientists. For example, the Plant Biology Program at Washington University described
by Walbot is heavily oriented toward plant taxonomy in developing
countries. (11) Many researchers from those countries participate
in the program; one returned to Peru and led an expedition that discovered 10,000 new varieties of potatoes, including five that are resistant to potato blight. Blase cited the example of high-lysine varieties of a crop found in a remote valley in Ethiopia which are being
grown in the U.S. (12) Witunski stated that U.S. universities were
running out of places to do new field experiments in the geological
sciences; cooperative research could be beneficial to both the U.S. and

the LDCs. (13) U.S. engineering schools might learn much from collaboration with LDCs on small-scale energy source R&D.

The potential benefits to the U.S. reach beyond the universities. Programs that move developing countries beyond subsistence agriculture can fuel demand for U.S. farm and manufactured products. Some of what is learned overseas could be useful in connection with domestic appropriate technology programs (National Center for Appropriate Technology, NSF, Department of Energy) and with efforts to develop more resource-conserving technologies and lifestyles. The information-gathering aspects of international involvements should not be overlooked, nor the political benefits that accrue from fruitful collaboration with citizens of other countries. The mutual benefit rationale for U.S. participation should receive more attention than it has in the past.

The rationale of mutual benefits to the U.S. from international S&T cooperation is acquiring growing support and credence. Increasing S&T capability in the LDCs will diminish the need for U.S. technical assistance and increase the possibilities for collaboration. Moravcsik has pointed out that mutuality need not be interpreted narrowly; we need not get the same *type* of benefits as the LDCs. (14)

Still the issue of public and congressional support for international technical assistance and cooperation projects is ever present. Sometimes these involvements can work against certain U.S. interests. Benefits and costs both to U.S. universities and to the larger society need to be continuously evaluated.

Bilateral vs. Multilateral Involvements

This issue of bilateral vs. multilateral involvements arises often in connection with debates on foreign aid. Multilateral involvements may be more politically acceptable to some LDCs than bilateral involvements. On the other hand, there is much more experience with bilateral programs and a general feeling that multilateral programs are more bureaucratic and less efficient in accomplishing objectives. U.S. universities and university personnel will probably continue to be called upon to interact in both these areas. We would expect the major demands for university involvement to continue to be in bilateral programs, but we feel that growing participation in multinational programs is a desirable goal for U.S. institutions and individuals to strive for.

A case deserving particular attention by the U.S. in connection with UNCSTED preparations is the United Nations University (UNU). Although we have had some negative comments and skepticism ex-

Policy Issues and Options   231

pressed about the UNU by a few U.S. professors, our impression is that it represents an innovative mechanism for involving university personnel from around the world in collaboration on helping to solve key development problems. It seems possible that the developing countries will press for U.S. support for the UNU at UNCSTED. We have not examined its work, such a study needs to be made. Several U.S. universities are participating in UNU programs; some financial support from the U.S. would seem to be called for.

Expanded U.S. University Involvement

Political interests undoubtedly will be foremost in the consideration of issues pertaining to UNCSTED. An expanded program of U.S. university involvement might go part of the way in meeting LDC demands for assistance in building an indigenous science and technology base. It might also serve to take some of the focus off other demands related to the New International Economic Order (NIEO), a code of conduct for multinational corporations, and access to proprietary technology. We have not sounded out the academic community on this matter, but there are certainly some who feel that the NIEO and a code of conduct are in U.S. as well as LDC interests; others would strongly disagree.

The issue of building an indigenous science and technology base appears to be a central one in terms of LDC thinking. However, whether an expanded program of U.S. assistance and cooperation in such an effort, with significant U.S. university participation, is greeted at UNCSTED with enthusiasm by the LDCs remains to be seen. U.S. involvement could be viewed as detrimental to the efforts of LDCs to build their own capability and they could push primarily for resource transfers from the developed to developing countries coupled with more cooperation among the LDCs themselves. An expanded program of U.S. technical assistance and cooperation also runs the risk that some activity would ultimately be harmful to business and labor in the U.S. Political issues are clearly of central concern in preparing for UNCSTED and need, and will undoubtedly get, careful consideration by many of the participants.

U.S. Minorities and International Cooperation

In many instances U.S. universities have greater numbers of foreign students enrolled in science and technology programs than U.S. minorities and women. They are subject to criticism by U.S. citizens and political leaders for not doing more for minorities at home. Foreign students coming to the U.S. are politically sensitive to this situation.

Wilburn has stressed the opportunity to link international development activity to domestic missions of U.S. universities in order to generate strong support for these activities in the future. (15) U.S. domestic needs are pressing in education; resources are scarce. People from other countries can help us with our problems here. According to Wilburn, some 30 to 50 percent of the science and mathematics faculty members at some of our predominantly black colleges are from Asia.

There is a relationship between strengthening the role of minorities in science in the U.S. and strengthening international science. According to Wilburn, the UNESCO science education materials used in many developing countries are better than those used in the public schools in Washington, D.C. He asks: Can we use scarce resources to teach Arabic to U.S. students and faculty who go overseas when students in U.S. urban school systems have pressing educational needs? (16) He replies that strengthening the participation of minority U.S. students in science and technology not only keeps potentially good students from being wasted, it also would appear to be an essential element in justifying, and generating political support for, international S&T activity by the U.S.

Interest in the potential role of predominantly black universities in S&T for development is growing, particularly with regard to programs in Africa and the Caribbean. Four 211(d) grants have been made to these institutions in the health sciences. Howard University has the largest foreign student enrollment in the U.S. A conference to explore the role of the predominantly black, 1890 land-grant institutions in S&T for development was planned for fall, 1978. These steps would appear to strengthen the political base for support of international activity in the U.S. as well as to make new, underutilized resources available for international S&T cooperation.

WHY UNIVERSITIES?

What Can Universities Do?

There appears to be skepticism in some quarters about the role of U.S. universities in meeting U.S. overseas objectives. This may stem in part from the opposition of students and faculty to U.S. policy in the late sixties in connection with Vietnam. Yet, our examination

Policy Issues and Options 233

of past U.S. university involvement indicates that it has been extensive, with universities and faculty undertaking a wide variety of activities over the past 20 years. Although we have not been able to quantify that involvement in the aggregative sense, it seems clear that universities are high on the list of sectors involved in international activity. Although independent evaluation of their involvements has not been extensive, it is clear that they have contributed to institution-building efforts and to educating many students from LDCs. It also seems likely that given the opportunity to contribute to a focused effort to help build an LDC indigenous S&T base, they would respond strongly and positively.

The rationale for existence of the university includes the passing on of knowledge (teaching), the generation of new knowledge (research), and, increasingly, service. Building an indigenous S&T base clearly has elements of all three associated with it and, in this sense, appears to fit well with the mission of the university.

The question then becomes the extent to which the teaching, research, and service a U.S. university might engage in is relevant or useful to LDCs. If one considers the role of a university to serve only the citizens of a particular state or a particular clientele, then its utility will be somewhat limited. However, today many universities and faculty see their role in much broader perspective - the university becomes international in scope and outlook. Such an attitude accounts for the widespread participation of U.S. universities in international matters.

In a statement from "Canadian Higher Education and International Development Cooperation, 1975-1980," approved by the board of directors of the Association of Universities and Colleges of Canada, the issue is addressed as follows:

> There is a gap between traditional university teaching and research largely directed toward the understanding of development needs, and activities and research directed to the immediate solution of the problems of developing countries. The Canadian community of universities cannot set aside its primary responsibilities which are to the Canadian people in order to direct its energies to the solution of practical problems, particularly, those of other countries. However, it is a role universities will wish to play in international development cooperation.

The report concludes by stating that the Canadian community of universities recommends:

1. that Canada reaffirm its policies of educational assistance to and cooperation with developing countries;

2. that Canadian universities be encouraged to have as one of their objectives, international development cooperation;

3. that the potential of Canadian universities to contribute to programs of international development cooperation at home and abroad be more fully recognized by the Canadian government;

4. that Canadian universities be assisted financially to meet their international development cooperation goals and obligations ideally by increased subsidies through the Canadian International Development Authority (CIDA) and the International Development Research Centre (IDRC);

5. that a better coordination of Canadian efforts in educational development cooperation be ensured by establishing more efficient consultative links between the federal and provincial governments and Canadian universities. (17)

Are Universities Relevant to Development Needs?

If development needs include building an indigenous S&T base centered on science and technology as practiced in the U.S., then universities clearly are relevant and can contribute to meeting those needs. Hazeltine stated at the Project Workshop that U.S. universities are more flexible, deal with a wider range of students, and use better pedagogy than alternative models LDCs might follow; for example, they can respond to the need for both science-based engineers and planner-managers. (18) Gomez reinforced this view by stating that the U.S. educational system is the best model. (19) Many developing governments and individuals would seem to agree; U.S. universities attract more foreign students than any other country. However, it is important that LDC institution-building and institution-strengthening efforts adapt U.S. models to their own needs and situations.

A more difficult question concerns the university's relevance to meeting basic needs and appropriate technology. Some observers feel that Western universities have little if anything to contribute and may in fact provide a model that is inimical to development. To them, appropriate technology is outside the normal scope of university concerns. If any Western educational institutions are relevant, they are likely to be below the university level. Furthermore, it may be that other models, e.g., China, are more relevant. (20)

Although U.S. universities cannot make a strong case for being relevant to basic needs and appropriate technology, it is our belief

Policy Issues and Options

that future involvement may very well be productive. Critics of university involvement generally seem to have been products of European universities. Yet U.S. universities are in many ways more flexible and have more of a service orientation than European universities. There is beginning to develop some appropriate technology activity in U.S. universities and community colleges. Such activity is likely to grow in the future.

How Do Universities Compare With Other Institutions?

In terms of helping to build an indigenous S&T base, U.S. universities have both advantages and disadvantages over other U.S. institutions - private industry, government, research institutes, private voluntary organizations. Universities and university personnel may have less trouble functioning politically overseas than other organizations; small, one-on-one programs and involvements may have more opportunities to last than larger, more visible efforts. Universities and faculty have extensive overseas contacts, including former students. Universities can provide integrated resources for teaching, research, and service activity; many possess agricultural, scientific, and engineering expertise.

Disadvantages include less contact with the productive sector than private industry and, in some cases, less flexibility in mobilizing resources than a research institute. In the case of the international agricultural research centers (IRRI, CIMMYT), it appears that universities in both the U.S. and the LDCs are not as effective for mobilizing such efforts and producing results as these centers. However, the university's ability to consider social and economic as well as scientific and technological factors might have been helpful in this connection.

Does U.S. University Resource Base Activity Help Development?

The "brain drain" continues to be an important issue. We mentioned in chapter 4 that at the 1979 UNCTAD-V, there will be discussion of a resolution calling for the developed countries to pay compensation for LDC students who go abroad to study and do not return; the issue will probably also arise at UNCSTED.

Charles Kidd has recently examined the "brain drain" issue in depth, in a parallel study to ours as part of U.S. UNCSTED preparations. (21) Kidd's study should be consulted for more information and analysis of this issue. We briefly summarize our own views as follows.

Whereas the overall percentage of foreign students in the U.S. universities is very small (less than 2 percent) compared with universities in some European countries and Canada (which is moving to limit foreign enrollments), (22) it is quite high in certain fields, e.g., around 40 percent (1975-76) in graduate engineering programs. Movement to restrict U.S. enrollments of foreign students would not appear to be in the short-term U.S. interest, particularly in programs that depend heavily on those students for research output, who pay their own way, or have it paid for by home country sources. However, building an indigenous LDC S&T base may suffer in many countries if large percentages of LDC students in science and technology do not go back. Changes in immigration regulations in the early 1970s have sharply reduced the immigration of scientists and engineers to the U.S. Kidd has summarized steps the LDCs can take to help ensure that their students will return home. (23) In our opinion, the primary long-term challenge would seem to be for the LDCs to focus more heavily on building and relying on their own institutions, while trying to move away from heavy dependence on U.S. institutions for educating S&T students. However, U.S. policies in international S&T can help support such a focus by placing more emphasis on activity within the LDCs themselves.

A related issue concerns whether or not support for international, development-focused research and teaching activity in the U.S. aids in S&T for development and contributes to building an indigenous S&T base in LDCs. AID's 211(d) program was specifically designed to enhance the capability of U.S. universities to contribute to development activity; Title XII institution-strengthening grants are to accomplish the same purpose in food and agriculture. We have acquired few well documented cases of 211(d) and other resource base activity in the course of this study and cannot answer the question definitively. Some programs that seem to provide positive results are the Georgia Tech small industries program and the INTSOY program at the University of Illinois. We would contend that efforts to carry out R&D in the U.S. relevant to development problems is an untapped, fertile field. (24) Contributions will be forthcoming if human and financial resources can be put to work.

Finally, we expect the flow of S&T students from LDCs to U.S. universities to continue for some time. Teaching and research that is focused on development problems in U.S. universities and that involve LDC students, contribute directly to building an indigenous S&T base for the many who return to significant positions in government, industry, and universities in their home countries.

BUREAUCRATIC VS. PROFESSIONAL APPROACH

Small vs. Large Projects

An issue that continually arises in the literature and in current thinking concerns whether support should be made available for small or large projects. Small, one-to-one projects seem to be favored in many areas of science although they are perhaps somewhat less appropriate for engineering and agriculture. They have the advantage of longer-term continuity, surviving political upheaval better, and of having the personal touch which builds more lasting relationships. They are more likely to be useful in attracting a variety of involvements on the part of a broader segment of the university science and technology community than would large projects. They are less likely to be dominant or disruptive or to overwhelm the abilities of the LDCs to develop their own capability.

On the other hand, larger projects may be more pertinent to certain areas such as agriculture and to institution building. Consortia of U.S. universities helping to build institutions in Algeria, Saudi Arabia, and Nigeria facilitate the recruitment of enough U.S. faculty to go abroad and still keep the resources of any one U.S. university from being too heavily strained. Mixes of needed skills - in engineering schools, technical institutes - can be provided. The demand for such involvements, particularly by the OPEC countries, appears to be increasing. In concentrating resources on strengthening a small number of U.S. universities for development work rather than scattering resources, the resource base type of involvement may be an efficient use of funds.

Thus there would appear to be a need for both small- and large-scale involvements, although there are perhaps more opportunities currently for U.S. faculty to get involved in the latter. Opportunities are lacking for a broader segment of U.S. scientists and technologists to become involved in S&T for development activity. Spending funds on a small grants program may generate more involvements than spending them on a large institution-building project; or, at least, the involvements will be of a different kind. The dearth of independent evaluation, or in some cases any evaluation of many U.S. university involvements makes judgments difficult. Glyde's analysis of institutional links between research institutions in Thailand and the United Kingdom indicates that the level of "link success" was greater when smaller numbers of people participated. (25) However, this was only one kind of involvement and the data were very limited.

Bureaucratic vs. Professional Approach

An issue that arises in the AID-GAO exchange (see chapter 1) and in other areas concerns the extent to which university involvement in development activities should be under bureaucratic control. On the one hand, there is a need for government bureaus and agencies to try to insure that program objectives and legislative mandates are met. On the other hand, university personnel are professionals who wish to be treated as such. (26) Somehow the proper balance must be struck, and some feel that AID has leaned too much toward the bureaucratic side. Other agencies such as NSF may be more congenial places for professional type of activity. This is one of those issues that will probably never be resolved to everyone's satisfaction. However, if it is desired to mobilize more U.S. university scientists and engineers to help build an indigenous science and technology base, it may be necessary to more heavily emphasize the professional approach than has been the case in the past to correct what we perceive to be an imbalance. As Moravcsik has noted, there is virtue in pluralism; there should be many parallel sources differing in style, personnel, and "taste" for obtaining support for LDC-related activities. (27)

Peer Review

Much university involvement in international activity appears not to have evolved through open, competitive, peer review processes. This point has been made several times before. It may be argued that the peer process is wasteful in terms of targeting resources to meet specific development needs. Yet, much activity in the U.S. university community is based upon peer review, and more open competitiveness in international work seems to be called for, particularly if it can be done in a way so as to get a greater variety of U.S. institutions and individuals involved in LDC activity. Peer review for a joint research project might involve LDC reviewers from countries not involved.

Peer review, narrowly defined, has its limitations. Rao has provided the following analysis:

1. On programs involving LDC institutions, the peer reviews of the kind now used by NSF and the government agencies would not work too well if applied uniformly to all the different kinds of possible involvements of U.S. and LDC universities.

2. For collaborative research, review criteria, especially of LDC institutional capabilities, will need to be more flexible.

Policy Issues and Options

There is an element of teaching and learning by both sides in the collaboration. It is unlikely that there will be perfect equality in competence between collaborators.

3. On institution-building projects, the traditional peer review is practically useless. Who are peers in this kind of work? We need to devise other methods of evaluation of the capabilities of U.S. institutions and LDC institutions. In many cases, one may have to go to sole sources. This is because we don't have in the U.S. such a large number of institutions <u>interested and capable</u> of working with LDC institutions.

4. On selection of individuals, we must use wider nets for getting people committed to LDC development. It is too easy to develop cliques, inner circles, etc. <u>Joint</u> visiting committees, <u>joint</u> selection committees, etc. are some of the kinds of things we should experiment with. (Joint: use both LDC and U.S. members on such committees.)

What I am saying is that different evaluation criteria and selection procedures should be used for different kinds of involvements, and we shouldn't lock ourselves into the one method of peer review now used extensively by U.S. agencies. (28)

EVALUATION

Need for Independent Evaluation

We have been struck in our efforts to obtain information for this project by the lack of independent evaluation of many university involvements. The field seems to have been badly underevaluated. Sometimes we came across final progress reports or studies by the sponsoring agency, but it is very difficult to find evaluations that can be considered truly independent. Project directors and agency sponsors have biases such that their evaluations cannot be totally conclusive, although they are, to some extent, helpful and can shed considerable light on what happened. But if large, future program involvement is contemplated, it seems essential that independent evaluation be built into such efforts.

Defining Objectives and Methods of Evaluation

Criteria for successful performance of U.S. universities have been discussed in chapter 5. These criteria need more attention than they have thus far received in this report.

Some individuals feel that specific, quantifiable goals need to be set for certain kinds of activity, such as the number of LDC faculty trained and the size of the graduating class achieved in an institution-building involvement. What those faculty and students do with the training they receive is the problem of the host (LDC) country. Others believe that the U.S. shares this responsibility with the LDC.

There is a need to focus upon methods of performing evaluations of U.S. university performance. Old methods will not suffice for international S&T activity and new methods need to be developed. For example, we feel that our evaluation of specific collaborative projects would have been greatly enhanced if we could have obtained the views of both U.S. and LDC participants or if the evaluations were carried out jointly by a team from both the U.S. and LDCs. There may be some projects, however, for which joint evaluation or *any* evaluation may not be politically feasible.

At the Project Workshop, Lucas stated that the term evaluation seemed to be used primarily in the sense of whether individual projects met project goals, such as training a specific number of students. She called for more fundamental evaluation to support policy decisions and to design effective future university programs; this evaluation should be comparative and address such questions as (1) would the effect have come about if the university had never been involved in the first place, and (2) what is the effectiveness of various projects designed to address the same problems. Lucas proposed that NSF's research and evaluative capability be linked with AID to carry out evaluations during the course of, as well as at the end of, the project, measuring the effectiveness of AID's technical assistance activities involving universities. (29) In the discussion which followed, some of the difficulties associated with performing such evaluations were highlighted (see Appendix A).

AID is evidently under considerable pressure from the General Accounting Office and the Congress to carry out evaluations of its programs including more independent assessments. Our impression, based upon our experience in connection with this project, is that more efforts at program evaluation are needed. We have had some discussions with AID officials who indicate that they heavily emphasize evaluation now. Our impression, admittedly superficial, is that AID's evaluations are primarily "in-house" and that it is difficult for the public to gain easy access to them. We feel that more independent

Policy Issues and Options   241

evaluation of certain kinds of involvements, carried out in an open, accessible way, can do much to improve the quality of future U.S. university involvements in S&T for development.

LEGISLATIVE OPTIONS

In this section, we sketch some of the major legislative options open to U.S. policy makers for expanding international S&T involvements and briefly discuss their possible impact on U.S. universities and LDCs. We are particularly concerned about their effect on building an indigenous LDC S&T base and on the reaction of LDCs to such changes in connection with the 1979 UNCSTED. The options we outline are not necessarily exclusive, two or more might be implemented.

Foundation for International Technological Cooperation

Based upon our study, the Foundation for International Technological Cooperation appears to be a key step if it is desired to implement an expanded program of international S&T. The functions of the foundation, which is now in the planning stages, are summarized in figure 1.2. We have discussed the status of the FITC and issues concerning it in chapters 1 and 5.

There are several organizational options for creating such a foundation:

1. Establish an independent agency somewhat parallel to the National Science Foundation.

2. Incorporate the functions of such a foundation within NSF.

3. Create the foundation within AID.

4. Establish the foundation through new legislation such as Senate Bill 2420.

5. Create the foundation by a reorganization proposal.

Table A.3 summarizes the views of our workshop participants on the first four of these organizational options. There was strong support for the independent agency concept.

The primary reasons for the independent agency outcome seems to be concern that AID leans too much toward the bureaucratic as opposed to the professional model, AID has restrictions against involvements with AID-graduate countries as well as a lack of focus on science and technology, and the NSF has primarily domestic mandates. On the other hand, politically it may be easier not to create a new entity that could cause jurisdictional disputes. In 1979, the administration chose to propose the establishment of FITC in the International Development Assistance Act of 1979. At the same time, a reorganization proposal was put forward which would give FITC separate but equal status in a new International Development Cooperation Administration (IDCA).

It is anticipated that U.S. universities would be a major participant in the work of the FITC. If adequately funded, it could be a source of renewed support and interest by the universities in international S&T cooperation. Therefore, it behooves the universities to express their support for it as the land-grant colleges and universities did to Congress in connection with Title XII.

It is our impression that the FITC will be one of the major new initiatives, if not the major new initiative put forward by the U.S. at UNCSTED. We feel that it has the potential for being well received by the LDCs, particularly if it has adequate funding authority. However, we see pitfalls. The LDCs may push primarily for multilateral involvements and resource transfers. The FITC seems more oriented toward bilateral activity. Furthermore, support must be won for the FITC in the Congress, a formidable task for any program requiring new money these days. Segal commented at the Project Workshop that new U.S. initiatives may fail because of lack of prior consultation with the LDCs and because of the inability of the U.S. to follow through on its commitments. Efforts are required to try to ensure that this does not happen. (30)

International Education Act of 1966

As far as we know, this act is still on the books. All it needs is funding authority. It directly affects U.S. universities. There has been much written about educating U.S. students to the realities of international development and the global implications of science and technology, as well as developing curricula relevant to S&T for development for the many foreign students in our own universities and our own students. Funding the act could have symbolic as well as real significance for the LDCs at UNCSTED, signaling to them the importance the U.S. places on international education.

Policy Issues and Options 243

Expanded Title XII Authority

The option involving expanded Title XII authority has been brought to our attention on several occasions. Currently, Title XII derives from a 1975 amendment to the Foreign Assistance Act (FAA) focused specifically on preventing famine and establishing freedom from hunger. It would seem to be difficult to amend the title to include nonfood and nonagriculture areas within its present context. However, this could presumably be accomplished by a new amendment to other parts of the FAA.

At present, Title XII programs are just getting underway. It may very well be that the involvement of U.S. universities through BIFAD is a major innovation which greatly strengthens the effectiveness of U.S. universities in contributing to AID's objectives. However, there is too little experience yet with Title XII to reach any conclusions one way or the other. One concern is that it appears to be evolving in a way that emphasizes large, coordinated projects and might not be suitable for many of the smaller, one-to-one projects that seem promising.

The development of the Title XII-BIFAD organization was greatly aided by the efforts of the land-grant colleges and state universities to win support for the program in Congress. Title XII restrictions exclude many other universities as prime contractors on collaborative research support grants and on institution-strengthening grants, the two key new elements as far as U.S. universities are concerned. If the Foreign Assistance Act were amended to extend the Title XII concept to other fields, e.g., engineering and science, it would be important that restrictions on eligible participants be removed because of the much broader university constituency in these fields. Whether the land-grant colleges and state universities would support the lifting of such restrictions, after their yeoman efforts to get Title XII passed, is questionable.

We do not feel that expansion of Title XII and lifting restrictions would have much impact or generate much enthusiasm by the LDCs at UNCSTED. The program has been slow in getting started and is probably not highly visible, even though potentially very important.

Funding Authority for 211(d)

AID's 211(d) authority is still on the books although the agency seems to be restricting new funding to support predominantly black universities and has been spending funds at considerably below the maximum funding authority of $10 million per year in fiscal years

1977, 1978, and 1979. This resource base grant program has many advantages and disadvantages and has been subject to certain criticisms (see chapter 1). Yet recent NAS reports have called for its reactivation and strengthening. (31) It appeared intact in the recently proposed Humphrey legislation (Senate Bill 2420).

Increased funding for 211(d) programs is an option to consider, particularly if some of the previous objections to the program can be overcome. To do so, it might be desirable to have a more open, competitive grant process, broader distribution of funds, more careful, built-in, independent evaluation, and a more integrated approach to using 211(d) resource bases along with other types of involvement, e.g., collaborative research and LDC-oriented curriculum development.

Such an expansion might be particularly useful for supporting U.S. university participation in engineering and science; the agricultural colleges have the Title XII program. Continued attention to strong resource base support for U.S. university involvement in international S&T seems called for to develop experiences more relevant to home-country needs for the many foreign students who will continue to come to the U.S. to study, as well as to strengthen U.S. resources for collaborative international S&T research. The Glyde study indicates that one factor in promoting successful links between developed and developing country institutions is core funding and a mandate for selected developed country institutions to become involved in LDC problems. (32) However, acting against such an expansion is the low level of use of 211(d) currently within AID and a feeling expressed by some AID officials that AID has developed all of the university resource base capability that it needs.

Unless a major expansion of 211(d) funding and use is contemplated, we feel a 211(d) initiative would have little if any impact at UNCSTED. Furthermore, the focus of the legislation on U.S. resource base development as opposed to LDC development would probably work against it as a major initiative. It also suffers from being an "old" program at a time when new initiatives may be required. However, some rejuvenation and expansion might be desirable as a supplement to the FITC initiative.

International Activity of the National Science Foundation

The National Science Foundation's authority to provide support for international science and technology activities would seem to be sufficient to justify an expansion of its international S&T role, subject to

Policy Issues and Options   245

the approval of the secretary of state. However, in the past, the
agency appears to have been reluctant to do so; its main international
cooperative science activities have been with the developed countries
(see Appendix D). The time may be ripe for such an expansion of
LDC-related activity; new programs would need to have the approval
of the National Science Board and the appropriate congressional committees. Initiatives that NSF might take include an expanded SEED
program, more intensive cooperative science activities with scientifically developing AID-graduate countries, and intensive short courses
for LDC science students in the U.S. that are oriented toward development needs of their own countries.

NSF has been a major source of support for U.S. universities.
Programs are generally governed by what we have termed a professional model, i.e., they are open, competitive, and peer reviewed.
Such a model seems particularly appropriate for involving a broader
segment of the U.S. academic community in international scientific
and technological cooperation than has been possible to date. If the
FITC is implemented, then expansion of NSF activities would seem
less compelling; however, if the FITC concentrates primarily on technology as opposed to science, then NSF expansion in international scientific cooperation seems like an attractive option. If the FITC is not
implemented, then the role of NSF could become even more significant,
particularly in collaborative R&D.

Some countries might find it more politically acceptable to work
with NSF than with AID in strengthening their indigenous S&T base.
NSF has the advantage of being a science and technology agency; its
reputation within the international scientific and technological community may be reasonably strong. However, it is not clear if NSF really
wants to divert significantly from its primarily domestic mandate or
if, having made up its mind to do so, it could garner the necessary
congressional support. Strong indications from the U.S. academic
science community that they wish NSF to assume such a role may be
a necessary but not sufficient condition to bring this about. We would
guess that the expansion of NSF's international S&T activity with developing countries would be well received at UNCSTED, particularly
if it were coupled with the FITC initiative.

Domestic Mission-Oriented Agencies

Authority for domestic, mission-oriented agencies has been provided recently in the U.S. Departments of Agriculture (USDA) and
Energy (DOE). In the latter, the Anti-Nuclear Proliferation Act of
1978 provides some funding for international S&T activity in small-

scale, alternative energy sources and to study the feasibility of a scientific Peace Corps. There are other U.S. government agencies that are involved and might be further involved in international activity, such as NASA, NOAA, DOI, HEW, DOT, EPA, DOC, DOL, etc. Current funding is often "pass-through" money from AID and subject to AID's mandate; direct funding for these agencies might give them more flexibility.

Expansion of international authority of mission-oriented agencies appears to be occurring although funding commitments to date are small. Such expansion conforms to some extent to bureaucratic reality - these agencies exist and have a natural tendency to want to grow and take on more activity. However, the result is a large number of activities that are difficult to keep track of and coordinate, and thus it is difficult to avoid duplication. Furthermore, the lack of a central focus is a minus when there is need to assemble data to document U.S. international S&T involvement, as is now the case.

U.S. universities interact heavily with these mission-oriented agencies. The universities are likely to welcome an expansion of their international activities provided that the funds for it are "add-on" and not diverted from domestic programs.

At UNCSTED, we would guess that an expanded program of mission-oriented activity would in itself not generate a great deal of enthusiasm among LDCs. We worry particularly about the U.S. coming in with a fragmented series of specific mission-oriented proposals, without a uniform, somewhat coordinated approach. However, as a supplement to the FITC, expanded mission-oriented activity, as in the case of NSF, could be a real plus.

The New International Communication Agency

The International Communication Agency incorporates the old U.S. Information Agency and the Bureau of Educational and Cultural Affairs in the State Department. The Fulbright program is supported from ICA funds; educational and cultural exchanges are an important part of its mandate.

One way to strengthen the indigenous S&T base in LDCs would be to earmark a larger percentage of funds for faculty and student exchange in S&T fields. Such an option should certainly be considered. The disadvantages are (1) the possible weakening of the cultural and non-S&T educational focus if total budgets are not increased, and (2) the possible political disadvantages of working through ICA as opposed to NSF or the new FITC. However, if both the quality and amount of Fulbright fellowship opportunities for U.S. university personnel in

LDCs were increased, such a step might be a useful supplement to other actions.

International Organizations

An alternative to be considered is expanded U.S. involvement in and funding for international (and regional) organizations. The U.S. might make a commitment to increase its contributions to selected international organizations with an S&T mandate. It might also choose to increase the involvement of U.S. scientists and technologists, including U.S. university personnel, in the multilateral programs of the UN specialized agencies, the UN University and other international bodies.

We see domestic U.S. political problems with such an alternative. Our impression is that support for international programs is not popular with the Congress; it will be difficult enough to gain support for new international S&T initiatives, let alone those that require increased funding for international organizations. U.S. university involvement in these organizations per dollar spent is likely to be less than through bilateral programs. Furthermore, though we have seen no studies that compare the two approaches, some feel that a dollar spent bilaterally is more effective than a dollar spent multilaterally in S&T for development work.

On the other hand, we would guess that the LDCs at UNCSTED might react negatively if the vast majority of proposed U.S. initiatives were bilateral, with control primarily resting with the U.S. The United Nations and its specialized agencies are important to the LDCs; they may very well press for financial support from the U.S. for activities that they have more control over. The Canadian International Development Research Centre provides for considerable input to project definition by developing countries and may prove to be an attractive model for them.

The situation would seem to call for some alternatives that involve the U.S. in expanded multilateral activity. We are not prepared to offer many such initiatives; we have not emphasized multilateral involvements in this study. Furthermore, it seems desirable that such possibilities be explored with the LDCs in some detail prior to UNCSTED. Such an approach might emphasize collaborative funding of efforts with OPEC nations, which are becoming a significant factor in funding for international development involvements.

7 Summary, Conclusions, and Recommendations

SUMMARY

Objectives

The purpose of this investigation was to examine the past, present, and future roles of U.S. universities in helping build an indigenous science and technology base in developing countries. Among the desired project outputs were an analysis of key policy issues and options, as well as definition of mechanisms for future U.S. university involvement in international S&T cooperation. These objectives were achieved by an examination of the previous legislative mandate for U.S. university involvement, and by studying past U.S. university international participation and current thinking in three fields: engineering, agriculture, and science. The data base for this analysis consists of grantee and contractor reports, some agency evaluations, published articles, reports, and books, and information supplied by knowledgeable practitioners. The deficiencies in the data base include lack of independent project evaluations and lack of inputs from LDC representatives. A Project Workshop held on July 13-14, 1978, provided a review of the Draft Study Paper and many constructuve suggestions for the final report.

The Legislative Mandate

Federal legislation pertinent to U.S. university involvement in S&T for development serves to: (1) set policy and directions; (2)

Summary, Conclusions and Recommendations

create government agencies with international missions, e.g., the Agency for International Development (AID), and provide international mandates for primarily domestic agencies, e.g., the National Science Foundation (NSF) and the Departments of Energy (DOE) and Agriculture (USDA); (3) provide specific programs directed at U.S. universities, e.g., AID's 211(d) and Title XII programs; and (4) provide for and regulate student and faculty educational exchanges. Trends in such legislation over the past 20 years were reviewed and key legislative programs were analyzed.

Foreign assistance legislation is currently focused on fulfilling AID's "New Directions" policy - meeting basic needs and assisting the poorest of the poor. This policy, as currently interpreted by AID, differs from one directed at helping to strengthen LDC S&T infrastructure and does not support activity in AID-graduate developing countries. The educational exchange thread in past and current legislation is also not directed toward S&T for development and has suffered from lack of support.

Two trends that could provide a more S&T-focused orientation are the broadening of domestic agency mandates and expansion of their funding for international activity, and the creation of a separate foundation or agency with a specific international S&T mandate. The latter idea had been called for in the early 1960s and revived periodically but never implemented. The proposed Foundation for International Technological Cooperation now under consideration by the U.S. government would address the central focus of UNCSTED - science and technology for development - and provide expanded opportunities for U.S. university involvement.

Engineering

U.S. engineering schools and faculty have a long history of involvement in international activity; including institution building, cooperative R&D, U.S. resource base development, and education and training. A 1976 report of the National Academy of Sciences and National Academy of Engineering summarizes 11 major program areas in which U.S. engineering schools can render development assistance, including research, curriculum improvement, industrial extension services, continuing education, and engineering technology. Some of these recommendations are reinforced in the recent National Academy study for the U.S. UNCSTED preparations which proposes a number of specific initiatives involving U.S. universities, including joint research on soils and water management, the industrialization process, and small-scale technologies based on renewable energy sources, as well as revival and expansion of AID's 211(d) program.

Our own report documents two large institution-building projects of the 1960s (the Indian Institute of Technology, Kanpur, India, and the College of Engineering, Kabul, Afghanistan), and several other past involvements, including the active Georgia Tech 211(d) Small Industries Program which contains elements of all four major types of U.S. university involvement, and the Rural Industry Technical Assistance Program, started by Prof. Morris Asimow of UCLA.

Large-scale consortia of universities are currently helping to build institutions of higher education in Saudi Arabia and Algeria; the latter, the $129 million, ten-year, INELEC project, also involves engineering technology colleges and private industry. Other consortia projects are being planned. An American Society for Engineering Education (ASEE) survey of ASEE member institutions reveals a shift in funding for their international activity from the U.S. government to foreign governments, particularly OPEC countries.

Engineering education in the U.S. is primarily concerned with supporting the modern industrial sector and rests on a sound scientific base. Many LDCs seem to want this kind of engineering education; they also need planner-managers, industrial engineers, engineering technicians, and science and technology policy analysts. Workshop participants felt that U.S. engineering education provided a good model. Institution-building efforts abroad have followed the science-based engineering pattern and appear in some instances to have been successful in helping to build an indigenous S&T base of this kind, although such a judgment must be tempered by lack of independent evaluation. U.S. engineering education is less likely to be relevant to basic needs and appropriate technology (AT), although the Georgia Tech and RITA programs appear to have had some success in this regard and some engineering colleges are getting into AT activity.

The engineering education profession seems to have difficulty gaining support for international efforts compared with the success of agricultural schools in connection with Title XII legislation. The important present and potential impact of U.S. engineering education on LDC science and technology infrastructure is indicated by the fact that about 25 percent of all foreign students in the U.S. major in engineering and that somewhere between 29 and 50 percent of all graduate engineering students in the U.S. in 1975-1976 were foreign students, mostly from LDCs. These numbers suggest that (1) new efforts might be made to develop courses, curricula, research projects, and special programs (e.g., summer institutes, industry training) relevant to international development for LDC engineering students in the U.S., and (2) increased emphasis on efforts to strengthen LDC institutions might have as a long-term objective the reduction of such heavy dependence of LDCs on engineering programs in the U.S.

Agriculture

U.S. state universities and land-grant colleges have played and will continue to play a central role in U.S. involvement in international agricultural development. They have done so through an extensive institution-building program in the late 1950s and the 1960s, and are getting geared up for renewed involvement in AID's new Title XII program. Title XII gives these universities a significant role in program definition as well as assurance of long-term support for international collaborative research and for strengthening their institutional capacity for international work. Their involvement is assured because of heavy emphasis on food and nutrition in legislation and programs concerned with technical assistance and cooperation. An important rationale behind past involvements has been the belief that the U.S. land-grant university is a relevant and transferable concept for LDC institutions.

Our report traces the history of U.S. university involvement from President Truman's 1949 Point Four declaration to the present. There are a significant number of cases in which U.S. universities and U.S. university personnel appear to have contributed to building an indigenous LDC S&T base for agricultural development. Among the cases examined in this report were the involvements of six U.S. agricultural colleges in helping to build nine agricultural universities in India over a 20-year period from 1952 to 1972; there was a considerable expansion of staff and facilities at the Indian institutions. The University of Illinois' INTSOY program of cooperative international R&D and related activities in soybeans has extensive linkages to many LDC institutions and international agricultural research centers. The University of Wisconsin's Land Tenure Center focuses on the relationship between land tenure and agricultural development. From June 1977 to 1978, Michigan State University's College of Agriculture and Natural Resources enrolled almost 300 students representing 59 countries; 93 percent were graduate students.

Our report analyzes some of the evaluations of early U.S. agricultural university involvements and traces the rocky history of the AID-university relationship, culminating in the establishment of the Board for International Food and Agricultural Development under the recently established Title XII program which gives U.S. land-grant universities important new inputs to U.S. technical assistance and cooperation activity in food and agriculture. Past evaluations suffer from being too general; in addition, we were not able to locate independent evaluations of specific projects during the study. Although U.S. university assistance is an important element, it is likely to be of secondary importance to the LDCs' own efforts to build and sustain

an agricultural base that will meet the food and nutrition needs of
their people. Critical factors for the success of a particular U.S.
university involvement include proper selection of U.S. university
personnel, adequate planning involving both LDC and U.S. participants, the commitment of the U.S. institution to support international
agricultural activity, and recognition of the interdisciplinary nature
of many projects. The current focus in agriculture seems to have
shifted from institution building to cooperative research.

The agriculture chapter contains discussions of 11 issues pertinent to U.S. university involvement in international agricultural development; the discussion of each issue contains a review of current
thinking and a set of questions. Among the issues examined are: (1)
the conflict between AID's "New Directions" policy and the present
focus of most U.S. agricultural colleges on large-scale farming; (2)
potential negative impact on women of changing agricultural practices;
(3) the degree to which the "land-grant" model is relevant or adapt-
able to LDC needs; (4) issues related to implementation of Title XII,
including university eligibility and lack of LDC planning inputs; (5)
attitudes of the public and state legislatures; (6) benefits to the U.S.
and U.S. universities from international program participation; (7)
the contributions of science and technology to LDC agricultural development, as well as negative impact; and (8) the relevance of the experience of the predominantly black, 1890 land-grant colleges with
poor farmers in the rural U.S. South to agricultural problems in
Africa and the Caribbean.

The Title XII program appears to have significant potential for
facilitating the involvement of U.S. state universities and land-grant
colleges in international agricultural development. Two collaborative
research programs involving consortia of U.S. agricultural colleges
-- one in sorghum and millet, the other in small ruminant animals --
were scheduled to get underway in October 1978.

Science

We include in science both <u>basic</u> and <u>applied</u> science but restrict
our exploration to the <u>natural</u> as opposed to the <u>social</u> sciences. Science involvements run the gamut from institution-building projects
to faculty exchanges. There appears to be more emphasis on small-
scale, individual-to-individual projects than in engineering or agriculture, although involvements in the latter two fields often have science underpinnings. The report briefly describes a variety of science involvements to illustrate the range of activity; science education programs are considered in some detail.

Summary, Conclusions and Recommendations

Bilateral involvements that seem to have been successful in achieving some or all of their stated objectives include: 1) the National Academy of Sciences-Brazil chemistry project, in which full-time, U.S. postdoctoral research associates went to Brazilian universities for an extended period to help build graduate chemistry programs, working with Brazilian faculty and part-time U.S. senior chemistry professors; 2) science education programs sponsored by the Ford Foundation and the National Science Foundation which included the adaptation of the biological sciences curriculum study materials to developing country conditions; and 3) NSF's SEED program which provides funds for U.S. university scientists and engineers to conduct research or teach, or both, in a participating developing country. Multinational involvements relevant to science include the Regional Scientific and Technological Program of the Organization of American States (OAS), the international scientific research projects (International Biological Project, Global Atmospheric Research Project, etc.), and collaboration between MIT's International Nutrition Policy and Planning Program and the new UN University.

Current thinking of several individuals on science involvements is presented and discussed. Prominent among these is Michael Moravcsik, whose detailed studies and recommendations on science development are summarized. Moravcsik favors many small involvements, coupled with widespread mobilization of both U.S. and LDC scientists, and mechanisms for better matching of LDC students to U.S. graduate science programs, such as the Physics Visiting Committee Project.

We have analyzed past involvements and current thinking to identify criteria for success or failure and key issues in science involvement. The latter include inadequate funding and continuity for projects, the size and scope of projects, and the balance to be struck among basic science versus applied science versus technological development. Some observers feel that any science of high quality is worthwhile in an LDC; others believe that scarce scientific resources need to be targeted toward development needs. The international centers of research excellence (e.g., the International Center for Insect Physiology and Ecology, the International Rice Research Institute) seem to straddle this argument somewhat and manage to perform both basic and applied research.

At present, there appear to be few opportunities for individual U.S. scientists to obtain support for LDC-focused research and other activity from U.S. government programs. Much funded international science activity takes place among the developed countries. The quality of Senior Fulbright-Hays fellowships and other opportunities for scientists to spend time in LDCs appear to be inadequate to attract them in sufficient numbers. Initiatives by individual scientists in international cooperative science activity can sometimes be significant, as was the case in the FORGE program and a Latin American molecular biology program. A competitive, small grants research program and improved study abroad opportunities are needed to involve more U.S. scientists in cooperative international activity.

Future Roles for U.S. Universities

We expect to see future U.S. university involvement in all four major categories; institution building, cooperative R&D, resource base involvement, and education and training. Rao defines three categories of LDCs in terms of the role and needs of their universities. The assistance and cooperation they might desire from U.S. universities varies widely and accounts in part for the diversity of involvements; the variety of interests and flexibility within the U.S. university community as a whole is another factor in bringing about this diversity.

We anticipate a shift away from large institution-building projects. A major exception to this trend is the current resurgence of involvement of U.S. engineering and engineering technology colleges in helping to build institutions in OPEC countries that provide the funds.

Interest in cooperative R&D activity appears to be increasing. Collaborative research projects in agriculture are a central element of the Title XII program. Limitations to collaborative research include the passage of laws and regulations by some LDCs restricting such activity and the difficulties associated with carrying out research of a truly collaborative nature. We also anticipate more emphasis on activity within the LDCs themselves and less on building resource bases in the U.S. However, this latter expectation is complicated in engineering by the larger percentage of foreign engineering graduate students currently desiring to study in the U.S. As a result, it seems essential to devote attention to making the experience of LDC engineering students more relevant to needs in their home countries. As a result, some revival of the 211(d) program and support for research at U.S. institutions focused on LDC problems may be in order.

Some mechanisms for an expanded program of U.S. university involvement include: an expanded NSF-SEED program; an open, peer-reviewed, competitive, small grants program for international collaborative R&D; a program to help build national and regional science and technology centers in LDCs; institutional grants to support interdisciplinary, development-oriented resource bases and LDC-focused research in the U.S.; new development-oriented involvement of U.S.

Summary, Conclusions and Recommendations 255

students and faculty in LDCs, e.g., a Scientific Peace Corps; and more and better fellowship opportunities. Multinational mechanisms included expanded involvement in and support for the United Nations University. A variety of approaches - some small, some large- should be encouraged. Bilateral programs are more developed; multilateral involvements need further consideration.

New links between U.S. universities and other U.S. organizations such as Volunteers in Technical Assistance (VITA), Appropriate Technology International, the Peace Corps, and industry may enhance the abilities of both parties to contribute to the S&T for development effort. New U.S. university consortia involving engineering and engineering technology colleges and private industry are active in building LDC institutions in Algeria and Saudi Arabia. Cooperation between U.S. universities and VITA (an S&T-oriented, private, voluntary organization) has been occurring for some time; increased cooperation in research and problem solving can combine VITA's outreach to LDC institutions and individuals with the university's substantial R&D capability and engineering and agricultural know-how. New University-Peace Corps arrangements might help recruit more volunteers with S&T capability who are sensitive to development problems and concerns. The university's traditional ties with the national laboratories might prove useful if the latter acquire a mandate to work on international problems, as appears to be happening in the small-scale, energy source field. University-industry collaboration is of considerable interest; U.S. universities can serve as linking organizations between LDCs and the U.S. private sector. Models of U.S. university-industry collaboration may be relevant to LDC needs. The role of professional societies in fostering international S&T cooperation and technology transfer is receiving increased attention.

Perhaps the most important mechanism as far as U.S. universities are concerned is the proposed Foundation for International Technological Cooperation (FITC). The functions of the FITC include emphasis on building LDC S&T infrastructure, on collaborative research, and on orienting activities of U.S. universities toward LDC problems. A planning office for the foundation has been set up, headed by a director with extensive experience in international university programs. The creation of the FITC would fill an important gap for U.S. universities in their efforts to become more involved in S&T for development.

There are several criteria upon which success or failure of U.S. university international involvements might be based. Some hold that meeting specific program objectives - e.g., number of participants trained, number of new curricula instituted - is sufficient and that what the LDC does with that capability is their own concern. Others feel that criteria must include the extent to which basic needs are met as a result of U.S. university activity. In many instances, data are lacking to establish whether these or other objectives have been achieved.

Some conditions for success of cooperative R&D activities include: 1) long-term, assured funding; 2) strong backing from a high university official; 3) dedicated, sensitive, capable personnel; 4) a "nonconventional" base within the U.S. university to avoid heavy teaching commitments and permit travel; and 5) a close connection between ongoing domestic activity at the U.S. university. The Georgia Tech Small Industries Program appears to satisfy these conditions.

There are many limitations to U.S. university involvement in S&T for development. They include: 1) an inadequate professional reward and supportive structure for U.S. faculty participation; 2) culture, language, and distance barriers; 3) lack of funding and program opportunities; 4) in some places, anti-U.S. sentiment which may work against large projects, but perhaps not against smaller involvements; 5) reluctance of U.S. personnel to work in certain countries for political reasons, e.g., human rights violations; and 6) laws and regulations in LDCs restricting foreign researchers. Some academic personnel feel that the universities should stick to what they do best, namely, teaching and research. Others feel that extension, outreach, and service are important; the agricultural colleges have had a long history of involvement in such activity. If an expanded program of university involvement is contemplated by the U.S., it should be kept in mind that an indigenous LDC S&T base ultimately means that the LDCs must rely primarily on their own skills and resources. Thus, more frequent situations in which U.S. or other developed country help is not wanted may be a healthy sign as well as a precursor to more truly cooperative activity.

We sketched three scenarios to illustrate and dramatize different future roles for U.S. universities: 1) a "status quo" scenario based upon minimal changes from previous years in both activity and budget; 2) an "all-out" science and technology scenario, which incorporates a large increase in funding and major involvement of U.S. S&T resources; and 3) a "modest increase" scenario which involves increases of about 25 percent of the "all-out" S&T scenario funding, coupled with a shift of science and technology activities and priorities. Of the three, U.S. actions in the "all-out" S&T scenario seem to us to be more likely to be greeted with favor by the LDCs at UNCSTED. However, given the current economic climate in the U.S., the "modest increase" scenario is more likely to occur than the "all-out" scenario.

Policy Issues and Options

We considered 23 policy issues under five major categories: 1) funding; 2) objectives; 3) why universities?; 4) bureaucracy vs. professional approach; and 5) evaluation. These issues emerge as being of

Summary, Conclusions and Recommendations

overall importance in considering the role of U.S. universities in S&T for development, based upon analysis of past involvements and the legislative mandate. Issues specific to engineering, agriculture, and science are discussed in chapters 2, 3, and 4, respectively.

Availability of funding will heavily influence the extent of U.S. university involvement. Funding sources include: the U.S. federal government, U.S. state and local governments, U.S. private industry, U.S. private foundations and development organizations, international organizations, OPEC and other foreign country governments, and foreign foundations. Foreign sources have increased while some private U.S. sources have grown less. An increase in U.S. government spending on S&T for development in connection with UNCSTED could stimulate new U.S. university interest; participation has been limited in the past because of lack of financial support, among other things. We have not had sufficient data on past spending upon which to base an estimate of future spending requirements. We guess, based on past UN guidelines, that an increase of the order of magnitude of hundreds of millions of dollars in new commitments might be needed to achieve a favorable response from the LDCs. However, if funds are primarily spent in the U.S. or funneled through U.S. universities to LDC counterparts, we would assume that the LDC response might be less favorable than if substantial funds were made available directly to LDCs or to international organizations such as the United Nations or its specialized agencies.

The above factors need to be considered in planning for the FITC. Workshop participants strongly supported the concept of the FITC as an agency independent of AID because of concern that AID leans too much toward the bureaucratic as opposed to the professional model, because of AID restrictions against involvements with AID-graduate countries, and because of AID's previous lack of focus on science and technology. Currently, it is difficult for new initiatives requiring funds to win approval in Congress. Because U.S. universities are likely to be major participants in FITC programs, they need to take an active part in defining these programs and voicing their support, as the land-grant colleges and state universities did in connection with the Title XII program.

The relationship between AID's basic-needs mandate and LDC S&T infrastructure building is an important one to consider. It can be argued that emphasizing meeting basic needs at the expense of S&T infrastructure serves to keep LDCs in a position of dependence on developed countries and makes it difficult for them to escape from poverty in the long run. Such an argument might very well be made at UNCSTED by developing countries. Conversely, it can be argued that building S&T infrastructure without helping to meet basic needs perpetuates LDC poverty and inequality, at least in the short run. At the Project Workshop, several participants felt that narrow interpretation of the basic-needs mandate by AID was severely limiting support for international S&T infrastructure activity involving U.S. universities. Others felt this was not the case. Our own view is that international S&T cooperation must strike a better balance between these two objectives.

Other policy issues discussed include the void that exists in U.S. policy with regard to the middle-income, AID-graduate countries that wish to participate in international S&T cooperation but fall in between the OPEC countries, with their ability to pay full costs, and the LDCs with low GNPs which fall within the AID mandate. Options for responding to these increasingly important countries include: 1) broadening AID's mandate; 2) focusing NSF's cooperative science program on these countries; and 3) incorporating AID-graduate country programs within the mandate of the FITC. The benefits to the U.S. from participation in international S&T activity are becoming increasingly apparent, strengthening the case for a "mutual-benefit" rationale for U.S. involvement.

We addressed the question of how U.S. universities compare with other U.S. institutions for S&T for development activity, the rationale for U.S. university participation, and the relevance of universities to development needs. Whereas, there are certain limitations to U.S. university involvement, we feel that a strong case can be made in its favor. We also addressed the manner in which U.S. government programs are carried out and discussed the need for more open, competitive, peer-review activity for collaborative R&D and other programs. The issue of small versus large projects was considered as was the need for a variety of approaches. Finally, we examined issues related to evaluation of U.S. university involvements. There is a lack of and need for independent evaluations and for methods to carry out such evaluations.

Eight specific legislative options were presented that might be considered by U.S. policymakers for expanding international S&T involvements involving U.S. universities. They include: 1) establish the Foundation for International Technological Cooperation; 2) fund the International Education Act of 1966; 3) expand Title XII authority to other than food and agriculture, and remove current restrictions on university participation; 4) increase the funding authority for 211(d); 5) expand the international activity of the National Science Foundation; 6) provide authority and funding for other, primarily domestic, mission-oriented agencies to expand their work in international S&T; 7) expand the S&T for development focus of the new International Communication Agency, and 8) expand funding for international organizations and support the United Nations University. These options are not necessarily exclusive. It seems to us that the key step to take is establishing the FITC but items 2), 4), 5), 6), and 8) would be desirable. Of these

Summary, Conclusions and Recommendations

latter five items, expanded funding for international organizations and support for the UN University are of particular importance to consider in detail because of the emphasis the LDCs may place on this approach to international S&T cooperation at UNCSTED.

CONCLUSIONS

Previous U.S. university involvement in international S&T activity has been extensive in the areas of institution building, cooperative R&D, resource base development, and education and training. In agriculture and engineering, U.S. universities and consortia of universities were involved in large institution-building projects in the 1960s. New institution-building projects involving engineering and engineering technology colleges have been funded by OPEC countries. In agriculture, the emphasis has shifted to collaborative research. In science, involvements have often been of the smaller variety and have been more successful where there already has been a local base. Since 1950, more than two million students from developing countries have come to the U.S. for higher education. Many have been in S&T fields and have returned home to significant positions in government, universities, and private and public enterprises.

There are a significant number of cases in which U.S. universities and U.S. university personnel appear to have contributed to building an indigenous S&T base in LDCs. Some examples are the NAS-SEED program, the building of engineering and agricultural institutions in India, the RITA and Georgia Tech collaborative programs focused on the development of small industries, the INTSOY program of the University of Illinois, and secondary science education programs sponsored by the Ford Foundation, to name just a few. This conclusion is based solely on our examination of project reports and some sponsoring agency evaluations. We have not obtained information directly from the LDC institutions involved, nor were we able to locate any independent evaluations.

There appears to be opportunities for future U.S. university involvement in all the categories is the first conclusion. However, in the future we expect to see some shift of emphasis away from large-scale, institution-building projects and toward cooperative research and development. In addition, more emphasis on education and training of LDC students within their own countries and less within the U.S. would seem to be called for to aid in building an indigenous S&T base. However, at the graduate level, the demand by LDC S&T students for education in the U.S. remains strong; new initiatives by the U.S. are needed to provide experiences for these students that are relevant to development needs of their home countries.

Foreign assistance legislation has shifted in recent years from emphasis on large, technical assistance projects to helping meet basic needs. Two AID programs relevant to universities are 211(d) and Title XII. The 211(d) program which provides support for U.S. university resource base development has a maximum annual funding ceiling of $10 million and is currently not being fully utilized; its revival and expansion could broaden the involvement of U.S. engineering and science programs in international S&T. The Title XII program which is just getting underway provides new opportunities primarily for land-grant universities with teaching, research, and extension capability in food and agriculture to participate in decision making, design, and implementation of long-term collaborative research programs. Other universities, federal laboratories, etc., may also be involved in providing services through Title XII.

The past and current legislative mandate for U.S. university involvement appears inadequate if it is decided to adopt a policy of greatly expanded efforts to help build an indigenous S&T base in LDCs in connection with U.S. initiatives at the 1979 UN Conference. The proposed Foundation for International Technological Cooperation appears to be an important mechanism in support of such a policy. Other steps accompanying the FITC that could be useful include expanding the international role of various federal agencies with primarily domestic missions, e.g., the National Science Foundation, the Department of Energy, and the Department of Agriculture.

Careful consideration must be given to objectives in defining success or failure for U.S. university involvement. Without the FITC, there is no legislation that appears to focus heavily on building an indigenous S&T base in LDCs. There is legislation that focuses on helping the poorest of the poor meet basic needs. Building an indigenous LDC S&T base does not insure that basic needs will be met; however, without such a base, LDCs may be kept in a position of dependence upon the developed countries. A broader interpretation of AID's basic-needs mandate would be needed to permit more emphasis on S&T infrastructure building in that agency.

U.S. universities and faculty are somewhat limited in what they can hope to accomplish. They are limited by distance, culture, human resources, available funding, and, in some cases, by lack of home (U.S.) institution interest or by not being wanted by the LDCs. U.S. university science, engineering, and agriculture programs are primarily relevant to the modern sector both in the U.S. and the LDCs; they have had little experience with what has been termed appropriate technology. However, we feel that they can develop the capability to work in appropriate technology, given the opportunity to do so. Another inherent limitation is that building indigenous capability largely means becoming more self-reliant; it may be that many things the U.S. would like to do the LDCs

Summary, Conclusions and Recommendations 261

may not want us to do. The U.S. role can at best be a supportive one. Our own study has not had sufficient input of LDC thinking.

Bilateral programs should contain funds for small, one-on-one projects as well as funds for institutional support. A variety of approaches are needed to meet a variety of situations. It is important that there be more open, competitive peer review of U.S. university involvement than there has been in the past to allow and encourage maximum participation, particularly in collaborative research programs.

Our study has not given much attention to multilateral involvements. U.S. university interaction with regional science and technology programs, regional universities like the Asian Institute of Technology, the World Bank, and the UN University all seem desirable and need further consideration, as does U.S. financial support for the UN University. Multilateral cooperation is likely to be an important topic at UNCSTED.

New links between U.S. universities and other U.S. organizations such as VITA, ATI, research institutes, national laboratories, Peace Corps, and private industry should be explored, as well as links between different sectors of the university community, e.g., agriculture and engineering schools, engineering and business schools, engineering and engineering technology colleges. These links may enhance the ability of all parties to contribute to the development effort.

Interdisciplinary centers and programs in U.S. universities focusing on science and technology for development might provide useful inputs to building an indigenous LDC S&T base. A small number of such programs and curricula have begun to emerge, some as the result of 211(d) grants, others due to university-supported initiatives. These programs can help provide relevant educational experiences for the many LDC students studying in the U.S., as well as a strengthened U.S. base for participation in development-focused, cooperative R&D.

We have examined several broad sets of policy issues that are of importance in considering the role of U.S. universities in S&T for development. These issues include the following.

Funding

Availability and continuity of funding will heavily influence the future role of U.S. universities; lack of funding has probably been the principal obstacle to their expanded involvement in international S&T programs. Three scenarios were considered which incorporate three levels of U.S. international S&T spending relative to current levels: no change, large increase, modest increase. Of these, the large increase option seems most likely to strengthen the U.S. position at

UNCSTED, provided that attention is paid to multilateral involvements; it also seems least likely to win approval in the U.S.

The legislative option that seems most vital for an expanded U.S. effort in international S&T is the proposed FITC. We conclude that the FITC will be most effective in carrying out its mandate if it is an independent agency, outside of the current policy and bureaucratic structure of AID. This conclusion was shared by our workshop participants.

In 1979, the FITC went through two name changes and as of this writing, was called the Institute for Scientific and Technological Cooperation (ISTC). It was put forward in Title II of the International Development Assistance Act of 1979. At the same time, a proposal was made to reorganize the U.S. foreign assistance program by Executive Order of the President. The reorganization proposal would establish an International Development Cooperation Administration (IDCA) in which AID and ISTC would be two separate but equal program elements. It is possible that this arrangement will give FITC/ISTC some degree of independence of the type we feel is desirable. The verdict is not in yet however. Furthermore, current budgetary constraints indicate that funding for new involvements of U.S. universities in S&T for development from FITC/ISTC will be very modest, at least in the initial stages.

Comparative Advantage of Universities

U.S. universities have the advantages of extensive overseas involvement, LDC alumni and professional contacts in the hundreds of thousands, a growing international perspective, teaching, research, and service objectives (albeit primarily domestic in orientation) closely related to building S&T infrastructure, and an ability to function in situations where other U.S. institutions (e.g., multinational corporations, direct government programs) may not be able to function. However, they are less well equipped to deliver industrial know-how than private industry and less attuned to political considerations than a government agency.

Bureaucratic vs. Professional Approach

The university-AID relationship has had its ups and downs. There is a need for the agency to insure that program objectives and legislative mandates are met. On the other hand, university personnel are professionals who wish to be treated as such. If it is desired to mobilize more university scientists and engineers to help build an indigenous

Summary, Conclusions and Recommendations

S&T base, it seems likely that involvements in programs that relate to their primary professional goals will yield better results than involvements that are peripheral to those goals. In addition, a variety of programs and opportunities for involvement, both individual and larger scale, need to be available and widely publicized.

Evaluation

U.S. university involvement in S&T for development may be characterized as lacking an independent evaluative base. It is important that efforts be made to develop such a base for past and present activity. New projects and programs should also require independent evaluations. Mechanisms for performing these evaluations need to be developed; involvement of LDC researchers should be considered. Many evaluations need to be more open and accessible to researchers and to the public than past, in-house agency evaluations have been, in order to improve the quality of future programs and projects.

RECOMMENDATIONS

We divided this section into two parts. First, recommendations for future studies and related activity stemming from this study are presented. Second, we address the matter of U.S. initiatives for UNCSTED.

Future Studies and Related Activity

The present study represents a first effort to assess the role of U.S. universities in S&T for development. There is a need for a continuing effort that will go well beyond what we have been able to accomplish in a six-month period. The elements of such an effort are listed below, not necessarily in order of priority:

1. Perform a comprehensive state-of-the-art review and analysis of the role of the U.S. universities in international science and technology, employing "case study" techniques. Develop an annotated bibliography and machine readable outputs. Extend our current study to areas of medicine, health, nutrition, and population. As part of this overall effort, examine in detail elements of the data base that we were not able to scrutinize in the cur-

rent study, namely Ford and Rockefeller Foundation program evaluations, as well as program reports and evaluations performed by AID. To accomplish the latter will require extensive time spent at AID, examining document-center material and interviewing AID personnel.

2. Examine the role of U.S. universities from the LDC perspective. This could be accomplished through analysis of the role of the U.S. universities as reflected by LDC UNCSTED-country papers, some foreign travel to discuss a small number of specific cases with LDC personnel involved in those activities, discussions with U.S. embassy officials in selected countries, and examination of the foreign literature. A variant on this element would be to take a small number of specific cooperative programs and perform interviews and evaluations with both LDC and U.S. participants.

3. Develop a methodology for carrying out independent formative and summative evaluations of selected types of U.S. university involvements, such as cooperative research and development, and institution strengthening. The need for such methodology and evaluations has been a strong finding of our current research. Elements of this activity include understanding how evaluations of U.S. university international S&T programs (e.g., those supported by AID) are currently carried out, comparison of these methods with methods of evaluating other federal programs, consideration of the role of LDC investigators in the evaluation of cooperative projects and, application of new methods to a small number of international S&T projects.

4. Convene a conference on the role of U.S. universities in science and technology for development, with the specific purpose of obtaining the views of the LDC academic community. Use the current study as an initial working document. At least 50 percent of participants should be from LDCs. An alternative would be to focus the conference on the role of LDC as well as U.S. universities.

5. Examine the current nature of the educational experience that foreign graduate students are receiving in the U.S. in selected fields, e.g., engineering, science, and agriculture, and define specific ways in which that experience can be altered or enriched to be more relevant to development concerns in their home countries. Information should be sought on the nature of thesis projects that foreign students carry out in the U.S. A workshop in-

Summary, Conclusions and Recommendations   265

volving foreign graduate students should be a part of this activity. This element seeks to provide a firmer information base than is now available on the current experiences of foreign graduate students in the U.S. and to use this base to generate new initiatives that U.S. universities and funding agencies might undertake.

Other possible elements follow:

6. Examine the interaction between U.S. universities and international organizations.

7. Examine in detail U.S. university spending on S&T for development.

8. Analyze international cooperative R&D activity in some detail. Consider mechanisms for implementation and criteria for evaluation.

9. Carry out comparative case studies of experiences of other countries (e.g., Germany, France, Great Britain, Japan, the Soviet Union, China) in international S&T involving universities.

10. Examine in detail past activity and opportunities for future cooperation between U.S. universities and other U.S. organizations (e.g., VITA, private industry, national labs, Peace Corps, etc.) in international S&T.

Item 4 would be valuable to undertake prior to UNCSTED. Item 6, portions of item 2, and item 5 could also be relevant to UNCSTED.

U.S. Government Initiatives in Connection with UNCSTED

Based upon our study, we recommend that the U.S. government take the following initiatives in connection with UNCSTED. These initiatives are responsive to UNCSTED agenda item 2(d) which calls for new mechanisms for international science and technology cooperation; they also are of significance for expanding the U.S. university role in such cooperation.

1. Establish the Foundation for International Technological Cooperation.

2. Appoint a high-level committee, including advisors from developing countries, with the stated purpose of increasing U.S. emphasis on and involvement in multilateral mechanisms for international S&T cooperation, including support for the United Nations University.

3. Expand the international activities of the National Science Foundation, the Department of Agriculture, the Department of Energy, and other primarily domestic, mission-oriented agencies concerned with science and technology.

4. Increase activity and broaden the scope of the AID 211(d) program.

5. Fund the International Education Act.

6. Provide improved opportunities for U.S. S&T faculty members to study and work in LDCs through the NSF-SEED program and the Fulbright-Hays program.

7. Establish a Scientific Peace Corps.

Appendix A

WORKSHOP ON THE ROLE OF U.S. UNIVERSITIES IN
SCIENCE AND TECHNOLOGY FOR DEVELOPMENT:
MECHANISMS AND POLICY OPTIONS
(UNCSTED PROJECT)

held at

Washington University
St. Louis, Missouri

July 13 and 14, 1978

Workshop Objectives and Mechanics

The workshop convened at 9 A.M. on July 13, 1978, at the offices of the Center for Development Technology (CDT) on the Washington University campus. The 17 workshop participants are listed in Fig. A.1. Although the group was small in number, as required by the grant, a wide range of university subject matter interests, and both government and nongovernmental organizations were represented.

Robert Morgan described the background of the project, which was carried out in response to an NSF program solicitation with the following objectives:

> Projects supported under this solicitation should contribute directly to U.S. preparation for the UNCSTED. The anticipated contribution is two-fold: (1) the production of Study Papers for use by the U.S. delegation to UNCSTED that address issues and options related to conference agenda items and (2) the promotion and improvement of consensus or clarification of important differences on these issues.

Merton R. Barry, Director
International Engineering Programs
University of Wisconsin
439 Engineering Research Building
1500 Johnson Drive
Madison, Wisconsin 53706

Melvin Blase, Director
International Programs
University of Missouri
Columbia, Missouri 65201

W.D. Euddemeier, Director
International Agricultural Programs
113 Mumford Hall
University of Illinois
Urbana, Illinois 61801

William Eilers
Department Director and UNCSTED
  Coordinator
Office of Science and Technology
Agency for International Development
Washington, D.C. 20523

Thomas Fox
Executive Director
VITA
3706 Rhode Island Avenue
Mt. Ranier, Maryland 20822

Mario Gomez
Professor of Mechanical Engineering
Washington University, Box 1185
St. Louis, Missouri 63130

Barrett Hazeltine
Associate Dean
Professor of Engineering
Brown University
Providence, Rhode Island 02912

Franklin Long
Program in Science, Technology, and
  Society
Cornell University
Ithaca, New York 14850

Barbara Lucas
Division of Policy Research and
  Analysis
National Science Foundation
Washington, D.C. 20550

Hugh Miller
Executive Director
Office of the Foreign Secretary
National Academy of Engineering
2101 Constitution Avenue, N.W.
Washington, D.C. 20418

K.N. Rao
Center for Policy Alternatives
MIT
Cambridge, Massachusetts 02139

Aaron Segal
Division of International Programs
National Science Foundation
Washington, D.C. 20550

Virginia Walbot
Assistant Professor of Biology
Washington University, Box 1137
St. Louis, Missouri 63130

Robert M. Walker
McDonnell Professor of Physics
Washington University, Box 1105
St. Louis, Missouri 63130

William Wight
Office of International Science
AAAS
1776 Massachusetts Avenue, N.W.
Washington, D.C. 20036

Adolph Wilburn
Council for the International
  Exchange of Scholars
1 Dupont Circle
Washington, D.C. 20036

Michael Witunski
Staff Vice President
McDonnell Douglas Corporation
P.O. Box 516
St. Louis, Missouri 63166

Washington University UNCSTED
  Project Personnel:

Robert P. Morgan, Principal
  Investigator
Ellen Irons, Research Assistant
Eduardo Perez, Research Assistant
Theodore Soule, Research Assistant
Ava Fried, Staff Associate

Center for Development Technology
Department of Technology and Human
  Affairs
Box 1106, Washington University
St. Louis, Missouri 63130

Fig. A.1. Workshop Participants.

Appendix A

An integral part of the Washington University UNCSTED project is, according to NSF guidelines, "a small workshop at which the Study Paper will be discussed and reviewed by peers." "Following the submission of the Study Paper," (which was done on June 1, 1978) "a workshop involving a small number of persons (approximately 10 to 20) who are professionally recognized peers in the subject area is to be convened by the project director. At the workshop, the Study Paper which has been distributed in advance will be discussed. From inputs obtained from this workshop, the original draft study paper will either be revised or supplemented with additional information."

In the invitation letters to the conference, participants were asked to read the Draft Study Paper, be prepared to comment briefly on a specific chapter, send any written comments on the report prior to the workshop, and be prepared to suggest three new initiatives for international S&T cooperation involving U.S. universities at the Friday afternoon (July 14) session. Some participants provided prior written comments while others did so at the workshop; reports and other printed materials were also provided to fill information gaps.

The following summary of presentations and remarks by individuals can in no way do justice to the quality, scope, and content of the discussions; the workshop was a very constructive, stimulating, and rewarding experience for us. A complete recording of workshop discussions was made and is on file at the Center for Development Technology, Washington University. The workshop summary and the tapes were utilized extensively in preparing the final report, which attests to the value of the workshop to our project. The feedback we have had from the participants indicates that they found the workshop to be a valuable experience as well.

Summary of Remarks (prepared by Robert P. Morgan)

Robert Morgan explained the purpose of the UNCSTED study project which was to examine the past, present, and future role of U.S. universities in helping to build an indigenous science and technology base in developing countries. This study is relevant to UNCSTED agenda item 2 (see Fig. A.2) which is concerned with institutional arrangements and new forms of international cooperation in the application of science and technology. Item 2d calls for the strengthening of international cooperation among all countries and the design of concrete new forms of international cooperation in the fields of science and technology for development. The importance of this topic was stressed by V.J. Ram, technical advisor to UNCSTED Secretary-General da Costa, at the American Society for Engineering Education annual meeting in

UN CONFERENCE ON SCIENCE AND TECHNOLOGY FOR DEVELOPMENT (UNCSTED)*

1. Science and technology for development:

    (a) The choice and transfer of technology for development;

    (b) Elimination of obstacles to the better utilization of knowledge and capabilities in science and technology for the development of all countries particularly for their use in developing countries;

    (c) Methods of integrating science and technology in economical and social development;

    (d) New science and technology for overcoming obstacles to development.

2. Institutional arrangements and new forms of international cooperation in the application of science and technology:

    (a) The building up and expansion of institutional systems in developing countries for science and technology;

    (b) Research and development in the industrialized countries in regard to problems of importance to developing countries;

    (c) Mechanisms for the exchange of scientific and technological information and experience significant to development;

    (d) The strengthening of international cooperation among all countries and the design of concrete new forms of international cooperation in the fields of science and technology for development;

    (e) The promotion of cooperation among developing countries and the role of developed countries in such cooperation.

3. Utilization of the existing United Nations system and other international organizations to implement the objectives set out above in a coordinated and integrated manner.

4. Science and technology and the future:

    Debate on the basis of the report of a panel of experts to be convened on this subject.

---

* Proposed by Economic and Social Council and approved by UN General Assembly on December 21, 1976.

Fig. A.2. Agenda.

Appendix A   271

Vancouver on June 22, 1978. It is Dr. Ram's opinion that the real
action at the conference will be on finding better and more effective
instruments of international and regional cooperation in S&T - both
advanced country-developing country and developing country-developing country - and that the conference will develop a better appreciation
and greater possibilities for such cooperation, both within and outside
of the UN system. The role of education and human resources development is also being stressed, as indicated by a statement of the Second
Preparatory Committee for the Conference on February 3, 1978, which
indicates that systems for education and training within the LDCs are
of particular concern, as well as "appropriateness of programs for
education and training of personnel from developing countries in
developed countries, migration of talent and skills for developing countries, need for real concern about R&D needs of developing countries."

Morgan briefly described the study plan for the project and steps
that were taken to collect and evaluate data. He defined the scope of
the study and some terms that were used, and briefly discussed the organization of the report. He pointed out limitations of the study which
included: (1) difficulties in locating and/or acquiring evaluations of
specific university involvements, (2) lack of input by individuals and
other sources from developing countries, and (3) omission of engineering technology college activity in the Draft Study Paper. The UNCSTED
study team is gathering information on this latter activity for the final
report. Several useful suggestions were made by participants for acquiring additional information. However, Morgan indicates that lack
of time and resources would limit what could be accomplished between
July 14 and September 1, 1978, the final report deadline.

Engineering

Barrett Hazeltine, reflecting on his Zambia experience sponsored
by the SEED program, indicated that Zambia and countries like Zambia
wish to avoid undue dependence on any one country; as a result, university institution building heavily dominated by one advanced country is
probably on the wane. In considering cooperative research, it should
be kept in mind that Zambia has a two-tiered technological structure,
in which some industries are at least as modern as the U.S., while
others are nowhere near as developed. U.S. engineering schools can
cooperate with and relate to problems of the former; activities in small-
scale solar and wind energy might be new, useful contributions. Some
of this research can be done more effectively in the LDCs.

In the discussion that followed these remarks, participants pointed
out that large-scale institution-building efforts still exist (there is a new
$129 million, ten-year project to build electrical engineering and engineering technology university capability in Algeria); that institution

building need not be "bricks and mortar" (in sub-Saharan Africa, many university professors are expatriates who contribute to institution building); that there is growing input of OPEC funds and programs to U.S. universities; and that there is competition among developed countries for these funds.

The participants discussed the sizeable number of foreign graduate engineering students in the U.S. A participant stated that we need foreign graduate students and cannot run our graduate academic and research programs without them. Subsidization of these students, particularly by large state universities which charge low tuition, are viewed by some as a contribution to international development, by others (including some state university presidents and legislators) as a drain on scarce state resources.

Hazeltine stated that U.S. engineering schools can offer LDCs help in building engineering education institutions based on the U.S. model; he feels that model is more flexible, deals with a wider range of students, and uses better pedagogy than alternative models. LDCs also need science and technology planners who understand broad issues such as choice of technology; U.S. universities may be more acceptable providers of this training than industry and governmental personnel. LDCs will get this help from somewhere – it is in their national interest to do so and it is in ours to help. Two types of engineering graduates appear to be needed: (1) those that are science based and (2) those who are planner-managers. According to Hazeltine, most LDC graduates will be employed as planner-managers. U.S. engineering faculty may want to relate more to the first of these two categories. In the discussion, the need for more practical education at the technology program and pre-college level was stressed. Countries vary greatly in their needs, their demands, and their ability to pay for technical assistance.

Merton Barry stated that the draft study paper adequately covers the principal kinds of engineering involvements but that some things are missing. He stressed the need to emphasize the software side of engineering as contrasted to the hardware. He singled out industrial engineers and science planners as being important; he quoted from an LDC government official at the 1978 World Congress on Engineering Education who said: "What I need in my ministry is people who can tell if we are being sold a bill of goods, who can understand contracts, etc." Not enough attention is paid to things that we have within our engineering schools that might be useful, including our ability to maintain equipment. The facilities of Wisconsin's Instrumentation Systems Center are being used to help the Singapore Institute of Industrial Research develop its metrology capability; this effort is supported by the United Nations Development Program.

Appendix A   273

Barry indicated that the report might have recognized and given more attention to early efforts by the Universities of Illinois and Wisconsin in India which laid the foundation for more extensive involvements of U.S. agricultural colleges in India and for the IIT/Kanpur program, respectively. Other programs that participants felt should be discussed or mentioned in the final report include various engineering technology programs, an in-teacher training program in India, the Middle-East Technical University in Ankara, Turkey, and continuing engineering education programs supported by UNESCO.

There was discussion and some debate about the role of engineering professional societies in technology transfer; the university's role as intermediary between private industry and LDCs and its role in performing international S&T in policy analysis were also discussed. The point was made that the engineering school's role in relation to other parts of the university - particularly the business school - needs to be reexamined for its contribution to S&T for development, and that new centers and other mechanisms are needed to couple LDC universities with their industrial sectors. One example cited both as a potential mechanism and source of technical assistance was the Polymer Processing Center at MIT; another was the Center for Development Technology at Washington University.

Mario Gomez stressed the differences in wealth, culture, education, and technological tradition among LDC countries. A major problem in many is lack of well-trained human resources. He feels that other developed countries are more successful than the U.S. in advising LDCs on where to send their students to study abroad; new U.S. mechanisms are needed. Comparative case studies of French, German, British, and Canadian experiences would have added useful information to the U.S. engineering cases in the report. Gomez cites the multinational program in metallurgy of the Organization of American States as being a good example of a successful regional collaborative research program. Other OAS programs include one on transfer of technology to Latin America and another involving graduate courses in Argentina and Mexico. Gomez stressed the following points: (1) the U.S. educational system is the best model - it and high technology developed as the U.S. grew out of underdevelopment, (2) it would be tragic for engineering research efforts here if we had no foreign graduate students, (3) it would be unwise to change the current emphasis on science-based engineering - LDCs need that versatility and flexibility, (4) business and industrial technology programs are also needed, and (5) traineeships in U.S. industry for foreign graduate students would be useful.

In the discussion following, participants stressed the need for science and technology policy studies and the role of engineering schools in performing such studies. One participant also mentioned that foreign governments are endowing chairs in U.S. universities - an interesting trend.

Agriculture

W. D. Buddemeier felt that the report should contain more information on university involvement with international agencies, such as UNDP and FAO, as well as work supported directly by foreign governments. He pointed out that agricultural colleges may be in a different situation than engineering and science programs in that the former have heavy experience in cooperative extension and getting research results out to users; they have close ties with agribusiness, farmers, and consumers. He feels that evaluation of programs is very important and needs to be done. Social scientists (economists and rural sociologists) are important in agriculture and should be covered in the final report.

Melvin Blase's remarks centered on three topics: (1) Title XII, (2) contributions of science to agricultural development, and (3) benefits to the U.S. He reviewed the history and current status of Title XII and discussed its policy implications. Some feel it is landmark legislation, of a significance approaching the Land-Grant Acts, which gives a fourth dimension to the U.S. agricultural university by legitimizing its long-term role in international development activity alongside of its traditional domestic missions of teaching, research, and extension. Title XII also is significant in that it provides U.S. universities, through the mechanism of the Board for International Food and Agricultural Development, with a decision-making role in the design of AID's international agricultural programs. Blase described the rationale for the Title XII program, its advocacy by universities to Congress, and some of the current issues in implementing the program, such as providing university matching funds and university eligibility.

Blase feels that the report needs to recognize to a greater extent the demonstrated contribution of science to agriculture in LDCs - in particular the work at international agricultural centers on high-yielding seed varieties - and its impact on agricultural policy and development. He also sees the need for more information on the Baldwin, CIC-AID evaluation study than we have provided, stating that it deals with responses such as changes in attitudes as a result of U.S. intervention.

Blase stated that a strong case can and should be made for the benefits to the U.S. from U.S. university involvements. For example, high-lysine corn varieties found in a remote valley in Ethiopia are being used here. International programs have generated new knowledge and have imported distant germ plasm to protect against both plant and animal diseases. He sees economic benefits to the U.S. in helping LDCs move from subsistence to a higher stage of development.

In the discussion that followed, participants considered the AID-university relationship under Title XII, possible state-level opposition to U.S. university international assistance and cooperation, the relationship of international agricultural research centers to other institutions in the LDCs in which they are located, and the responsiveness of U.S. universities to AID's "New Directions" policy.

Science

Robert Walker stated that pure science is tremendously important for developing countries; it is a fascinating activity and it changes boundary conditions in significant ways. He quoted from M. Moravcsik's book, Science Development, and from an article by the Indian scientist, H. Bhabba, to reinforce this view. LDCs need scientists who do first-class science and who are part of the international science community, with frequent contacts and leaves to developed country institutions. The line between basic and applied science is often fuzzy and a fruitless distinction to make. Although science is a long-term investment, it has a way of paying off in practical and unexpected ways. Walker cited pure research at the G.E. Research Lab which led to practical applications; also, an Indian cosmic ray physicist ran the Indian Satellite Television Experiment. It is difficult to interest U.S. physical scientists in LDC problems but may be easier in more location-dependent fields such as geology, botany, and oceanography.

Walker argued for the value of several initiatives, including: 1) programs of frequent visits by LDC scientists to the U.S., 2) efforts such as the Physics Visiting Committee Project, and 3) increasing contacts between developing and developed country scientists through lecture programs and overseas visits. Participants discussed the suitability of the NAS-Brazilian chemistry program as a model, and how much science poor countries with limited budgets could afford.

K.N. Rao, after discussing the importance of a scientific attitude toward development, suggested that the final report might be couched in a framework which would provide direct inputs to U.S. UNCSTED preparations. Universities are knowledge bases and generators of manpower; they have their limitations. Rao presented results from his recent Impact paper, in which he defines three categories of LDC universities and assesses their needs. U.S. universities can make useful contributions, depending upon the needs of LDC universities and their stage of development. Rao feels that a major U.S. contribution has been the education of large numbers of LDC students who came to the U.S. and then returned to important decision-making positions in their home countries. Support for special programs at U.S. universities to orient LDC students to development problems may be necessary. He described Ford Foundation efforts to assist LDC students (e.g., Colombian math Ph.D's) in the U.S. to return to their home countries.

Rao called for future initiatives and mechanisms including: 1) a small country-level grants program in support of local activities; 2) internships for LDC students in U.S. industry (France and Germany do this); 3) more industry-university cooperation; 4) collaborative research on problems of mutual interest - industrialization, urbanization, health delivery - problems from which, incidentally, we also stand to learn a lot in the U.S.; and 5) mechanisms for continuous interaction with LDC scientists and technologists, such as traveling lectureships for U.S. professors. It is hard to build up continuing collaborative relationships with LDC universities once these are broken by discontinuities in funding.

Although U.S. government support is desirable, initiatives can be taken by individuals, private foundations, and universities. The FORGE program and International Foundation for Science were discussed. According to Rao, such nongovernmental efforts must be fostered. The OAS and LDC national research councils for science and technology are also useful vehicles for quick, small grants programs.

Rao stated that it is desirable to increase participation of U.S. scientists in international research centers and other activities located in LDCs, and that it might be done through established mechanisms at NSF at relatively small cost. There is much more money available and demand for such participation in big international science programs located in developed countries than for participation in LDC-focused programs; some redirection is desirable. Significant new international programs concentrating on global problem areas - energy, desertification, ocean resources, environment - should be organized to permit vastly increased opportunities for U.S. and LDC scientists to participate. Experiences with programs such as the International Biological Year and International Geological Year would provide guidelines for new programs.

Satellite technology might be used to link LDC and U.S. universities. The latter idea generated some discussion of relative costs and benefits. Rao felt that it might be better to integrate the science and engineering sections of the report.

Virginia Walbot described the international involvement and significance of the plant biology program at Washington University. The program has 12 faculty members and 17 graduate students in plant toxonomy. Twelve of the graduate students are doing some of their research in foreign countries. Four of the Ph.D. candidates are older

Appendix A

foreign scientists who will head research institutes or botanical gardens. Much has been learned and will be learned in the future about plants in other countries that will be useful here in the U.S. However, she stated that there is little government support for such activity; support comes mainly from private sources, including the National Geographic Society; foreign students are supported by their own countries. Walbot feels that developing countries benefit from sending students to the plant biology program by learning about the need to conserve and expand the narrowing base of germ plasm in the world's collection of crops, by becoming sensitive to the impact of industrial development activity on flora and fauna, and by becoming aware of the economic botany and utility of plants for medical and other uses. One former Washington University plant biology foreign trainee returned to Peru and led an expedition that discovered 10,000 new varieties of potatoes, including five that are resistant to blight.

All countries have botanical gardens; many developing country scientists know much about plants that is unknown in the U.S. Therefore, there is a strong case for mutual benefit to be made here, both for the university scientists who participate and for their countries.

A debate ensued as to whether government funds in this area (systematic biology) were available; the consensus seemed to be that funds were scarce. Participants also discussed the appropriateness of various kinds of plant biology activity in LDCs. One developing country wanted assistance to become a world center of excellence in research on recombinant DNA in plants - something far from being accomplished in the U.S.

The Legislative Mandate

F.A. Long found the chapter on the legislative mandate to be a good, concise summary of past and recent activity. There have been fluctuations in both amount and expectation of support for university programs; such fluctuations may very well continue. The role of AID has clearly been quite important and despite complaints to the contrary, that agency has provided substantial support for U.S. university programs. A five-year AID 211(d) program grant was important to Cornell. NSF's mandate has been modest and narrowly directed. The new International Communication Agency and the Office of Education might be potential sources of additional support for international programs dealing directly with education.

Long felt that more could be said about the mandate of other agencies - the Office of Education, Environmental Protection Agency, Departments of Energy, Interior, and Commerce. They have international programs which deserve coverage in the final report; expansion of these programs might be desirable. The National Science Foundation has carried out a survey of S&T programs for international development in selected federal agencies which should contain relevant material. Some indication in the chapter of university strength and interest tied to specific mandates might be useful to government officials who read it. There was a brief discussion about whether in fact there is no legislative mandate for U.S. involvement in building an indigenous LDC S&T base, as is stated in the Draft Study Paper.

In turning to the future role of U.S. universities, Long stressed the university's role in collaborative policy studies, both analytical and comparative. The Council on S&T for Development is interested in this activity. He would like to see university input into the priorities for a new Foundation for International Technological Cooperation, with a substantial line item for policy studies. He also sees the universities as having a role to play as intermediaries between government and the private sector for technology transfer. The universities themselves might develop their own mandate for what they want to do; support need not come solely from the U.S. government - private foundations and foreign governments provide it also.

With large amounts of money, large external and internal programs would be generated; there could be substantial bilateral programs, new teaching efforts, a large program of summer short courses, etc. With smaller amounts of money, U.S. universities could do a better job of teaching LDC students about development problems. in their home countries, as well as teach our own students about the problems of international development; we could elevate the visibility of the subject through seminars like Cornell's World Food and Nutrition Seminar, visitors, and by having more university-to-university collaboration among scholars. In discussing the 1979 UNCSTED, Blase proposed that a facilitative mechanism be devised for building an international network of scholars to permit small, scholar-to-scholar programs and other activity, with the U.S. taking the lead. In this connection, Segal warned that such new initiatives have floundered in the past because of no prior consultation with, and poor reception by, LDCs as well as lack of real commitment in the U.S.

Hugh Miller singled out the executive branch as being an important element in establishing S&T programs through such mechanisms as bilateral agreements, a point neglected in the chapter. The legislation establishing Appropriate Technology International should be included in table 5 of the Study Paper. Furthermore, policy statements through clauses in annual foreign aid legislation (light-capital

Appendix A

technology, "poorest of the poor," no aid for AID-graduate countries) affect international S&T as well as foreign policy. Another significant piece of legislation under consideration in Title V of the new Foreign Relations Authorization Act is concerned with developing a specific mandate for creating a science and technology foreign policy for the United States. The bill, which passed the House in May 1978, requires the State Department to strengthen its science and technology policy activities. However, the outcome of the legislation, which would have an impact on the universities because of its impact on AID, is uncertain. (The bill subsequently was passed.)

A discussion ensued about the extent to which U.S. universities and university personnel are involved in UN programs, such as FAO and UNDP. Some universities have contracts. There is evidently debate within the UN about the extent to which universities should work with UN agencies. U.S. engineering educators have been working closely with UNESCO, which seems interested in expanding its links with U.S. universities.

Aaron Segal analyzed the legislative mandate and sees three key structural problems. First, the basic-needs policy of U.S. foreign aid is incompatible with much of U.S. science and technology. There are political factors in the U.S. working against changing that policy. Unless the Foundation for International Technological Cooperation is concerned with matters broader than basic needs, U.S. universities may not have much of a role. Several participants disagreed that universities cannot relate to basic needs, especially in agriculture; some specific examples were cited.

A second structural problem described by Segal is that although many federal agencies are active, they primarily receive "pass-through" money from AID. Several federal agencies have reduced their international involvement (USGS was cited as an example) due to shifts in allocation of AID "pass-through" funds as a result of the basic-needs mandate. Segal feels that they need their own funding authority. NSF has some program authority. They use their money to support unsolicited, single, discrete cooperative research proposals. NSF counterparts overseas find this to be an unsatisfactory, luxurious approach; they want long-term, problem-oriented, targeted, team research. NSF has gone to the National Science Board with options for expanded NSF involvement in new approaches to international scientific cooperation. The options include long-term, problem-oriented, cooperative research with scientifically advanced LDCs (e.g., India, Brazil, Mexico). They also include small-scale grants, support for faculty abroad, and curriculum development and short courses for foreign students in the U.S.; these activities are for the less scientifically advanced LDCs. However, even if the National Science Board supports one or more of these proposed initiatives, many obstacles to their implementation will remain. According to Segal, the Office of Management and Budget (OMB) has indicated that there will be no new R&D initiatives for FY 1980 (or at least, a very strong case would need to be made). Congress, which carefully scrutinizes and must approve the NSF budget, may also be an obstacle. If the U.S. university community wants such activity, they will need to make their views known both to the Congress and the administration.

A third structural problem identified by Segal concerns the 60 non-AID recipient developing countries. Countries that can pay full costs for NSF assistance can be accommodated but many other that can only pay for part of assistance rendered are excluded. It appears to be a funding and a mandate problem. AID has a small amount of seed money in its reimbursable programs' office for this purpose; program activity of this office may expand. Segal would like to see formulas developed for non-AID recipients which are less than fully reimbursable, that is, in which the recipients do not have to pay all of the costs.

A fairly lengthy discussion took place ranging over the following: 1) the importance of universities developing their own mandates and rationale for participation (e.g., BIFAD has done this, representing agricultural colleges); 2) the trend toward interdisciplinary research institutes in U.S. universities, staffed with full-time people; 3) the ability of universities, or lack of it, to relate to AID's basic-needs strategy; and 4) the negative attitudes of LDCs toward that strategy. Segal and Blase called attention to a void in U.S. aid policy. AID can only assist the poorest countries; OPEC countries can pay for help. We are not prepared to relate to needs of AID-graduate, middle-income, developing countries.

Future Roles of U.S. Universities

William Eilers reviewed prospects for "S&T for development" funding for the next two or three years, and factors that may cause AID's basic-need strategy to be revised. He then summarized the status of the Foundation for International Technological Cooperation; its functions are listed in figure 1.2. White House thinking is that the foundation will be the major new element of the FY 1980 AID budget request, with funding provided no later than October 1979,

Appendix A

and that $200 million in on-going AID program money will be diverted into the foundation, along with $50 million in new program funds. A planning office, headed by Ralph Smuckler of Michigan State University, to flesh out the program over the next six months; Smuckler reports to John Gilligan in the latter's capacity as head of the Development Coordinating Committee. The functions of the foundation were discussed (see figure 1.2), as were difficulties associated with setting up the foundation within the policy framework and current structure of AID. There was discussion of the need for diversity in U.S. technical assistance policy and in the new foundation, in order to respond to diverse needs in LDCs. The experience and status of Appropriate Technology International (ATI) were also discussed.

Eilers analyzed future roles for U.S. universities corresponding to categories in the Draft Study Paper and as they might relate to functions of the new foundation. He concurred with the view that there is a trend away from U.S. support for large-scale institution building and baseline support for regional centers, and toward more modest funding for collaborative research and institution strengthening. Cooperative R&D might be carried out in a low-profile way; an interesting area, for example, is small-scale renewable energy sources. Denver Research Institute and Georgia Tech programs make small grants with AID money to overseas institutions; such a mechanism seems useful for funding activity in LDCs.

Eilers discussed some of the difficulties with the 211(d) program; he feels gaps in U.S. university capability have been filled. Therefore for these and other reasons, including the need for AID regional bureau concurrence, it is hard to justify new 211(d) grants; they may be phased out. Efforts to further subsidize foreign students already in the U.S. do not seem practical, but addressing their needs with workshops, seminars, etc. may be possible. In turning to the report subsection on mechanisms, Eilers mentioned the role of individual entrepreneurs like Hollander (molecular biology) and Kelleher (FORGE). Sources of support for such efforts and problems associated with them were discussed. The NAS-Brazil chemistry program has had some success, but there were difficulties; mixed feelings about the program were generated in the National Academy of Sciences including concern about cost-effectiveness.

Eilers would like to see U.S. university personnel involved in the new foundation in every way possible. He likes the model of having in-country, LDC resident representatives of the foundation (like the Ford Foundation did), and cited the experience of Bruce Billings in Taiwan as a good one, in which Billings had the background to identify scientific and technological opportunities. Eilers feels this will be particularly effective if the foundation can move into non-AID, middle income countries, with people who can relate both to industry and government. Ford and Rockefeller Foundation experiences may be valuable here. Billings was categorized by one participant as a high-innovative content, entrepreneurial type. The university science and engineering community needs to assert itself within the foundation and on the board. Eilers does not think it will be possible to have participatory decision making of the BIFAD type; however, an advisory board which includes university scientists and engineers comparable to the Joint Committee on Agricultural Development or the Joint Research Committee in the Title XII programs would be useful and important. Small grants to individuals in developing countries are valuable and should be a part of the foundation program, perhaps delivered through individual U.S. university faculty. Participants mentioned the possibility of the foundation having the ability to pass through funds to NSF, and of using organizations like the Agricultural Development Council and the Council of Science and Technology for Development for various purposes. The point was made that universities should mobilize their resources to make inputs to the planning office of the foundation to insure a role for themselves in future decision making.

William Wight supported the early involvement of U.S. university personnel in planning for the new foundation as well as their strong participation on its board, as was and is the case with BIFAD. He feels the Draft Study Paper covers well the various types of U.S. university involvement in institution building. He would like to see more discussion of the problems associated with centers of excellence in LDCs and their negative impact in pulling away faculty from other institutions; more discussion of links between universities and the productive sector and more mention of the role of university faculty as consultants to LDC governments and industry. He supports small, one-on-one cooperative R&D projects with objectives set jointly by U.S. and LDC participants.

According to Wight, education and training of foreign students is of the most immediate and significant impact and deserves more attention in the report. The American Association for the Advancement of Science (AAAS) has held seminars with foreign graduate students to consider their needs. (Two such reports are available - one by AAAS, one by NAS.) The scientific subject matter in this country seems not to be the problem, rather it is the approach - the equipment that we have in the U.S. will not exist when they return home. Industrial programs in LDCs differ from R&D as conducted in the U.S. These differences need to be highlighted and some information provided that relates to home-country conditions. There also might be more stress on economic factors related to the sciences they study.

Appendix A

Wight feels the scenarios in the report need revision - they are too extreme. One should consider what can be done with relatively little money. The "appropriate technology" scenario seems inappropriate. Morgan responded that he anticipated adding a fourth scenario which would reflect modest additional resources. Segal expressed his views on various possible funding sources and how they might fare in the future, and suggested they might be added to the final chapter of the study. Wight would like to see use of bilateral programs such as SEED, FORGE, Fulbright. A discussion followed about SEED - there seemed to be some conflict and uncertainty about whether NSF and/or AID would continue the program, even though both agencies seem to feel it is worthwhile. Wight would like to see foreign graduate students do thesis research in their home countries, also, financial assistance should be tied to their returning to their home countries. There should be more links between universities and community colleges, the latter for training in administration and supportive infrastructure for science programs. Also, cooperative research may be affected by the passage of laws in LDCs regulating foreign researchers.

Long emphasized the need to work harder to see that LDC students are trained at home rather than in the U.S.; U.S. programs to support that goal might include help with institution building, more arrangements for on-leave people, and traveling, short courses. The UNCTAD-V agenda has an item concerned with compensation to LDCs for losses due to the "brain drain."

Adolph Wilburn stressed the opportunity to link international development activity to domestic missions of U.S. universities in order to generate strong support for these activities in the future. U.S. domestic needs are pressing in education and other sectors; resources are scarce. Can we use scarce resources to teach Arabic to U.S. students and faculty who go overseas when students in U.S. urban school systems have pressing educational needs? It may be that people from other countries can help us with our problems here. Some 30 to 50 percent of the science and mathematics faculty members at some of the 100 or so predominantly black colleges in the U.S. are from Asia (India, Taiwan, etc.). Wilburn would like the report to highlight our ability to strengthen LDC universities as universities first and not jump immediately into the more complex issues of their role in economic development.

Wilburn described the Fulbright program in some detail. He expressed concern that some LDC scholars do not get proper advice as to which universities in the U.S. are most relevant to their needs and interests. About 20 percent of the Fulbright scholars who go out from the U.S. are in science, engineering, and technology. More than 50 percent of those who come to the U.S. are in those areas. Some of the problems facing Fulbright scholars were discussed. In a recent study of former American Fulbright scholars to Latin American universities, it was found that 85 percent of the respondents maintain links of various kinds with their host institutions. The program would like to expand. There are 650,000 U.S. faculty; 7,000 are abroad at any time. Five hundred of the latter are in the Fulbright program; that number is down from 650. The quality of the terms of the grants is also down. The new, more visible position of the Fulbright program in the International Communication Agency (ICA) may help to some extent, but the case must be made to show the Congress that this program doesn't just benefit isolated individuals; it is important in increasing the quality of academic life in the U.S. and thus American students will benefit.

The potential role of the predominantly black, 1890 land grant institutions in development needs careful consideration. A conference will be supported in the fall, 1978 to enable these institutions to search their experiences for ways to make relevant inputs to the U.S. preparations for UNCSTED. The status of the "International Linkages in Higher Education" study was discussed as were ways to provide good experiences and institutional matches for foreign scholars coming to the U.S.

Further discussion took place on the relationship between strengthening the role of minorities in science in the U.S. and strengthening international science. Wilburn argued that the case needs to be made that it is good for science to increase the participation of minority students in science; potentially good scientists are being wasted. Furthermore, foreign students coming to the U.S. are politically sensitive to the fact that we are not doing all that we could do for our own minorities at home. The UNESCO science education materials used in many developing countries are better than those used in the public schools in Washington, D.C. We have a lot to learn from other societies in our international development involvements.

Michael Witunski emphasized several aspects of U.S. university involvement which need more emphasis in the report. These include extensive involvement of U.S. faculty with UNDP and World Bank in-country teams to perform preinvestment surveys and background studies. He stated that U.S. universities were running out of places to do geology field experiments in the geological sciences and that cooperative research could be beneficial to both the U.S. and LDCs. He cited Brazil as a possible example and the need for institutional mechanisms and support to facilitate such activity.

Appendix A   285

Witunski suggested several possibilities for U.S. university-industry collaboration. MNCs might be persuaded to use money they cannot bring back to the U.S. because of tax laws to support research in LDCs. U.S. community and technical colleges and MNCs might collaborate to provide packages which include sale of products to LDCs coupled with appropriate training courses at the colleges. Finally, he commented that it would be desirable for the new foundation to be able to accept funds from a variety of private sources.

Policy Issues and Options

Thomas Fox wants the Study Paper to be hard, direct, and sharp on basic needs and appropriate technology. He was distressed that the report seemed to accept the prevailing foreign aid philosophy of basic needs and appropriate (i.e., small-scale) technology as unshakable; it needs to be altered to include a broader range of technologies which may have little to do with scale. Unless that happens, Fox feels that the U.S. university role may be a very limited one.

For Fox, the gut issue is the relevance of U.S. universities to basic needs and appropriate technology -- the latter defined much more broadly than current, small-scale philosophy. He cited instances at a recent Department of Energy meeting in which suggestions for research on small-scale energy sources were not accepted because they were not seriously considered for application here in the U.S.; unless the latter happens, our assistance may have no credibility. He emphasized this point, as well as the importance of linking efforts to solve pressing domestic problems to international activity.

There are a wide range of opportunities available to involve U.S. universities in work related to developing countries; however, the linkages are absent. VITA represents one such linking organization; many VITA volunteers come from U.S. university campuses. Other useful links include sister-city programs, the Georgia Tech-University of Kumasi link, "switchboards" for properly placing visitors (the Institute for International Education is important but limited in this regard), and links with private voluntary organizations and other development organizations.

Fox went on to briefly describe several specific examples of VITA-university collaboration: portions of a Penn State thesis on alternative sources of cement in New Hebrides which came through the VITA network provided a quick useful response to a VITA requester; a Washington University-VITA intern is surveying VITA requesters in the wind power area as part of a student project with immediate utility as well as potential for longer-term research. Other long-term problems like low-cost stored solar energy for cooking often come to VITA's attention and could challenge university research teams. Fox feels these kinds of specific opportunities need more emphasis in the report.

A discussion ensued about how the report handled the concept of appropriate technology; some participants did not like the small-scale light-capital approach of the report and felt that other definitions of appropriate technology were more suitable.

Barbara Lucas stated that whereas the report did a good job in very strongly emphasizing the need for independent evaluation, the latter term as used in the report was primarily in the sense of whether individual projects met project goals, such as training a specific number of students. More fundamental evaluation is needed to support policy decisions and to design effective, future university programs. Such evaluations need to be comparative and to address questions such as (1) would the effect have come about if the university had never been involved in the first place, and (2) what is the effectiveness of various projects designed to address the same problems. Lucas suggested that linkages between a research and evaluative agency, such as NSF and AID, could provide formative evaluations to enable questions of effectiveness of technical assistance programs to be answered. She would like to see the final chapter of the report extended, with pros and cons considered in more detail.

A discussion of evaluation ensued. The Rockefeller and Ford Foundations performed evaluations of projects which may be available at the foundation offices. Although not as scientifically based as might be desirable, they should provide useful information. Multilateral evaluations to assess the impact of projects were suggested. Political obstacles to performing evaluations in LDCs were discussed. One participant expressed the view that impact can only be evaluated for a string of projects after 15 or 20 years, unless a project has specific, quantifiable goals. The difficulties associated with evaluation were set against the need for and desirability of having such information. University social and policy scientists may have a key role to play in such evaluations. Many other points and arguments were made in the discussion of evaluation. Currently, AID heavily emphasizes in-house evaluation; however, there is pressure from the GAO and the Congress for further evaluation, including more independent assessments.

Appendix A

Concluding Session

The concluding session at the workshop dealt with three topics: 1) suggestions for new U.S. initiatives for international S&T cooperation involving universities, 2) a ranking by participants of alternative legislative steps which might foster such cooperation, and 3) discussion of how the final report and its summary might best be focused. Some of the initiatives suggested include: 1) encourage universities, linked worldwide through an "open university" network, to propose radical new solutions in such areas as energy use; 2) involve university science and engineering faculty in predicting when significant technological advances (photosynthesis research, nitrogen fixation, etc.) might yield results to solve pressing world problems; 3) develop new programs for enriching the experiences of foreign students in the U.S. and career opportunities for U.S. faculty overseas consonant with a philosophy that international development work is an exciting and important new frontier; 4) propose a network of universities, on a worldwide basis, to facilitate the interaction of academic personnel at various levels (the UN University is one mechanism that might be considered for doing this), 5) increase the pool of international scholars and support development education, perhaps through the expanded programs of the International Communication Agency in the U.S.; 6) support the concept of the university as intermediary between government and MNCs; 7) establish regional centers in LDCs, e.g., Africa, with easy access to U.S. supplies, expertise, and information about graduate study, etc.; 8) establish a Technology Corps and a Scientific Peace Corps. The need for funding to sustain such initiatives was emphasized, as was the need for universities to get involved in the process of gaining political and financial support for these initiatives. The pros and cons of the Canadian International Development Research Centre as a model for the FITC was discussed; some inaccuracy in the Draft Study Paper in this regard was pointed out.

A discussion took place about possible outputs of the study. Some of the suggestions were: 1) produce a powerful executive summary setting forth what U.S. universities can do (identify opportunities); 2) produce an advocacy document for U.S. universities; and 3) prepare articles and monographs addressed to the U.S. academic community and to the international community. These things need not necessarily be done within the context of the report to NSF. The report might be used by others to make the case. The point was made that it seemed ironic that universities were trying to change things overseas when they were having difficulty garnering support here at home for international development activity.

A survey was taken of the participants in which they were asked to rank the legislative options listed in the Draft Study Paper that would facilitate U.S. university involvement in international S&T cooperation. Results are summarized in table A.1. The proposed new foundation for International Technological Cooperation was the top choice. Expanding international activities within NSF received many second and third rankings. There was very heavy support for having the foundation be an independent agency, somewhat parallel to the National Science Foundation.

TABLE A.1. Legislative Options to Facilitate Enhanced International Science and Technology Cooperation Involving Universities

	Number of 1st, 2nd, 3rd, 4th, 5th Place Choices				
	FIRST	SECOND	THIRD	FOURTH	FIFTH
Establish International Foundation for Technological Cooperation	10	2	1	-	2
Fund the International Education Act of 1966	-	1	-	2	1
Expand Title XII Authority of Nonfood and Nonagriculture Areas; Remove Current Restrictions on University Participants	1	2	2	1	3
Increase the Funding Authority for 211(d).	-	3	2	1	-
Expand International Activity of the National Science Foundation	1	4	6	-	1
Provide Authority for Primarily Domestic, Mission-Oriented Agencies to Work in International S&T	2	3	1	2	1
Expand the "S&T for Development" Focus of the New International Communication Agency	-	1	2	4	-
Expand Funding for International Organizations.	1	2	1	1	-

IF YOU FAVOR ESTABLISHING AN INTERNATIONAL FOUNDATION FOR TECHNOLOGICAL COOPERATION, IT SHOULD BE:

	Number of 1st, 2nd, and 3rd Choices		
	FIRST	SECOND	THIRD
An independent agency somewhat parallel to the National Science Foundation	12	1	-
Incorporated within NSF activities	1	5	2
Within AID	1	1	1
Within a reorganized development assistance agency as in Senate Bill 2420	1	4	4

Appendix B

The contribution of the United States to worldwide development would be enhanced by the establishment of an International Development Foundation (IDF), which would be a catalyst and coordinator of U.S. scientific, technical, and education activities related to development problems. This Foundation would be governed by a board of trustees with both public and private members, the latter in the majority.

The IDF might be established on a permanent basis by the Congress, with multi-year authorization for sustained work on the major development problems. Annual appropriations would be sought; the IDF should not be under pressure to obligate all of its funds on an annual basis.

The IDF would be autonomous, in the sense the National Science Foundation is autonomous, its Executive Director would be appointed by the President and confirmed by the Senate.

The purposes of the IDF would be:

- to expand knowledge of the nature of the development process;

- to facilitate the application of U.S. and international research competence to the search for solutions to critical scientific and technical problems of developing countries;

- to improve access to U.S. research and technical resources for developing countries;

- to facilitate the growth of institutional and individual capacity in developing countries for research and experimentation on development problems;

Appendix B   291

- to encourage technical cooperation by U.S. institutions with institutions in developing countries on topics of mutual interest such as food production, environmental quality and population; and

- to assist U.S. private and voluntary organizations and foundations to contribute effectively to international development.

In order to accomplish these objectives, the foundation would perform the following functions:

1. It should serve as a central source of knowledge concerning research needs and priorities on selected development problems.

2. The IDF should serve as coordinator and catalyst of research and development problems by government research facilities.

3. The Foundation would enhance the contribution of U.S. universities and private research and training facilities to the solution of key development problems.

4. It would encourage and support U.S. participation in international research and development programs on development problems.

5. It would improve access to U.S. training and research facilities by the developing countries.

6. It would help to build indigenous capacity for training, research, and experimentation through:

    - funding training in the U.S. and third countries for prospective local institution staff members;

    - funding research projects and competitions;

    - organizing research methodology workshops and sponsoring regional conferences on key problems;

    - promoting links between U.S. and international research efforts and local institutions;

    - strengthening indigenous training institutions and in-service training programs;

- making grants when necessary for equipment and furnishings; and

- supporting experimental and pilot projects.

7. It would help public and private foundations and voluntary organizations interested in development to become more effective. (1)

Appendix C

FORD FOUNDATION AND SECONDARY SCIENCE EDUCATION IN BRAZIL

This case study illustrates the strong impact of U.S. science curricula on LDC secondary science education.

The project was an interaction between the Ford Foundation and the Brazilian Institute for Education, Science and Culture (IBECC), a national Brazilian committee in Sao Paulo affiliated with UNESCO. One strong Brazilian individual, Isaias Raw, joined this committee in 1952 with the objective of changing not just the science curricula in Brazil's secondary schools, but also "the approach and the understanding of what science education meant." One of Raw's first activities was to produce science kits with low-cost laboratory equipment applicable to biology, chemistry, and physics. Although production started on a small-scale basis for a local science club, it soon blossomed into a major industry, putting thousands of kits into homes and schools.

Dr. Raw, and his activities through IBECC came to the attention of the Ford Foundation in 1960. After visits, discussions, and planning, in 1961 Ford awarded IBECC a grant of $125,000 for a three-year period. This grant enabled three approaches to be taken in introducing modern science teaching materials and practices: 1) distribution of kits, 2) upgrading of science teachers through summer institutes, and 3) translation and distribution of selected U.S. science curriculum materials.

The third task was taken on in a large-scale manner, enlisting both the public and private sectors of Brazil.

... texts were sold through commercial channels by contract between private publishers and IBECC and the University of Brasilia Press. At the beginning no private publisher would gamble on these publications, so arrangements were made with the University of Brasilia Press to advance the necessary capital, the guarantee of USAID to pay for 36,000 copies being the decisive factor in this arrangement. After first printing, royalty arrangements were made, with 10% being returned to IBECC, who then handled the royalty payments of about 5% to the copyright holders in the United States. (2)

The summer courses also involved use of the science curriculum studies. (In January 1962, IBECC brought 50 teachers together in a Physical Sciences Secondary Curriculum physics course and nearly 50 in a Biological Sciences Curriculum Study biology course.) Other teacher training courses in 1963 and 1964 also used the science curriculum materials extensively.

The transfer involved much more than translation; it included an entire adaptation to local conditions. Often the translators traveled to the U.S. to the homes of the specific curriculum projects for consultation. This involved a degree of commitment such that the Ford Foundation awarded a supplementary grant of $45,000 specifically to pay for translation and distribution of the U.S. science curriculum materials. By 1965 IBECC had produced over 140,000 copies of a generalized science text.

In 1964, the foundation extended its efforts to support the creation of a network of science teacher training centers (CECIs) through a $150,000 grant to the University of Recife to serve the northeastern part of Brazil. In 1965, the Ministry of Education set up six additional centers throughout the country. These were all modeled after IBECC.

Because of the new centers, a need arose for trained leaders of teacher training programs and the creation of teacher training manuals. In January 1966, IBECC received $86,000 from the Ford Foundation to serve this purpose. This specifically assisted preparation of teachers' guides to accompany the translated U.S. science curriculum materials.

At this point, IBECC gained new secure legal status and came to be known as FUNBEC (Fundacio Brasileira para o Densenvolvimento do Ensino de Ciencias).

By 1970, training had been proceeding regularly and 37 teachers guides had been planned, 8 were ready for the publisher, first drafts of 7 had been completed, 7 others were being written or translated. Ten guides had already sold nearly 3,000 copies.

During this period, the Ford Foundation made efforts to introduce an evaluation system into the program. In particular, K. N. Rao was concerned that: "Since these are translations of materials developed in the U.S., what are the special difficulties faced by Brazilian teachers and students in using these new texts?" He expressed further concern about the effect on students' performance on entrance exams to the universities and other areas. (3)

While negotiating for a further grant in May 1969, which would include funds for evaluation, Dr. Isaias Raw was abruptly retired from his post at the university. The attack was the culmination of years of conflict and rivalry between Raw and university colleagues hostile to his progressive efforts. Upon his dismissal, Raw left Brazil and came to the U.S.

Despite the importance of Raw's presence in establishing and maintaining FUNBEC, the institution had grown strong enough to readjust and continue after his departure. In October 1969, the Ford Foundation proceeded to negotiate a two-year grant of $194,000 to FUNBEC for continuation of activities and to begin evaluation. Myriam Krasilchik traveled to the U.S. to visit the curriculum development centers and the Educational Testing Service in Princeton. Upon her return, she set up programs evaluating the science materials and their impact. The foundation assisted her efforts in 1972 by providing consulting assistance and a grant to the BSCS project headquarters to facilitate collaboration between BSCS and FUNBEC. The BSCS report, "An Evaluation of Biologia," covered 300 classrooms. The effectiveness of the improved science curriculum on the students' education was quite unclear. It did determine, however, that the in-service teacher training workshops had a strong effect in increasing a teacher's ability to initiate and deal with class discussion. The report expressed that "this single shift in teacher behavior [was] a significant contribution for such a short in-service program."

Overall, in the evaluation by Robert H. Maybury, of all five of the Ford Foundation's involvements, the sentiment was stated thus:

> First and foremost there is a growing realization that reliable and significant measures of the effectiveness of innovations in science education for the school rooms of developing countries still escape the science educators. As the record shows, not one of these Ford Foundation projects has been able to demonstrate that effectiveness in any conclusive way. Fortunately, more sophisticated psychometric work is underway in these developing countries that may soon yield the longed-for reliable and useful indications of educational effectiveness of the instructional materials and methods in the sciences prepared in these projects. (4)

FORD FOUNDATION IN LEBANON AND OTHER ARAB COUNTRIES OF THE MIDDLE EAST

Two officers of the Ford Foundation office in Beirut started emphasizing the importance of improved science teaching in the late 1950s. They invited Professor Milton Pella, a science educator from the university of Wisconsin, to come to the Middle East to study difficulties in science teaching in various countries. Professor Pella, who had previously been a consultant to the Ford Foundation in this area regarding Turkey, visited the Middle East for several months in 1962. He prepared a report, returned to the Middle East by request of the Ford Foundation, and out of his second trip recommended that the University of Wisconsin help in developing a program of research in problems of science education in developing countries, and in training graduate level persons from Arab countries. In April 1963, the Ford Foundation granted $141,000 to the University of Wisconsin, over a five-year period, to become a resource base for a program of cooperation in science education with Arab countries.

Included in this grant were provisions for a program specialist in science education from the University of Wisconsin who would spend two years in Beirut, various stipulations about the specialist's actions after returning, and provisions for up to five science educators from Arab countries to be brought to Wisconsin for graduate study over five years. In 1966, the University of Wisconsin received a supplemental grant of $57,000 to assist the graduate students and defray unexpected costs of their training. By the end of 1969, five graduates had trained at the University of Wisconsin - one from Damascus, one from Jordan, and three from Beirut. One important stipulation in the grant was that each graduate return to teach in his own native country. All did return except the Jordanian who eventually ended in Lebanon, joining three others at the American University of Beirut (AUB). In 1969, through the Lebanese Science Advisory Committee, one of the Lebanese teachers negotiated a $59,000 grant for summer institutes designed to train Lebanese teachers. Later that year, financial difficulty at the AUB and threatened discontinuation of the University of Wisconsin/Lebanese graduates' salaries, prompted the Ford Foundation to issue a grant of $167,000 to AUB primarily to cover salaries, equipment and overhead.

The official Ford Foundation program at the University of Wisconsin ended in 1969. However, Pella found that through tight management, a balance of $80,000 remained. With that, Pella arranged with the foundation to continue the graduate program and cooperative research for another six years. As of 1974, a Lebanese mathematics teacher was finishing his doctoral program, and six science educators from the United Arab Republic were working under Pella's guidance.

Appendix C 297

THE UNITED NATIONS UNIVERSITY

The United Nations University was chartered in 1974 by the General Assembly to be an instrument for conducting research, postgraduate training, and dissemination of knowledge through centralized coordination of research and postgraduate centers and programs concerned with "pressing global problems of human survival, development and welfare." The university is an autonomous institution within the structure of the UN. It consists of a governing council, headquarters and field staff, program advisory committees, and the associated institutions and scholars that conduct its research and advanced training activities. These constituents perform four primary functions.

1. identifying major problems appropriate for the university to tackle,

2. organizing and supporting research,

3. strengthening individual and institutional capabilities to work on these problems,

4. disseminating the knowledge generated to international organizations, governments, scholars, policy makers, and the public.

Since its start in September 1975, the UN University has launched three major programs: World Hunger, Human and Social Development, and the Use and Management of Natural Resources. It has established institutional ties with the Central Food Technological Research Institute (CFTRI) in Mysore, India; the Institution of Nutrition of Central America and Panama (INCAP) of Guatemala City; the Nutrition Center of the Philippines (NCP), Manila; the Institute of International Affairs, Sopjia University, Tokyo; and recently with the International Nutrition Policy and Planning Program at MIT.

Funding comes from national governments; major contributions have been from Japan, Venezuela, Saudi Arabia, Ghana, Sweden, India, Austria, Norway, Netherlands, Holy See, and Libyan Arab Jamahiriya. The U.S. pledged $10 million, although as of spring 1978 this had not been paid. Total paid contributions to date are $76,414,066. The 1978 operating budget (as of March 1978) is $1,632,695.

In a February 1978 report by James M. Hester, rector, UN University, certain distinctive characteristics of the university were cited. Among other things the report maintains that the university:

1. Guarantees academic freedom,

2. Mobilizes advanced science and scholarship on a nongovernmental basis, as no other UN agency does,

3. Emphasizes strengthening scholarly abilities in developing countries,

4. Is governed by an international council of scholars and individual citizens, not by representatives of governments and national institutions,

5. Receives no funds from the UN or required payments from UN member states.

NATIONAL ACADEMY OF SCIENCE/ NATIONAL ACADEMY OF ENGINEERING/ NATIONAL RESEARCH COUNCIL ACTIVITIES

BOSTID (Board on Science and Technology for International Development) activities have been primarily through workshops and studies. Other programs have taken the form of joint panels and committees. Most interactions have been bilateral, although there are a few multinational exceptions. BOSTID has received funding primarily through AID, but some private funds have been contributed for activities in Brazil, Guatemala, Singapore, and Zaire. Some countries have used resources in their control, such as science bloc grants from AID, to assist BOSTID. The BOSTID report summarizing activities from 1970-1976 maintains that "BOSTID acts as a mechanism for communication among individual scientists and engineers without regard to politics." (5) (See table A.2 for BOSTID personnel deployment.)

The Office of the Foreign Secretary handles most of the international activities of the National Academy of Sciences, the National Academy of Engineering, and the National Research Council. Several international research programs are conducted by appropriate NRC divisions or by groups such as the Environmental Studies Board - a joint NAS-NAE board. However, the foreign secretary is a liaison member of all such groups. One working committee is the Advisory Committee on International Organizations and Programs (ACIOP).

Appendix C

TABLE A.2. U.S. University Personnel Involved in BOSTID Projects, 1970-76

BOSTID Members

Current	Past
7	13

This represents 51% of total board members since 1970.

Overseas Programs

Area	Number of people involved
Africa	32
Asia (excluding Taiwan)	75
Latin American	85
Central America	6
Middle East	4

Advisory Committee on Technology Innovation

Number of projects	Number of people
12	101

Special Studies/Advisory Panels

Number of projects	Number of people
18	114

Another committee formed in 1970 is the International Environmental Programs Committee. It advises the State Department in environmental matters, acts as the U.S. adhering committee for the Scientific Committee on Problems of the Environment of the International Council of Scientific Unions (ICSU), establishes bilateral workshops on environmental questions in the less developed countries, and provides a national clearinghouse on international environmental activities.

Other NAS activities have included:

1. ACIOP counseling the organizers of international conferences, all held within the U.S.

2. Conducting a joint ICSU-UNESCO study of the feasibility of a world science information system (UNISIST). The NAS foreign secretary was convener of the UNISIST Central Committee. The study recommended that it *was* feasible.

3. Establishing the NAS/CNPq chemistry project in Brazil.

4. Supporting ICIPE in Nairobi.

5. Supporting international research programs.

6. Advising ICSU on the role they should play in questions of human welfare in developing countries.

7. Administering international travel grants, including the Fulbright-Hays program. (The Senior Fulbright-Hays Program is no longer administered by NAS. It is now handled by the Committee for the International Exchange of Scholars, associated with the American Council on Education.) (6)

An interesting study by Michael Moravcsik in Thailand examined what role NAS reports and brochures played in developing countries. NAS study reports are meant to be readily accessible to all scientists and be useful in both general and specific areas. Moravcsik's results were based on 140 questionnaires returned out of 800 sent to members of the science and technology community in Thailand. Two questions and their respective answers were most revealing. (7)

Appendix C   301

1. Is your access to NAS studies:

zero	very poor	poor	adequate	good	excellent	no answer
33	11	25	8	6	1	16

2. Do you find these reports:

useless	little use	some use	definitely useful	indispensible	no answer
2	1	29	32	1	35

These results seem to indicate that while there is some problem with accessibility, those who read the studies had a reasonably favorable response.

HIGHLIGHTS OF FORD FOUNDATION SCIFNCE-RELATED INTERNATIONAL ACTIVITIES 1969-76

1. Strong support of International Centers of Excellence. Motivating force, with Rockefeller Foundation, behind IRRI, CIMMYT, IITA, CIAT.

2. Secondary science education improvement around the world. Much stress on adopting U.S. science curriculum studies.

3. Undergraduate and graduate education improvement worldwide. Successful examples cited were in support of graduate engineering and basic science programs at the Universidad Nacional Autonoma de Mexico (UNAM), Mexico City, and the Monterrey Institute of Technology.

4. Support of student and faculty exchange as a further means of strengthening educational capacity of LDC institutions.

5. Other research projects related to agriculture and food production.

As stated by the Ford Foundation's annual reports, the four overall areas of primary interest have been: education and research, agriculture, population, and development planning. (8)

ROCKEFELLER FOUNDATION INVOLVEMENTS: THE UNIVERSITY DEVELOPMENT PROGRAM

The Rockefeller Foundation has been heavily involved in international science and technology activity. The foundation was an early supporter of CIMMYT, the International Wheat and Maize Research Center, and has continued its support for public health related activity.

A 1972 report by Thompson describes and evaluates the foundation's University Development Program, which began in the early 1960s and focused on providing technical assistance for overall university development to a few selected LDC institutions. In Thailand three basic disciplines - medical and basic sciences, agriculture, and economics - were emphasized at three universities in Bangkok - Thammasat, Kasetsart, and Mahidol, respectively. At the Universidad del Valle in Cali, Colombia, a strong effort was made to help raise the level of engineering, economics and agricultural economics, university administration, the humanities, and basic sciences. Related strengthening programs took place at three East African universities, at the University of the Philippines, and at the University of Ibadan in Nigeria. These efforts involved U.S. university faculty and/or institutions to some degree. (9)

Thompson's report begins to address the issue of the effectiveness of these involvements. However, we know of no good, comprehensive evaluation that is available of the extensive Ford and Rockefeller Foundation's activities. Future research of this kind could shed valuable light on both LDC university-strengthening activity and the role of U.S. universities in that process.

Appendix D

A 1973 AID rough estimate of FY 1972 U.S. funding for S&T aid is reported by Maravcsik and summarized in table A.3. Moravcsik suggests that the last category, which includes research done in the U.S. that might also have incidental benefits for LDCs, should not be strictly counted as scientific assistance. (1) This leaves a total of $320 million for FY 1972, or about 1.3 percent of U.S. R&D expenditures (3 percent if category 3 is included).

A February 1978 preliminary analysis of selected R&D in the FY 1979 budget by AAAS and six other organizations provides some interesting information on spending for R&D. Total FY 1979 R&D spending outlays (excluding R&D facilities) are estimated to be $27.9 billion. Of this, $13.8 billion is for defense (49.5 percent), $3.4 billion for space (12.2 percent), and $10.7 billion for civilian R&D (38.3 percent). The total breaks down to $3.6 billion for basic research (12.9 percent), $6.6 billion for applied research (23.7 percent), and $17.6 billion for development (63.0 percent). The AID budget for FY 1979 for R&D is estimated at $76 million, or 0.27 percent of the total R&D budget. The AID figure represents a 53.5 percent increase over FY 1978. (2)

The United Nations has previously established targets for developed country spending on S&T for development at 0.05 percent of GNP in direct scientific and technical aid to LDCs and a shift of 5 percent of non-military R&D towards research related to LDC problems. In the U.S., FY 1979 R&D outlays are estimated at 5.7 percent of Total Budgeted Outlays and 1.2 percent of GNP. The UN target of 0.05 percent of GNP would be 5.17 percent of the U.S. R&D budget or $1.16 billion in direct aid. A shift of 5 percent of non-military R&D would be either $535 million or $705 million, depending upon whether or not space activities are included.

TABLE A.3. U.S. Spending on R&D for Development
Rough AID Estimates for FY 1972
(Reported by Moravcsik)

	Millions of Dollars
1. R&D explicitly for benefit of developing countries:	
a) AID	100
b) U.S. Contributions to Multi-national Organizations	70
c) U.S. Foundations	40
Subtotal	210
2. Other R&D in Developing Countries	
a) Through Other U.S. Governmental Agencies	90
b) Private Industry	20
Subtotal	110
3. Other R&D of Potential Short-Term Benefit to Developing Countries	
a) Governmental Agencies	500
b) Private Industry	30
Subtotal	530

TABLE A.4. National Science Foundation Budget Submission to the Congress for International Activities, FY 1979

INTERNATIONAL COOPERATIVE SCIENTIFIC ACTIVITIES PROGRAM SUBACTIVITY $10,600,000

Obligations by Program Element

Program Element	Actual FY 1977	Budget Request FY 1978	Current Plan FY 1978	Estimate FY 1979	Difference FY 1979/78
Cooperative Science	$5,054,517	$ 5,400,000	$ 5,530,000	$ 6,100,000	$570,000
Scientific Organizations and Resources	2,604,687	3,100,000	2,862,981	3,300,000	437,019
UN Conference on Science and Technology for Development	-0-			200,000	-1,000,000
International Travel	537,340	1,200,000	1,200,000	600,000	100,000
Dollar Support for Special Foreign Currency	289,544	500,000	500,000	400,000	79,750
		300,000	320,250		
Total	$8,486,088	$10,500,000	$10,413,231	$10,600,000	$186,769

SPECIAL FOREIGN CURRENCY PROGRAM ACTIVITY SUMMARY
FY 1979 PROGRAM TOTAL $6,000,000

Page	Subactivity	Actual FY 1977	Budget Request FY 1978	Current Plan FY 1978	Estimate FY 1979	Difference FY 1979/78
K-I	Research and Related Activities	$3,959,889	$4,900,000	$4,334,055	$4,900,000	$565,945
K-II	Science Information	443,537	1,100,000	1,100,000	1,100,000	-0-
	Total	$4,403,426	$6,000,000	$5,434,055	$6,000,000	$565,945

Total FY 1979 support for R&D at colleges and universities is estimated at $3.566 billion or 12.78 percent of total R&D spending. Of this total, $53 million is estimated to be provided by AID, or 1.49 percent of the total. The AID projected increase of 94.1 percent from FY 1978 to FY 1979 in constant dollars is the largest percentage increase by far of any government agency. Universities spent about half of AID's R&D funds in FY 1977 and 1978, with a somewhat higher percentage projected for FY 1979.

There are other federal agencies involved in international R&D and S&T. For example, the National Science Foundation 1979 budget submission to Congress contains $10.6 million for International Cooperative Scientific Activities which is subdivided as shown in table A.4. Fifty-eight percent of these funds are for Cooperative Science Programs, primarily with developed or wealthier developing countries. Thirty-one percent is for Scientific Organizations and Resources, the largest single part of which is the U.S. contribution to the International Institute for Applied Systems Analysis, a joint U.S.-USSR-supported undertaking (again, an activity not focused primarily on S&T for development). NSF also put $6 million in special foreign currency into research and related activities and science information. These totals ($10.6 M + $6 M) represent 1.76 percent of the total NSF budget. (3) This by no means exhausts the U.S. government agency spending on international science and technology. The "Quantification" study by Schlie supported by NSF should shed additional light on the matter.

U.S. university involvement is more extensive than the R&D figures indicate. For example, AID-financed university contracts and grants active during the period from 4/1/77 to 9/30/77 totaled $232.9 million (see table A.5). Of this total, 25.3 percent was for Technical Assistance to host countries, and presumably the rest (74.7 percent) went directly to U.S. universities. Of this latter figure, 32.0 percent went for research while 68.0 percent went for grant support, training, and technical services to AID. If we apply these percentages to the $53 million estimated for AID R&D to universities in FY 1979, then $166 million would be projected to be spent on U.S. universities in FY 1979 if we include all the AID categories. This number is a rough estimate which needs to be used with caution.

TABLE A.5. Statistical Summary: AID-Financed University Contracts and Grants Active During the Period 4/1/77 Through 9/30/77

Type of Activity	Number of Countries	Number of U.S. Universities	Number of Contracts/Grants	Amount in Dollars (a)
Technical Assistance to Host Countries				
Latin America	9	11	12	$ 5,683,373
Near East	7	12	12	12,985,293
Africa	13	14	20	25,260,112
Asia	2	11	12	15,124,811
Subtotal	31	48	56	$ 59,053,589
Grant Support	--	48	66	$ 58,281,774
Training	--	96	107	16,769,132
Central Research	--	55	111	55,587,145
Technical Services to AID	--	42	68	43,228,671
Subtotal	--	241	352	$173,866,722
TOTAL	--	289 (b)	408	$232,920,311

New Contracts/Grants and Amendments effective second half fiscal year 1977		
Contracts/Grants	82	Dollar Amount $16,095,127
Amendments	734	Dollar Amount 31,110,308
Total	816	$47,205,435

(a) Funds are cumulative and represent total dollars obligated over the life of the contract/grant.
(b) Column totals 289 but there are only 127 universities engaged in contract/grant activities, since some universities are operating in more than one area or type of service.

Source: AID Report, No. E840W42A.

Appendix E

BARBED WIRE FENCE-MAKING MACHINE

Since 1968, village entrepreneurs in Botswana have been using simple, handpowered machines (designed by VITA) to make chain link fencing for gardens, small livestock, etc. These machines are manufactured in two workshops in the country.

In late 1976, the Rural Industries Innovation Center (RIIC) of Kanye, Botswana, one of the two workshops manufacturing the chain link fence-making machines, came to VITA with a request for a similar machine to make barbed wire fencing. With the rapid expansion of agriculture and cattle raising, increasing amounts of barbed wire were needed for effective range enclosure projects, but few villagers could afford the high cost of imported barbed wire. A simple, low-cost machine to make barbed wire would, RIIC noted, provide barbed wire at a cost that villagers could afford and would create new businesses and jobs in the light-industries sector.

VITA worked with RIIC to define the economic and technical parameters for a viable barbed wire fence-making machine. A concise problem statement was prepared and, together with similar statements for other low-cost technologies, was submitted to two VITA volunteers on the engineering faculties of two universities, New Mexico State and Dartmouth.

The problem statements were then made available to engineering students looking for design project ideas and the barbed wire making machine was selected by a student at Dartmouth. After doing background research on the history of barbed wire fence technology, the student designed a machine based on the existing chain link fence

Appendix E

making machine now in widespread use in Botswana. A prototype was built and tested. After several modifications were made in the design, the prototype's performance and cost were subjected to a computer program, developed by the student, to test its economic viability given known price and demand structures in Botswana. Both physical and computerized testing indicated that the machine would fit ideally into the village context.

Copies of the design plan, which included step-by-step construction instructions, have been sent to the RIIC and to VITA affiliates in Tanzania and Upper Volta, where similar demand for low cost barbed wire exists. All three groups are now studying the design and hope to begin prototype construction shortly.

An abbreviated version of the plan will be made available to other interested groups in the Third World by VITA publications program. (1)

Notes

INTRODUCTION

1. V. J. Ram. (Remarks at Annual Meeting, American Society for Engineering Education, Vancouver, British Columbia, Canada, June 22, 1978).

2. National Science Foundation, Program Solicitation, Policy Related Studies on Science and Technology for Development, Washington, D.C., November 1977.

3. From resolution adopted by the Second Preparatory Committee for UNCSTED, Geneva, Switzerland, Feb. 3, 1978.

4. Michael J. Moravcsik, "Science and Developing Countries" (paper presented for the U.S. National Science Foundation, October 1977), p. 1. See also Derek J. DeSolla Price, "The Difference between Science and Technology" (Address at the International Edison Birthday Celebration, Thomas Alva Edison Foundation, February 1968).

5. See, for example, N. Jequier, Appropriate Technology: Problems and Promises (Paris: OECD, 1976).

6. United Nations, World Plan of Action for the Application of Science and Technology to Development (New York: United Nations, 1971).

7. UNESCO, Committee on Science and Technology for Development, <u>Quantification of Scientific and Technological Activities Related to Development</u> E/C. 8/44, September 29, 1976.

8. Jean Wilkowski (Statement at Second Preparatory Committee Plenary Session, U.N. Conference on Science and Technology for Development, January 27, 1978).

9. Joao da Costa (Statement by the Secretary of the Conference at the Opening Meeting of the Second Session for the U.N. Conference on Science and Technology for Development A/CONF. 81/PC/L. 2, United Nations, January 24, 1978).

10. See note 1.

CHAPTER 1

1. U.S., Congress, House Committee on International Relations and Senate Committee on Foreign Relations, <u>Legislation on Foreign Relations through 1977</u> (Washington, D.C.: Government Printing Office, 1978), vol. 1.

2. U.S., Congress, House, Committee on Education and Labor, <u>A Compilation of Federal Education Laws</u> (Washington, D.C.: Government Printing Office, 1971), p. 377.

3. Ibid, p. 593.

4. <u>Legislation on Foreign Relations</u> 3:166.

5. <u>Congressional Quarterly Alamanac</u> 17 (1961): 294.

6. <u>Congressional Quarterly Almanac</u> 33 (1977): 360.

7. Trend Analysis, <u>Congressional Quarterly Almanac</u> 16-33 (1960-77).

8. <u>Congressional Quarterly Almanac</u> 22 (1966): 399.

9. Ibid.

10. <u>Congressional Quarterly Almanac</u> 25 (1969): 435.

11. U.S., Congress, House, Committee on International Relations, <u>New Directions in Development Aid</u>, Excerpts from the Legislation (Washington, D.C.: Government Printing Office, 1977).

12. <u>Congressional Quarterly Almanac</u> 33 (1977): 360.

13. <u>New Directions in Development Aid</u>, pp. 4, 6.

14. Statement of the Hon. Clarence D. Long before the Subcommittee on International Development, Institutions and Finance of the Committee on Banking, Finance and Urban Affairs, U.S. House of Representatives, March 14, 1978.

15. U.S., Congress, <u>International Development and Food Assistance Act of 1975</u>, 94th Cong. 1st sess.

16. <u>Congressional Quarterly Almanac</u> 17 (1961): 293.

17. <u>Congressional Quarterly Almanac</u> 19-25 (1963-69).

18. <u>Congressional Quarterly Almanac</u> 26 (1970): 989.

19. U.S., Congress, Senate, Committee on Foreign Relations, <u>Foreign Assistance Legislation, Fiscal Year 1972</u>, Hearings (Washington, D.C.: Government Printing Office, 1971).

20. <u>Congressional Quarterly Almanac</u> 27 (1971): 387.

21. U.S., Congress, <u>Congressional Record</u>, 95th Cong. 2d sess. 1978, vol. 124, no. 5, January 25, International Development Cooperation Act of 1978, Senate Bill 2420.

22. The Brookings Institution, "Interim Report: An Assessment of Development Assistance Strategies," mimeographed (Washington, D.C., October 1977).

23. <u>U.S. Code</u>, supplement 2, 2 (1970): 1239.

24. <u>U.S. Code</u>, 1976 Edition, vol. 10, Title 42, Chapter 16, Section 1862(b), pp. 1128-1129.

25. <u>U.S. Code</u>, supplement 2, 2 (1970): 1239.

26. U.S. Code, Congressional and Administrative News, 94th Cong., 2d sess. 1976, vol. 2 (St. Paul, Minn.: West Publishing Co.) p. 90 STAT 2054.

27. U.S., National Research Council, Commission on International Relations, Supporting Papers, World Food and Nutrition Study, vol. 5, National Academy of Sciences (Washington, D.C.: NAS 1977).

28. George Waldman to Eduardo A. Perez, July 27, 1978.

29. Congressional Quarterly Almanac 17 (1961): 294.

30. U.S., Congress, Foreign Assistance Act of 1966, 89th Cong., 2d sess.

31. U.S., Congress, Foreign Assistance Act of 1968, 91st Cong., 2d sess.

32. U.S. GAO, Report of the Comptroller General of the United States, "Strengthening and Using Universities as a Resource for Developing Countries," 1976, Front Cover.

33. Ibid, p. 2.

34. Agency for International Development, "AID Comments on the GAO Report 'Strengthening and Using Universities as a Resource for Developing Countries,'" mimeographed (Washington, D.C.: AID, July 1976).

35. Agency for International Development, A Directory of Institutional Resources Supported by 211(d) Grants: U.S. Centers of Competence for International Development (Washington, D.C.: AID, 1975).

36. Ervin J. Long to Robert P. Morgan, April 5, 1978.

37. U.S., National Research Council, The Role of U.S. Engineering Schools in Development Assistance (Washington, D.C.: NAS, 1976); U.S. Science and Technology for Development: A Contribution to the 1979 U.N. Conference, National Academy of Sciences (Washington, D.C.: NAS, May 1978).

38. Agency for International Development, BIFAD, The First Year: A Progress Report (Washington, D.C.: AID, 1977), p. 2.

39. Ibid., p. 3.

40. Agency for International Development, Report to the Congress on Title XII: Famine Prevention and Freedom From Hunger of the Foreign Assistance Act of 1961 As Amended (Washington, D.C.: AID, 1978).

41. BIFAD, The First Year: A Progress Report, p. 40.

42. Ibid., p. 18.

43. Title XII -- Famine Prevention and Freedom From Hunger.

44. Eduardo A. Perez, The Role of U.S. Universities in International Agricultural Development (M.S. thesis, Department of Technology and Human Affairs, Washington University, St. Louis, 1978).

45. U.S., Congress, House, A Compilation of Federal Education Laws, p. 377.

46. Ibid., p. 381.

47. Ibid., p. 383.

48. Ibid., p. 394.

49. U.S., Congress, Mutual Security Act of 1960, P.L. 86-472, 86th Cong., 2d sess.

50. Legislation on Foreign Relations Through 1977 3:166.

51. Ibid., p. 167.

52. Fred H. Harrington, "International Linkages in Higher Education: Feasibility Study," Draft Final Report, February 1978.

53. Charles V. Kidd, "Manpower Policies for the Use of Science and Technology for Development" (Washington, D.C.: George Washington University, October 1978), p. 211.

54. Aaron Segal, Remarks at Workshop on Role of U. S. Universities in Science and Technology for Development, Washington University, St. Louis, Mo. July 13-14, 1978.

55. Kidd, "Manpower Policies," chap. 1.

56. See note 54.

57. Melvin Blase, Remarks at Workshop on Role of U. S. Universities in Science and Technology for Development, Washington University, St. Louis, Mo. July 13-14, 1978.

58. Franklin A. Long, Remarks at Workshop on Role of U. S. Universities in Science and Technology for Development, Washington University, St. Louis, Mo. July 13-14, 1978.

59. See note 54.

60. Harrington, International Linkages.

61. The Brookings Institution, "Interim Report: An Assessment of Development Assistance Strategies."

62. U. S. Office of Science and Technology Policy Statement, Washington, D. C., May 1978.

CHAPTER 2

1. National Research Council, The Role of U. S. Engineering Schools in Development Assistance (Washington, D. C.: NAS, 1976).

2. Charles L. Miller and Frederick J. McGarry, The M. I. T. Inter-American Program in Civil Engineering: Research Report R64-36 (Cambridge, Mass.: Department of Civil Engineering, School of Engineering, MIT, 1964).

3. J. Giral et al., "Appropriate Technology for Chemical Industries in Developing Economies," <u>Report on Foreign Area Fellowship Program Summer Research Training Project</u> (held at National Autonomous University of Mexico, July-August, 1972) (St. Louis, Mo.: Center for Development Technology, 1972).

4. Institute of International Education, <u>Open Doors</u> (New York: IIE, 1978), p. 24.

5. Institute of International Education, <u>Open Doors</u> (New York: IIE, 1974), p. 5.

6. Institute of International Education, <u>Open Doors</u>, 1978, p. 23.

7. Ibid., p. 26.

8. Engineering Manpower Commission, Engineers Joint Council, Tabular Summary (undated).

9. National Center for Education Statistics, <u>Digest of Education Statistics, 1976 Edition</u> (Washington, D.C.: U.S. Government Printing Office, 1977), p. 125.

10. National Science Foundation, "Scientists and Engineers from Abroad: Trends of the Past Decade, 1966-1976," <u>Reviews of Data on Science Resources</u>, NSF 77-305 no. 28 (Feb. 1977).

11. Kidd, "Manpower Policies," p. 62.

12. Gordon S. Brown and H.E. Hoelscher, "Open Forum: Are We Mistraining Our Foreign Graduate Students?" <u>Engineering Education</u> 61, no. 3 (Dec. 1970): 272-275.

13. Personal correspondence from Ms. Bonnie Kienitz, Coordinator, Wisconsin-Monterrey Tec Program, April 26, 1978.

14. Fred H. Harrington, "International Linkages in Higher Education: Draft Final Report," Feb. 1978.

15. Henry R. Glyde, "Institutional Links in Science and Technology: The United Kingdom and Thailand," <u>International Development Review</u> (Focus) January, 1973, pp. 7-12.

16. Morris Asimow and John S. McNown, "Engineering Education and International Development," Journal of Engineering Education 56 (November 1965): 65-70.

17. Neil Boyle, "An Evaluation of the Rural Industrialization Technical Assistance (RITA) Program," Northeast Brazil, 1962-1968 (Report to the World Bank, 1976).

18. Ibid., p. 30.

19. Personal communication to Robert P. Morgan, March 21, 1978.

20. Georgia Institute of Technology, Engineering Experiment Station, Employment Generation Through Stimulation of Small Industries, 211(d) Grant, Annual Reports, 1974-1978, Atlanta, Ga.

21. Ibid., 5th Annual Report (Final), p. 4.

22. "Report of Comprehensive Review of Georgia Tech 211(d) Grant," submitted to AID March 17, 1977.

23. T. C. Clark, Chairman, Grant Review Team, letter of transmittal accompanying 211(d) Annual Report, 1978.

24. Consortium members: Carnegie-Mellon University, Education Development Center, Georgia Institute of Technology, Illinois Institute of Technology, Lehigh University, North Carolina State University at Raleigh, Purdue University, Rice University, Stevens Institute of Technology, University of Cincinnati, University of Notre Dame, and Washington University.

25. Education Development Center, Kabul Afghan-American Program, Final Report (Newton, Mass.: EDC, 1973).

26. Ibid., p. 5.

27. Merton R. Barry, Written Communication in Connection with Project Workshop, July 13-14, 1978.

28. Kabul Afghan-American Program, Final Report, p. 95.

29. Education Development Center, Kanpur Indo-American Program, 1962-1972 (Newton, Mass.: EDC, 1972).

30. California Institute of Technology, Carnegie-Mellon University, Case Western Reserve University, Massachusetts Institute of Technology, The Ohio State University, Princeton University, Purdue University, University of California, and University of Michigan.

31. EDC, Kanpur Indo-American Program, p. v.

32. See note 27.

33. EDC, Kanpur Indo-American Program, p. 35.

34. Ibid., p. 36.

35. Ibid., p. 26.

36. Ibid., p. 27.

37. Robert P. Morgan, "International Directions for Engineering Education in the United States" (Paper presented at the Conference on Engineering for International Development, Estes Park, Colorado, August 27 - September 1, 1967).

38. Merton R. Barry, AID-Wisconsin Engineering Education Project in India, Final Report (Madison, Wis.: The University of Wisconsin, Engineering Experiment Station, 1967).

39. Merton R. Barry, Engineering Degree Development Program in Singapore, 1966-1967: Final Report to the Ford Foundation (Madison, Wis.: The University of Wisconsin, Engineering Experiment Station, 1977).

40. Merton R. Barry, Remarks at Workshop on Role of U.S. Universities in Science and Technology for Development, Washington University, St. Louis, Mo., July 13-14, 1978.

41. Personal correspondence from Ms. Bonnie Kienitz, Coordinator, Wisconsin-Monterrey Tec Program, April 27, 1978.

42. The Technology Adaptation Program, Massachusetts Institute of Technology. Technology Adaptation Program, Massachusetts Institute of Technology (Cambridge, Mass.: TAP, 1976).

43. AID Panel, "Comprehensive Review MIT 211(d): The Adaptation of Industrial and Public Works Technology to the Conditions of Developing Countries," mimeographed (undated).

44. Members are: The University of Alabama; University of Michigan; University of Rochester; Princeton University; California Institute of Technology; Colorado School of Mines; Massachusetts Institute of Technology; Milwaukee School of Engineering; Wentworth Institute of Technology.

45. Engineers Council for Professional Development, 43rd Annual Report, 1976 (back cover).

46. Ibid.

47. K. N. Rao, "Technical Education in the Developing Countries," The Ford Foundation, 1965; "Training and Employment of 'Middle Level' Technicians in Latin America," The Ford Foundation (undated), "The Education and Training of Chemical Technicians," discussion paper submitted to Commonwealth Conference on the Education and Training of Technicians, Huddersfield, England, 1966.

48. Rao, "Technical Education in the Developing Countries."

49. K. N. Rao to R. P. Morgan, July 17, 1978.

50. Wentworth Institute of Technology, "Summary Statement on Foreign Projects," June 1978.

51. Education Development Center, "INELEC: The Institute, Handbook II" (Newton, Mass.: EDC, March 1977).

52. Lawrence L. Barrell, "The Role of Engineering Technology Schools in International Development" (Paper presented at Annual Meeting of American Society for Engineering Education, Vancouver, B.C., Canada, June 20, 1978).

53. T. N. Soule and R. P. Morgan, "Summary of ASEE International Activities Survey," Draft Report, August 1978.

Notes 321

54. General Accounting Office, <u>Report of the Comptroller General of the United States: Strengthening and Using Universities as a Resource for Developing Countries</u> (Washington, D.C.: U.S. Government Printing Office, 1976).

55. Barrett Hazeltine, "Strategies for Engineering Schools in Developing Countries," <u>Technos,</u> January-March 1977, pp. 16-26.

56. See note 40.

57. Barrett Hazeltine, Remarks at Workshop on Role of U.S. Universities in Science and Technology for Development, Washington University, St. Louis, Mo. July 13-14, 1978.

58. L. P. Grayson, "The Design of Engineering Curricula," <u>UNESCO Studies in Engineering Education</u>, no. 5 (7 Place de Fontenoy, 75700 Paris, France, 1977).

59. UNESCO, <u>Final Report: International Conference on the Education and Training of Engineers and Higher Technicians</u> (New Delhi, India: April 20-26, 1976).

60. Kidd, "Manpower Policies," p. 129.

61. Brown and Hoelscher, "Open Forum."

62. Mario Gomez, Remarks at Workshop on Role of U.S. Universities in Science and Technology for Development, Washington University, St. Louis, Mo., July 13-14, 1978.

63. See note 57.

64. See note 62.

65. See note 57.

66. National Science Foundation, "Scientists and Engineers from Abroad."

67. K. N. Rao to R. P. Morgan, March 14, 1978.

CHAPTER 3

1. Kenneth W. Thompson and Barbara R. Fogel, <u>Higher Education and Social Change: Promising Experiments in Developing Countries</u>, Vol. 1: <u>Reports</u> (New York: Praeger, 1976).

2. For a more comprehensive history of early stages of involvement of U.S. universities in international agricultural programs see T. Keith Glennan and Irwin T. Sanders, <u>The Professional School and World Affairs: Report of the Task Force on Agriculture and Engineering</u> (New York: Education and World Affairs, 1967), pp. 21-68; Edward W. Weidner, <u>The World Role of Universities</u> (New York: McGraw-Hill, 1962), pp. 153-173; and Richard A. Humphrey (ed.), <u>Universities ... and Development Assistance Abroad</u> (Washington, D.C.: American Council on Education, 1967).

3. Michigan State University, Institute of Research on Overseas Programs, <u>The International Programs of American Universities: An Inventory and Analysis</u> (East Lansing: Michigan State University, 1958).

4. Ibid., p. 31.

5. East-West Center, Institute of Advanced Projects, <u>The International Programs of American Universities: An Inventory and Analysis</u> (East Lansing: Michigan State University, 1966).

6. Kathleen M. Propp et al., <u>AID: University Rural Development Contracts 1951-1966</u> (Urbana: University of Illinois, 1968), p. 1.

7. William N. Thompson et al., <u>AID: University Rural Development Contracts and U.S. Universities</u> (Washington, D.C., 1968), p. 15.

8. Propp et al., <u>AID: University Rural Development Contracts 1951-1966</u>.

9. Thompson et al., <u>AID: University Rural Development Contracts and U.S. Universities</u>.

10. Propp et al., <u>AID: University Rural Development Contracts 1951-1966</u>.

11. William N. Thompson et al., AID: University Rural Development Contracts 1951-1966, p. 7.

12. Ibid., p. 9.

13. Ibid., p. 11-12.

14. Ibid., p. 14.

15. Ibid., p. 15.

16. See note 2.

17. National Research Council, World Food and Nutrition Study: The Potential Contributions of Research, National Academy of Sciences (Washington, D.C.: NAS, 1977).

18. Irene L. Gomberg and Frank J. Atelsek, "International Scientific Activities at Selected Institutions, 1975-76 and 1976-77" (Washington, D.C.: American Council on Education, January 1978).

19. National Research Council, World Food and Nutrition Study.

20. National Research Council, Commission on International Relations, Supporting Papers: World Food and Nutrition Study, vol. 5, National Academy of Sciences, (Washington, D.C.: NAS, 1977).

21. Ralph W. Cummings, Jr., Food Crops in the Low-Income Countries: The State of Present and Expected Agricultural Research and Technology (New York: The Rockefeller Foundation, 1976).

22. Ibid.

23. National Center for Education Statistics, Digest of Education of Statistics, 1976 Ed., U.S. Department of Health Education and Welfare (Washington, D.C., 1976).

24. Alfred C. Julian and Robert E. Slattery, (eds.), Report on International Educational Exchange: Open Doors 1975/6 - 1976/7, Institute of International Education (New York: IIE, 1977).

25. Agency for International Development, A Directory of Institutional Resources Supported by Section 211(d) Grants: U.S. Centers of Competence for International Development (Washington, D.C.: AID, 1975).

26. The authors gratefully acknowledge helpful discussions with Melvin Blase and W. D. Buddemeier in connection with preparation of this section.

27. "BIFAD: The First Year: A Progress Report," (Washington, D.C.: Agency for International Development, November, 1977).

28. National Research Council, Commission on International Relations, Supporting Papers.

29. Ibid., p. 103.

30. Agency for International Development, Report to the Congress on Title XII-Famine Prevention and Freedom From Hunger of The Foreign Assistance Act of 1961 as Amended (Washington, D.C.: AID, 1978).

31. Agency for International Development, Guidelines for the Preparation and Submission of Proposals for Matching Formula Title XII University Strengthening Grants (Washington, D.C.: AID, 1978).

32. National Research Council, Commission on International Relations, Supporting Papers.

33. Ibid.

34. Ibid.

35. Ibid.

36. Report of the Comptroller General of the United States, Strengthening and Using Universities as a Resource for Developing Countries (Washington, D.C.: U.S. Government Printing Office, 1976).

37. International Soybean Program, "The Sri Lanka Soybean Development Program, Interim Report of Progress, March 1, 1975 - September 30, 1976" (Paper submitted to Food and Agriculture Organization of the United Nations, University of Illinois College of Agriculture at Urbana-Champaign, October 1976).

38. See note 3.

39. Ibid.

40. H. Read, Partners with India: Building Agricultural Universities (Urbana: University of Illinois, 1974).

41. Roland R. Renne, Agricultural Universities in India (Urbana: University of Illinois, 1974), pp. 1-7.

42. Ibid., p. 4.

43. Ibid., p. 3.

44. Land Tenure Center, The Land Tenure Center Annual Report, 1976-1977 (Madison, Wis.: University of Wisconsin, 1978).

45. Land Tenure Center, "A Brief Description of the Land Tenure Center Program, University of Wisconsin," mimeographed, May 1977, p. 1.

46. Read, Partners with India.

47. University of Illinois, "A Brief Outline Summary of Progress 1973-1978: The International Soybean Program (INTSOY)," mimeographed (Urbana, Ill.: University of Illinois, 1978).

48. Brookings Institution, Interim Report: An Assessment of Development Assistance Strategies (Washington, D.C., October 1977), pp. II-27.

49. College of Agriculture and Natural Resources, Michigan State University, "Highlights -- International Activities: College of Agriculture and Natural Resources (June 1977 - June 1978)."

50. Ibid., p. 6.

51. J. F. Metz to R. P. Morgan, April 12, 1978.

52. Kenneth L. Turk, The Cornell-Los Banos Story: Two Decades of Cooperation in Agricultural Education and Research Between Cornell University and the University of the Philippines (Ithaca, N.Y.: Cornell University, 1974).

53. Richard N. Adams and Charles C. Cumberland, "United States University Cooperation in Latin America," Institute of Research on Overseas Programs (East Lansing, Mich.: Michigan State University, 1960).

54. Edward W. Weidner, <u>The World Role of Universities</u> (New York: McGraw-Hill, 1972).

55. Ibid., p. 163.

56. Ibid., p. 171.

57. Ibid., p. 172.

58. John W. Gardner, <u>AID and the Universities</u> (New York: Education and World Affairs, 1964).

59. Ibid., p. xiii.

60. Ibid., p. 2.

61. Ibid., p. 3.

62. Ibid., p. 7.

63. Ibid., pp. 4-5.

64. Ibid., p. 10.

65. Ibid., p. 17.

66. Ibid., p. 20.

67. Ibid., p. 21.

68. Ibid., p. 24.

69. Ibid.

70. Ibid., p. 40.

71. Ibid., p. 45.

72. Ibid., p. 47.

73. Education and World Affairs, The University Looks Abroad: Approaches to World Affairs at Six American Universities.

74. Ibid., p. 266.

75. Ibid., p. 271.

76. Ibid., p. 272.

77. Ibid., p. 278.

78. Ibid., p. 284.

79. Keith T. Glennan and Irwin T. Sanders, The Professional School and World Affairs: Report of the Task Force on Agriculture and Engineering (New York: Education and World Affairs, 1962).

80. Richard H. Wood, U.S. Universities: Their Role in AID Financed Technical Assistance Overseas (New York: Education and World Affairs, 1968).

81. Agency for International Development-University Cooperation in Technical Assistance, Building Institutions to Serve Agriculture, Committee on Institutional Cooperation (Purdue University, Lafayette, Ind., 1968).

82. Ibid., p. 229.

83. Ibid.

84. Ibid., p. 44-45.

85. Ibid., p. 4.

86. Ibid.

87. Some studies that deal indirectly with this question are: National Research Council, World Food and Nutrition Study; National Research Council, Commission on International Relations; and Kenneth W. Thompson and Barbara R. Fogel, Higher Education and Social Change: Promising Experiments in Developing Countries, Volume 1: Reports (New York: Praeger, 1976).

88. U.S., Congress, House, Committee on International Relations, New Directions in Development Aid, excerpts from the legislation (Washington, D.C.: U.S. Government Printing Office, 1977), pp. 1-2.

89. Agency for International Development, Guidelines for the Role and Function of the Joint Committee on Agricultural Development (Washington, D.C.: AID, 1978).

90. Douglas Ensminger, Statement made to St. Louis Forum, 1979 U.N. Conference on Science and Technology for Development, U.S. Department of State and the National Research Council, January 23, 1978.

91. Buford L. Nichols, Statement submitted to the St. Louis Forum, 1979 U.N. Conference on Science and Technology for Development, U.S. Department of State and the National Research Council, January 23, 1978.

92. Frances Moore Lappe and Joseph Collins, Food First: Beyond the Myth of Scarcity (Boston: Houghton Mifflin, 1977).

93. Irene Tinker and Michele Bo Bramsen, (eds.), Women in World Development (Washington, D.C.: Overseas Development Council, 1976).

94. Development Associates, Inc., "A Seven Country Survey of the Roles of Women in Rural Development," a report prepared for the Agency for International Development (Washington, D.C., 1974).

95. John L. Fisher, Summary Report on the Conference on Women and Food, unpublished report, conference sponsored by Consortium for International Development held at Tucson, Arizona, January 9-11, 1978.

96. Dale E. Hathaway, "Applying American Science and Technology in Developing Countries," Higher Education in the World Community, Stephen K. Bailey (ed.) (Washington, D.C.: American Council on Education, 1977), p. 69; reprinted by permission.

97. Ralph H. Smuckler, U.S. Cooperation with Emerging Centers for Science and Technology in Low and Middle Income Countries Including Regional Aspects" (Draft paper presented to the Association of University Directors of International Agriculture, Logan, Utah, June 21-22, 1978), p. 4.

98. Ibid., pp. 5-6.

99. Lowell H. Watts, "Linkage Between Science and Technology Bases and the Agricultural Producer in LDC's" (Paper presented at the Annual AUSUDIAP Conference, Logan, Utah, June 21-23, 1978), p. 9.

100. Agency for International Development, Report to Congress on Title XII.

101. H. F. Massey (Statement submitted to the St. Louis Forum, 1979 U.N. Conference on Science and Technology for Development, U.S. Department of State and the National Research Council, January 23, 1978).

102. Agency for International Development, Report to Congress on Title XII.

103. Ibid., pp. 22-23.

104. Ibid., p. 20.

105. Morris D. Whitaker and E. Boyd Wennergren, "U.S. Universities and the World Food Problem," Science 194, no. 4264 (October 29, 1976).

106. T. Kelley White, "Science and Technology: Institutional Development and the U.S. University" (Statement submitted to St. Louis Forum; 1979 U.N. Conference on Science and Technology for Development, U.S. Department of State and National Research Council, January 23, 1978).

107. LaVern A. Freeh, "University of Minnesota, Suggested Policy and Guidelines Statement Relating to the University's World-Wide Mission and Responsibility," Draft Proposal, March 20, 1978.

108. Julian and Slattery, Report on International Educational Exchange.

109. Hathaway, "Applying American Science and Technology in Developing Countries;" Thompson and Fogel, Higher Education and Social Change; and Leopoldo S. Castillo, "Graduate Training in the United States as Seen by a National From a Developing Country," Journal of Dairy Science 51, no. 2 (1968): 237-242.

110. Hathaway, Higher Education in the World Community, reprinted by permission.

111. Ibid., p. 71, reprinted by permission.

112. J. Collom, H. Matteson, and L. Zuidema, "The University Programming Role in AID Participant Training: Its Conduct and Support" (Position paper of the Association of U.S. University Directors of International Agricultural Programs, 1978).

113. Statement made at Workshop on the Role of U.S. Universities in Science and Technology for Development. (For a summary of this workshop, see Appendix A.)

114. Agency for International Development, Benefits to the United States from Technical Assistance Activities Abroad, Some Case Studies (Washington, D.C.: AID, 1972).

115. Melvin Blase, Remarks at Workshop on the Role of U.S. Universities in Science and Technology for Development, St. Louis, Mo., July 13-14, 1978.

116. See note 90.

117. See note 90.

118. Robert McNamara, Address to World Bank Board of Governors, September 25, 1972.

119. National Research Council, World Food and Nutrition.

Notes 331

120. T. T. Williams, "Strategies for Science and Technology Transfer for the Benefit of Disadvantaged Populations" (Proceedings of 14th Annual Conference, Association of U.S. University Directors of International Agricultural Programs (AUSUDIAP), Logan, Utah, June 21, 1978).

121. Ibid., p. 25.

122. Agency for International Development, Report to the Congress on Title XII, p. 14.

CHAPTER 4

Science

1. Michael J. Moravcsik, "Science and the Developing Countries," (A contribution to the U.S. Country Paper for the UNCSTED Conference, October, 1977), pp. 2-3.

2. Carl Djerassi, "A Modest Proposal for Increased North-South Interaction Among Scientists," Bulletin of the Atomic Scientists, February 1976.

3. Jerrole Meinwald et al., "Chemical Ecology: Studies from East Africa," Science 199 (March 1978): 1167.

4. W. A. Copeland, "Iran's Pahlavi University: A Decade of Cooperation with the University of Pennsylvania," International Educational Cultural Exchange 7 (Summer 1971): 27-33.

5. Ibid.

6. Department of Educational Affairs, Inter-American University Cooperation, Pan American Union (Washington, D.C.: OAS, 1968).

7. Ibid., p. 40.

8. See, for instance, National Science Foundation Annual Report (Washington, D.C.: U.S. Government Printing Office, 1965-1966).

9. See also A. V. Baez, "Innovation in Science Education - Worldwide" (Paris: UNESCO Press, 1976).

10. <u>National Science Foundation Annual Report</u> (Washington: U.S. Government Printing Office, 1970).

11. <u>National Science Foundation Annual Report</u>, 1966-1970.

12. <u>National Science Foundation Annual Report</u>, 1968.

13. Department of Educational Affairs, <u>Inter-American University Cooperation,</u> p. 62.

14. Telephone conversation between Ellen Irons and K. N. Rao, May 9, 1978.

15. J. L. Morrill and K. N. Rao, "Science as Inquiry - Improvement of Secondary Education - Science and Mathematics in Latin America," (Discussion Notes, Ford Foundation, November 1965), pp. 2-3.

16. Robert H. Maybury, <u>Technical Assistance and Innovation in Science Education</u> (New York: John Wiley and Sons, 1975).

17. Telephone conversation between Ellen Irons and Dr. Richard Tolman, May 11, 1978.

18. Arnold B. Grobman, "Evaluation Abstracts: Factors Influencing International Curricular Diffusion," <u>Studies in Educational Evaluation</u> 2, no. 3 (Winter 1976): 231.

19. See note 14.

20. See note 17.

21. National Research Council, BOSTID, <u>Programs of the Board on Science and Technology for International Development: Summary of Activities, 1970-76</u> (Washington, D.C.: NAS, 1977), pp. II-57.

22. K. N. Rao, "Financing Graduate Education and Research in Science and Engineering in Latin America" (Mimeographed draft paper for The First Pan-American Conference on Post-Graduate Education in Engineering, Caracas, August, 1967).

23. See note 14.

24. Kidd, "Manpower Policies."

25. Barrett Hazeltine, Remarks at Workshop on the Role of U.S. Universities in Science and Technology for Development, Washington University, St. Louis, July 13-14, 1978.

26. National Research Council, BOSTID, Programs of the Board on Science and Technology for International Development, pp. II-46.

27. Carl Djerassi, "A High Priority? Research Centers in Developing Nations," Bulletin of Atomic Scientists 24, no. 1 (1968): 22-27.

28. National Research Council, BOSTID, Programs of the Board on Science and Technology for International Development, pp. II-46.

29. CIMMYT, International Maize and Wheat Improvement Center, Londres, Mexico, 1971.

30. Department of Educational Affairs, Inter-American University Cooperation, p. 2.

31. National Science Foundation Annual Report 1976 (Washington: U.S. Government Printing Office, 1976), p. 109.

32. Department of Educational Affairs, Inter-American University Cooperation, p. 137.

33. National Science Foundation, "Cooperative Science Programs in Latin America," descriptive leaflet (Washington, D.C.: NSF, 1973).

34. National Science Foundation, "Links That Connect a Hemisphere," Mosaic 8, no. 8 (November/December 1977): 44-50.

35. National ResearchCouncil, BOSTID, Programs of the Board on Science and Technology for International Development.

36. See, for instance, National Science Foundation Annual Reports, 1963-1976.

37. Mahindra Naraine, "Science for Progress: OAS Involvement in Science in Latin America," Nature 267 (May 1977): 298-299; reprinted with permission.

38. Mario Gomez, Remarks at Workshop on Role of U.S. Universities in Science and Technology for Development, Washington University, St. Louis, July 13-14, 1978.

39. Telephone conversation between Ellen Irons and Dr. Michael Greene, May 11, 1978.

40. See note 37.

41. See note 39.

42. Office of Program and Methodology, A Directory of Institutional Resources, U.S. Centers of Competence for International Development, January, 1975 (Washington, D.C.: AID, 1975).

43. Ibid., p. 52.

44. Center for International Studies, "Massachusetts Institute of Technology International Nutrition Policy and Planning Program, Spring, 1978," Draft paper, mimeographed.

45. National Research Council, BOSTID, Programs of the Board on Science and Technology for International Development.

46. National Science Foundation Annual Report 1968.

47. David Pines, "On Building U.S. and World Science through International Scientific Exchange," New U.S. Initiatives "In International Science and Technology" Workshop Reports, conducted by Denver Research Institute, Center for Public Issues (University of Denver, August 1977) pp. 163-164.

48. Ibid.

49. A. Julian and R. Slattery (eds.) Open Doors 1975/6 - 1976/7, (New York: IIE, 1978).

50. Council for International Exchange of Scholars, Annual Report to the Board of Foreign Scholarships: 1976-1977, 11 Dupont Circle, N.W., Washington, D.C. 20036.

51. NSF Annual Reports 1970-76.

52. U.S., Congress, Senate, Foreign Assistance Authorization. Hearings before the Committee on Foreign Relations, 1977 (Washington, D.C.: U.S. Government Printing Office, 1977), pp. 52, 40.

53. Ibid., p. 40.

54. Mutual Security Act of 1960, PL 86-472.

55. East-West Resource Systems Institute, "The Interests and Activities of the East-West Resource Systems Institute, East-West Center," (Honolulu, Hawaii: East-West Center, January 1978).

56. Annual Report, 1977, The International Association for the Exchange of Students for Technical Experience (Zurich, Switzerland: IAESTE, October 1977).

57. National Research Council, "Stewards for International Exchange: The Role of the National Research Council in the Senior Fulbright-Hays Program 1947-1975," National Academy of Sciences (Washington, D.C.: NAS, 1976).

58. Adolph Wilburn, Remarks at Workshop on Role of U.S. Universities in Science and Technology for Development, Washington University, St. Louis, Mo., July 13-14, 1978.

59. Estimated from data in note 57.

60. U.S., Congress, House, Committee on International Relations, Science, Technology, and American Diplomacy, vol. 2 (Washington: U.S. Government Printing Office, 1977), p. 889.

61. Ibid., p. 900.

62. Ibid., derived from chart, p. 958.

63. Ibid., p. 977.

64. Ibid., p. 975.

65. National Science Foundation, "Small Projects; Large Impacts," Mosaic 8, no. 6 (November/December 1977): 38-43.

66. Charles V. Kidd, "An Evaluation of the NSF-SEED Program - Scientists and Engineers in Economic Development" (Report to NSF, mimeographed, November 1977).

67. W. Eilers and A. Segal, Remarks at Workshop on Role of U.S. Universities in Science and Technology for Development, Washington University, St. Louis, Mo., July 13-14, 1978.

68. Kidd, "An Evaluation of the NSF-SEED Program," p. 1 (Recommendations).

69. Ibid., pp. 7-8 (Recommendations).

70. Ibid., p. 9 (Recommendations).

71. National Science Foundation Annual Report, 1969, p. 104.

72. Ibid., p. 218.

73. Michael J. Moravcsik, "The Physics Interviewing Project," International Educational and Cultural Exchange, Summer 1972, p. 16.

74. Earl Callen and Michael Scandron, "The Physics Interviewing Project: A Tour of Interviews in Asia," Science 200 (June 2, 1978): 1018-1022.

75. Michael J. Moravcsik, Science Development: The Building of Science in Less Developed Countries, PASITAM (Bloomington, Ind., 1975).

76. Ibid., p. 174.

77. K. N. Rao, "University Based Science and Technology for Development: New Patterns of International Aid," Impact of Science on Society 28, no. 2 (1978): 119; © UNESCO, 1978. Reproduced by permission of UNESCO.

78. Ibid., p. 119.

79. Ibid., pp. 124-125.

80. Michael Greene, Physics in Latin America: Peru and Chile (College Park: University of Maryland, 1971).

81. Ibid., p. 7.

82. H. Harry Szmant, "Foreign Aid Support of Science and Economic Growth," Science 199 (March 1978): 1181-1182.

83. Carl Djerassi, "A Modest Proposal for Increased North-South Interaction among Scientists," Bulletin of the Atomic Scientist 32 (February 1976): 58.

84. K. W. Thompson and B. R. Fogel, Higher Education and Social Change: Promising Experiments in Developing Countries, vol. 1, Reports (New York: Praeger, 1976).

85. F. H. Harrington, "International Linkages in Higher Education: A Feasibility Study," Draft Final Report, February 1978.

86. L. Erk, "A Seminar in Training for Development," (Paper presented by workshop sponsored by the AID/NAFSA Liaison Committee, July 14, 1977). Information on AAATDC available from AAATDC, the University of Michigan, Ann Arbor, Mich.

87. R. M. Walker, Remarks at Workshop on Role of U.S. Universities in Science and Technology for Development, Washington University, St. Louis, Mo., July 13-14, 1978.

88. F. A. Long, Remarks at Workshop on Role of U.S. Universities in Science and Technology for Development, Washington University, St. Louis, Mo., July 13-14, 1978.

89. K. N. Rao, Remarks at Workshop on Role of U.S. Universities in Science and Technology for Development, Washington University, St. Louis, Mo., July 13-14, 1978.

90. V. Walbot, Remarks at Workshop on Role of U.S. Universities in Science and Technology for Development, Washington University, St. Louis, July 13-14, 1978.

91. H. R. Glyde, "Institutional Links in Science and Technology: The United Kingdom and Thailand," International Development Review/Focus 1 (1973): 7-12.

92. Telephone conversation between Ellen Irons and Michael Moravscik, May 9, 1978.

93. Ibid.

CHAPTER 5

1. H. R. Potter, in Institution-Building: A Model for Applied Social Change, D. W. Thomas et al., eds., (Cambridge, Mass.: Schenkman Publishing Co., 1972).

2. Education Development Center, Annual Report 1977, 55 Chapel Street, Newton, Mass. 02160.

3. B. Hazeltine, Remarks at Workshop on the Role of U.S. Universities in Science and Technology for Development (St. Louis, Mo.: Washington University, July 13-14, 1978).

4. K. N. Rao, "University Based Science and Technology for Development: New Patterns of International Aid," Impact of Science on Society 28, no. 2 (April-June 1978); © UNESCO 1978, reproduced by permission of UNESCO.

5. K. N. Rao, "Away from the Metropolis: The Role of the Engineer and Technologist in Regional and Provincial Development" (Address at 1st International Congress on Educational Research in Superior Technical Education, Durango, Mexico, August 2-5, 1978).

6. H. H. Szmant, "Foreign Aid Support of Science and Economic Growth," pp. 1173-1182.

7. Carl Djerassi, "A Modest Proposal for Increased North-South Interaction Among Scientists," Bulletin of Atomic Scientists (February 1976), pp. 56-60.

8. D. Hathaway, in Higher Education in the World Community, Stephen K. Bailey, ed., (Washington, D.C.: American Council on Education, 1977).

9. D. Delasanta, E. Perez, and J. Byram, Statement prepared in connection with St. Louis Forum on UNCSTED Conference, January 23, 1978.

10. Examples of such institutions in LDCs are cited in Nicolas Jequier, Appropriate Technology: Problems and Promises (Paris: O.E.C.D., 1976).

11. C. Barker to R. P. Morgan, May 30, 1978.

12. R. P. Morgan, "An International Development Technology Center," Journal of Engineering Education 60, no. 3 (November 1969): 247-249.

13. National Academy of Sciences, U.S. Science and Technology for Development: A Contribution to the 1979 U.N. Conference (Washington, D.C.: NAS, May 1978).

14. M. Moravcsik to R. P. Morgan, June 19, 1978.

15. GAO Report, "Strengthening and Using Universities as a Resource for Developing Countries," May 5, 1976.

16. National Research Council, The Role of U.S. Engineering Schools in Development Assistance, National Academy of Sciences (Washington, D.C.: NAS, 1976).

17. National Academy of Sciences, U.S. Science and Technology for Development (Washington, D.C.: NAS, 1978), ch. 2.

18. R. P. Morgan, "An International Development Technology Center"; and "International Directions for Engineering Education in the United States" (Paper presented at Conference on Engineering for International Development, Estes Park, Colorado, August 27 - September 1, 1967).

19. M. Moravcsik, "Science and the Developing Countries" (Contribution to the U.S. Country Paper for the UNCSTED Conference, October, 1977).

20. Institute for International Education, Open Doors: 1975-1976, 1976-1977 (Washington, D.C.: IIE, 1978).

21. National Center for Education Statistics, Digest of Education Statistics, 1976 Edition (Washington, D.C.: U.S. Government Printing Office, 1977), p. 125.

22. Lynda Woodcock, "Influx of Foreign Students an Issue in Canada," Chronicle of Higher Education, March 13, 1978, p. 3.

23. Kidd, "Manpower Policies."

24. J. Mullin, Remarks at Session on UNCSTED Conference, Annual Meeting, American Society for Engineering Education, Vancouver, B.C., Canada, June 22, 1978.

25. Adolph Wilburn, Remarks at Workshop on the Role of U.S. Universities in Science and Technology for Development, Washington University, St. Louis, July 13-14, 1978.

26. L. Middleton, "Welcome Cools for Iranians on Many Campuses," Chronicle of Higher Education 16, no. 20 (July 24, 1978): 9-10.

27. National Academy of Sciences, U.S. Science and Technology for Development.

28. C. V. Kidd, "An Evaluation of the NSF-SEED Program"; and B. Hazeltine, "Strategies for Engineering Schools in Developing Countries," TECHNOS, January-March, 1977, pp. 16-26.

29. Kidd, "An Evaluation of the NSF-SEED Program," p. 2.

30. Szmant, "Foreign Aid Support of Science and Economic Growth," pp. 1173-1182.

31. Ibid., p. 1180.

32. M. Moravcsik, Science Development: The Building of Science in Less Developed Countries.

Notes 341

33. William Eilers, Remarks at Workshop on the Role of U.S. Universities in Science and Technology for Development, Washington University, St. Louis, July 13-14, 1978.

34. Anti-Nuclear Proliferation Act of 1978, N.R. 8638.

35. National Academy of Sciences, U.S. Science and Technology for Development (Washington, D.C.: NAS, 1978), p. 34.

36. Educational Development Center, Kanpur Indo-American Program 1962-1972.

37. "African Design Team Returns," IE&R Newsletter 2, no. 1, Purdue University (September 1978).

38. F. H. Harrington, International Linkages in Higher Education: Feasibility Study, Draft Final Report, February 1978.

39. From brochure on Council on Science and Technology for Development, 2010 Massachusetts Avenue, Washington, D.C. 20036, February 1978.

40. Melvin Blase, Remarks at Workshop on the Role of U.S. Universities in Science and Technology for Development, Washington University, St. Louis, July 13-14, 1978.

41. K. N. Rao, Remarks at Workshop on the Role of U.S. Universities in Science and Technology for Development, Washington University, St. Louis, July 13-14, 1978.

42. Merton Barry, Remarks at Workshop on the Role of U.S. Universities in Science and Technology for Development, Washington University, St. Louis, July 13-14, 1978.

43. "The United Nations University: The Second Year, 1976-1977," adapted from Annual Report of the Council of the University to the General Assembly of the United Nations, 1977.

44. Walter Shearer-Izumi, "The Natural Resources Program at the United Nations University," Science 198 (Dec. 2, 1977): 896-897.

45. Mario Gomez to R. P. Morgan, August 1978.

46. E. B. Hartman and R. P. Morgan, eds., <u>Proceedings of Conference on University Education for Technology and Public Policy</u> (St. Louis, Mo.: Washington University, December 1976).

47. American Association of Community and Junior Colleges and the National Liaison Committee on Foreign Student Admissions, <u>The Foreign Student in United States Community and Junior Colleges</u> (New York: College Entrance Examination Board, 1978).

48. Lawrence L. Barrell, "The Role of Engineering Technology Schools in International Development" (Paper presented at Annual Meeting of American Society for Engineering Education, Vancouver, B.C., Canada, June 20, 1978).

49. Education Development Center, "INELEC: The Institute, Handbook II," (Newton, Mass.: EDC, March 1977).

50. From "VITA and Its Work with Universities" (Statement provided to Robert P. Morgan by Lauryl Druben, Volunteers in Technical Assistance, 3706 Rhode Island Avenue, Mt. Rainier, MD 20822, August 1978).

51. M. Witunski to Robert P. Morgan in connection with Workshop on the Role of U.S. Universities in Science and Technology for Development, Washington University, St. Louis, July 13-14, 1978.

52. K. N. Rao, Remarks at Workshop on Role of U.S. Universities in Science and Technology for Development, Washington University, St. Louis, July 13-14, 1978.

53. Hugh Miller, Remarks at Workshop on Role of U.S. Universities in Science and Technology for Development, Washington University, St. Louis, July 13-14, 1978.

54. "The International Development Institute," National Academy of Sciences, July 1971.

55. Brookings Institution, "An Assessment of Development Assistance Strategies: Interim Report," Washington, D.C., 1977.

56. John Walsh, "New Institute Passes First Test in Congress," <u>Science</u> 204 (April 27, 1979): 385-388; and Institute for Technological Cooperation, "Congressional Presentation," Washington, D.C., Feb. 23, 1979.

57. John W. Gardner, <u>AID and the Universities</u> (New York: Education and World Affairs, 1964).

58. William Eilers, Remarks at Workshop on Role of U.S. Universities in Science and Technology for Development, Washington University, St. Louis, July 13-14, 1978.

59. Michael Witunski, Remarks at Workshop on Role of U.S. Universities in Science and Technology for Development, Washington University, St. Louis, July 13-14, 1978.

60. Louis Berlinguet, letter to R. P. Morgan, March 23, 1978; <u>The Role of Canadian Universities in International Development</u>, Interim Report on behalf of the Association of Universities and Colleges of Canada and the Royal Society of Canada (Ottawa, July, 1977); <u>Canadian Higher Education and International Development Cooperation, 1975-1980</u>, as approved by Board of Directors, AUCC; and <u>Survey of Programmes of Cooperation Established Between Canadian Universities and Foreign Institutions: 1976</u>, Association of Universities and Colleges of Canada, February 1977.

61. Barbara Lucas, Remarks at Workshop on Role of U.S. Universities in Science and Technology for Development, Washington University, St. Louis, July 13-14, 1978.

62. Liam Finn, Remarks at Session on Technology and International Development: The Role of Engineering Schools, ASEE Annual Meeting, Vancouver, B.C., Canada, June 20, 1978.

63. Wilburn, Remarks at Workshop.

64. Moravcsik, <u>Science Development</u>.

65. D. Goulet, <u>The Uncertain Promise</u> (New York: IDOC/North America and Overseas Development Council, 1977), pp. 81-82.

66. W. D. Buddemeier to R. P. Morgan in connection with Workshop on Role of U.S. Universities in Science and Technology for Development, Washington University, St. Louis, July 13-14, 1978.

67. Henry Glyde, "<u>Institutional Links in Science and Technology</u>," 60 pp., no date. (Author is currently with Department of Physics, University of Ottawa, Ottawa, Canada).

68. Ibid., pp. 44-46.

69. Edward Cornish, The Study of the Future (Washington, D.C.: World Future Society, 1977), p. 111.

CHAPTER 6

1. Aaron Segal, Remarks at Workshop on Role of U.S. Universities in Science and Technology for Development, Washington University, St. Louis, July 13-14, 1978.

2. U.N. World Plan of Action on Science and Technology for Development (New York: United Nations, 1971); and R. Clarke, The Great Experiment: Science and Technology in the Second U.N. Development Decade (New York: United Nations, 1971).

3. M. Moravcsik, Science Development, p. 180.

4. William Eilers, Remarks at Workshop on Role of U.S. Universities in Science and Technology for Development (St. Louis, Mo.: Washington University, July 13-14, 1978).

5. M. S. Wionczek, "Some Questions for the World Jamboree," Bulletin of Atomic Scientists, December 1977, p. 32.

6. I. Adelman and C. T. Morris, Economic Growth and Social Equity in Developing Countries (Stanford, Ca.: Stanford U. Press, 1973).

7. Robert P. Morgan, "Technology and International Development: New Directions Needed," Chemical and Engineering News 55 (November 14, 1977): 31-39.

8. Thomas H. Fox, Testimony before the Subcommittee on Domestic and International Scientific Planning, Analysis and Cooperation of the House Committee on Science and Technology, July 25, 1978.

Notes 345

9. Ibid., pp. 2-3.

10. Clarence D. Long, Statement before the Subcommittee on International Development Institutions and Finance by the Committee on Banking and Urban Affairs, March 14, 1978.

11. Virginia Walbot, Remarks at Workshop on Role of U.S. Universities in Science and Technology for Development, Washington University, St. Louis, Mo., July 13-14, 1978.

12. Melvin Blase, Remarks at Workshop on Role of U.S. Universities in Science and Technology for Development, Washington University, St. Louis, Mo., July 13-14, 1978.

13. Michael Witunski, Remarks at Workshop on Role of U.S. Universities in Science and Technology for Development, Washington University, St. Louis, Mo., July 13-14, 1978.

14. Michael Moravcsik to Robert P. Morgan, June 19, 1978.

15. Adolph Wilburn, Remarks at Workshop on Role of U.S. Universities in Science and Technology for Development, Washington University, St. Louis, July 13-14, 1978.

16. Ibid.

17. "Canadian Higher Education and International Development Cooperation, 1975-1980," As approved by the Board of Directors, Association of Universities and Colleges of Canada, p. 4.

18. Barrett Hazeltine, Remarks at Workshop on Role of U.S. Universities in Science and Technology for Development, Washington University (St. Louis, July 13-14, 1978).

19. Mario Gomez, Remarks at Workshop on Role of U.S. Universities in Science and Technology for Development, Washington University, St. Louis, July 13-14, 1978.

20. N. Jequier, *Appropriate Technology: Problems and Promises*, (Paris: OECD, 1976); and J. Van Brakel, *Chemical Technology for Appropriate Development* (Delft University Press, 1978).

21. Charles Kidd, "Manpower Policies."

22. Lynda Woodcock, "Influx of Foreign Students an Issue in Canada," Chronicle of Higher Education, March 13, 1978, p. 3.

23. Kidd, "Manpower Policies," pp. 232-241.

24. We hypothesized that U.S. universities were an underutilized resource in S&T for development; yet we concluded their involvements have been extensive. We have not been able to test this hypothesis adequately with the time, resources and data available to us. One Workshop participant felt we should state that only some universities have been extensively involved; clearly, some have been more involved than others. Our feeling is that the hypothesis is correct and not necessarily at odds with the conclusion.

25. Henry Glyde, "Institutional Links in Science and Technology."

26. Harold L. Wilensky, "The Professionalization of Everyone?" The American Journal of Sociology 52, no. 2 (September 1964): 137-158.

27. Moravcsik, personal communication.

28. K. N. Rao to Robert Morgan, July 21, 1978.

29. Barbara Lucas, Remarks at Workshop on Role of U.S. Universities in Science and Technology for Development, Washington University, St. Louis, Mo., July 13-14, 1978.

30. Aaron Segal, Remarks at Workshop on Role of U.S. Universities in Science and Technology for Development, Washington University, St. Louis, Mo., July 13-14, 1978.

31. U.S. Science and Technology for Development, National Academy of Sciences, 1978; and National Research Council, The Role of U.S. Engineering Schools in Development Assistance (Washington, D.C.: NAS, 1976).

32. Glyde, "Institutional Links in Science and Technology," p. 43.

APPENDIX B

1. All of Appendix B is taken verbatim from "Interim Report: An Assessment of Development Assistance Strategies," The Brookings Institution, submitted to the Department of State in accordance with grant no. 1722-720235, October 6, 1977.

APPENDIX C

1. Robert H. Maybury, Technical Assistance and Innovation in Science Education (New York: John Wiley and Sons, 1975).
2. Ibid., pp. 31-32.
3. Ibid., p. 41.
4. Ibid., p. 209.
5. National Academy of Sciences, Program of the Board on Science and Technology for International Development: Summary of Activities, 1970-1976, Washington, D.C., 1977.
6. National Academy of Sciences, Annual Reports, 1969-1975.
7. Personal communication: Ellen Irons and Michael Moravcsik, May 9, 1978.
8. Ford Foundation, Annual Reports, 1969-1976.
9. Kenneth W. Thompson, "Higher Education for National Development: One Model for Technical Assistance," Occasional Paper No. 5, International Council for Educational Development, 522 Fifth Avenue, New York, NY 10036.

APPENDIX D

1. "Intersociety Preliminary Analyses of R&D in the FY 1979 Budget," AAAS and six other societies, Washington, D.C., Feb. 1978.

APPENDIX E

1. Information for Appendix E was supplied by Lauryl Druben, Volunteers in Technical Assistance (VITA), 3706 Rhode Island Avenue, Mt. Rainier, Md. 20822.

Bibliography

DEVELOPMENT ASSISTANCE POLICIES, PROGRAMS
AND MECHANISMS: OVERVIEW

Acton, M. Science and Technology for International Development: A Selected List of Information Sources in the U.S. and Bibliography of Selected Materials. Ann Arbor: AID, 1975.

Agency for International Development. Report to the Congress on Title XII: Famine Prevention and Freedom from Hunger of the Foreign Assistance Act of 1961. Washington, D.C.: AID, 1977.

_____. A Directory of Institutional Resources: U.S. Centers of Competence for International Development. Washington, D.C.: AID, 1975.

_____. The Role of Voluntary Organizations in International Assistance: A Look at the Future. Washington, D.C.: AID, 1974.

_____. Benefits to the United States from American Technical Assistance Activities Abroad: Case Studies. Washington, D.C.: Brookings Institution, 1970.

Asher, Robert E. Development Assistance in the Seventies: Alternatives for the United States. Washington, D.C.: Brookings Institution, 1970.

Berg, R. J.; Kitchell, R. E.; Schwab, P.; and Swan, A. Report of Review Team on Institutional Grants Program (211D): AID. Washington, D.C.: AID, 1973.

Blase, Melvin G. Institution Building: A Source Book. Washington, D.C.: AID, 1973.

Brookings Institution. An Assessment of Development Assistance Strategies: Interim Report. Washington, D.C.: Brookings Institution, 1977.

Brown, Lester R. with Eckholm, Erik P. By Bread Alone. New York: Praeger, 1974.

Canadian International Development Agency. Canada: Strategy for International Development Cooperation, 1975-1980. Ottawa: Canadian International Development Agency, 1975.

Clarke, Robin. The Great Experiment: Science and Technology in the Second U.N. Development Decade. New York: United Nations, 1971.

da Costa, Joao. Statement by the Secretary of the Conference at the Opening Meeting of the Second Session for the U.N. Conference on Science and Technology for Development. United Nations, New York, 1978.

Derge, David R. and Souder, Donald L. Institution Building and Rural Development: A Study of United States Technical Assistance Projects. Bloomington: Indiana University, 1968.

Erb, Guy F., and Kallab, Valeriana (eds.). Beyond Dependency: The Developing World Speaks Out. Washington, D.C.: Overseas Development Council, 1975.

Ford Foundation. International Programs of the Ford Foundation. New York: Ford Foundation, 1973.

Gant, George F. "The Institution Building Project." International Review of Administrative Science 32 (Nov. 3, 1966).

Gardner, John W. AID and the Universities. New York: Education and World Affiars, 1964.

General Accounting Office, Comptroller General. Federal Program Evaluations: July 1, 1975 through June 30, 1977. Washington, D.C.: U.S. Government Printing Office, 1978.

Glyde, Henry R. "Institutional Links in Science and Technology: The United Kingdom and Thailand." International Development Review 15, no. 1 (1973): 7-12, Focus section.

Goulet, Denis. The Cruel Choice: A New Concept in the Theory of Development. New York: Atheneum, 1973.

_____. The Uncertain Promise. Washington, D.C.: IDOC/North America and ODC, 1977.

Hammond, George S., and Todd, W. Murray. "Technical Assistance and Foreign Policy." Science 189 (September 26, 1975): 1057-1059.

Hansen, Roger D., and the staff of the Overseas Development Council. The United States and World Development: Agenda for Action. New York: Praeger, 1976.

Haq, Mahbub ul. The Poverty Curtain: Choices for the Third World. New York: Columbia University Press, 1976.

Harrington, Fred Harvey. "International Linkages in Higher Education: A Feasibility Study." Draft Final Report. Washington, D.C.: February 1978.

Hayden, R. "MUCIA: The Consortium Approach to International Cooperation." International Educational and Cultural Exchange 9 (1973): 57-63.

Hayes, Samuel P., Jr. Evaluating Development Projects. Paris: UNESCO, 1966.

Howe, James, and the staff of the Overseas Development Council. The U.S. and the Developing World: Agenda for Action 1974. New York: Praeger, 1974.

Jequier, Nicolas (ed.). Appropriate Technology: Problems and Promises. Paris: OECD, 1976.

Kidd, Charles. "Manpower Policies for the Use of Science and Technology for Development." Washington, D.C.: George Washington University, October 1978.

MUCIA. MUCIA Program of Advanced Study in Institution Development and Technical Assistance Methodology. Annual Reports, 1974/75 and 1975/76. Ann Arbor: AID R&D Report Distribution Center.

_____. Private Voluntary Organizations and Appropriate Technology: A Report Prepared for the Agency for International Development. Ann Arbor: AID, 1976.

_____. Expanding the Roles of Private Humanitarian Agencies in Development Assistance: Manpower and Management Issues. Bloomington: MUCIA, 1976.

National Research Council. U.S. Science and Technology for Development: A Contribution to the 1979 U.N. Conference. Washington, D.C.: U.S. Government Printing Office, 1978.

_____, Board on Science and Technology for International Development. Programs of the Board on Science and Technology for International Development: Summary of Activities, 1970-1976. Washington, D.C.: National Academy of Sciences, 1977.

_____. The International Development Institute. Washington, D.C.: NAS, 1971.

National Science Foundation. "Scientists and Engineers From Abroad: Trends of the Past Decade, 1966-75." Reviews of Data on Science Resources 28 (February 1977).

Nau, Henry R. (ed.). Technology Transfer and U.S. Foreign Policy. New York: Praeger, 1976.

OAS. "Progress Through Partnership, Partners of the Americas." Washington, D.C.: OAS, 1977.

Paddock, W. and Paddock, E. We Don't Know How: An Independent Audit of What They Call Success in Foreign Assistance. Ames: Iowa State University Press, 1973.

Posz, Gary S.; Sjun, Jong; and Storm, William B. Administrative Alternatives in Development Assistance. Cambridge, Mass.: Ballinger Publishing Co., 1973.

Propp, Kathleen M.; Guither, H. D.; Regnier, E. H.; and Thompson, W. N. AID-University Rural Development Contracts 1951-1966. Urbana: University of Illinois, 1968.

Rabinowitch, Victor, and Rabinowitch, Eugene (eds.). Views of Science, Technology, and Development. New York: Pergamon Press, 1975.

"Reader's Guide to Focus: Technical Cooperation." International Development Review (1973/1).

Rockefeller Foundation. President's Ten Year Review and Annual Report. New York: Rockefeller Foundation, 1971.

Sewell, John W. The United States and World Development: Agenda 1977. Washington, D.C.: ODC, 1977.

Sommer, John G. Beyond Charity: U.S. Voluntary Aid for a Changing Third World. Washington, D.C.: ODC, 1977.

_____. "U.S. Voluntary Aid to The Third World: What Is Its Future?" Development Paper No. 20. Washington, D.C.: ODC, 1975.

Thomas, D. Woods; Potter, Harry R.; Miller, William L.; Aveni, Adrian F. (eds.). Institution Building: A Model for Applied Social Change. Cambridge: Schenkman Publishing Co., 1972.

Thompson, Kenneth W. Foreign Assistance: A View From the Private Sector. London: University of Notre Dame Press, 1972.

"Transcript of St. Louis Forum on the 1979 UNCSTED Conference." Washington University, St. Louis, Mo., January 23, 1978.

UNESCO. Bilateral Institution Links in Science and Technology. Prepared by Daisy Loman. Paris: UNESCO, 1969.

_____. "Quantification of Scientific and Technological Activities Related to Development." Paris: UNESCO, 1976.

United Nations. Multinational Corporations in World Development. New York: United Nations, 1973.

_____. World Plan of Action for the Application of Science and Technology to Development. New York: United Nations, 1971.

_____. Science and Technology for Development: Proposal for the Second U. N. Development Decade. New York: United Nations, 1970.

U.S. Congress. Legislation on Foreign Relations Through 1977. 3 vols. Washington, D. C.: U. S. Government Printing Office, 1978.

_____. Inter-American Relations. Washington, D. C.: U. S. Government Printing Office, 1973.

_____. House. New Directions in Development Aid: Hearings. Washington, D. C.: U. S. Government Printing Office, 1977.

_____. Foreign Relations Authorization for Fiscal Year 1978: Hearings. Washington, D. C.: U. S. Government Printing Office, 1977.

_____. Science, Technology and American Diplomacy: An Interaction of Science and Technology with U. S. Foreign Policy. 3 vols. Washington, D. C.: Government Printing Office, 1977.

_____. Interagency Coordination of Federal Scientific Research and Development: The Federal Council for Science and Technology: Hearings. Washington, D. C.: U. S. Government Printing Office, 1976.

_____. Senate. "A Bill to promote the foreign policy, security, and general welfare of the United States by assisting the peoples of the world in their efforts toward economic development by establishing the International Development Cooperation Administration." Introduced by H. H. Humphrey, January, 1978.

_____. Foreign Assistance: Hearings. Annual. Washington, D. C.: U. S. Government Printing Office, 1961-1977.

_____. Legislation on Foreign Relations: Hearings. Annual. Washington, D. C.: U. S. Government Printing Office, 1961-1977.

_____. Foreign Policy Choices for the Seventies and Eighties: Hearings. 2 vols. Washington, D. C.: U. S. Government Printing Office, 1976.

Bibliography   355

_____. Foreign Assistance Authorization: Examination of U.S. Foreign Aid Programs and Policies. Washington, D.C.: U.S. Government Printing Office, 1975.

_____. Foreign Assistance Legislation, Fiscal Year 1972: Hearings. Washington, D.C.: U.S. Government Printing Office, 1971.

Ward, Barbara; Runnalls, J.D.; and D'Anjou, L. The Widening Gap: Development in the 1970's. Report on the Columbia Conference on International Economic Development, February 15-21, 1970. New York: Columbia University Press, 1971.

Weiss, Charles, Jr. "A Proposed Institute for Development Science," Paper presented at the Conference on The Potential Contribution of Individual Scientists to the Solution of Development Problems, USAID, Washington, D.C., March 4, 1971.

Wilkowski, Jean. Statement at Second Preparatory Committee Plenary Session, U.N. Conference on Science and Technology for Development, January 27, 1978.

Wionczek, M.S. "Some Questions for the World Jamboree." Bulletin of Atomic Scientists, December 1977, p. 32.

INTERNATIONAL EDUCATION: OVERVIEW

Adams, Richard N., and Cumberland, Charles C. United States University Cooperation in Latin America. East Lansing: Michigan State University, 1960.

Agency for International Development. "AID Comments on the GAO Report Strengthening and Using Universities as a Resource for Developing Countries." Mimeographed. Washington, D.C.: AID, 1976.

Altbach, Philip G. (ed.). University Reform: Comparative Perspectives for The Seventies. Cambridge, Mass.: Schenkman, 1974.

_____. Comparative Higher Education. Washington, D.C.: American Association for Higher Education, 1973.

_____, and Kelly, David H. Higher Education in Developing Nations: A Selected Bibliography 1969-1974. New York: Praeger, 1974.

American Association of Community and Junior Colleges and the National Liaison Committee on Foreign Student Admissions. The Foreign Student in United States Community and Junior Colleges. New York: College Entrance Examination Board, 1978.

Ashby, Eric. "The Structure of Higher Education: A World View." Occasional paper no. 6. New York: International Council for Educational Development, 1973.

Association of Universities and Colleges of Canada/Royal Society. "The Role of Canadian Universities in International Development." Ottawa: Association of Universities and Colleges of Canada, 1977.

Association of Universities and Colleges of Canada. Survey of Programmes of Cooperation Established Between Canadian Universities and Foreign Institutions. Ottawa: Association of Universities and Colleges of Canada, 1977.

Atelsek, Frank J., and Gomberg, Irene L. Foreign Area Research Support Within Organized Research Centers at Selected Universities. Washington, D.C.: American Council on Education, 1976.

Bailey, Stephen K. (ed.). Higher Education in the World Community. Washington, D.C.: American Council on Education, 1977.

Brembeck, C. S., and Hanson, J. W. (eds.). Education and the Development of Nations. New York: Holt, Rinehart, and Winston, 1966.

Buarque, C. "Role of the University: Criticisms and Hopes." Universities Quarterly 27 (Summer 1973): 346-355.

Butts, R. Freeman. American Education in International Development. New York: Harper and Row, 1963.

Canter, Jacob. "Our Common Challenge in International Exchanges." Junior College Journal 37 (February 1967): 14-16.

Collins, Charles C. "Exporting the Junior College Idea." Junior College Journal 38 (April 1968): 10-14.

Bibliography

Conference on University Contracts Abroad. Addresses and Summary of Proceedings. Annual, 1955-1960. Washington, D.C.: American Council on Education.

Coombs, Philip H. The World Educational Crisis: A Systems Analysis. New York: Oxford Press, 1968.

_____. The Fourth Dimension of Foreign Policy: Educational and Cultural Affairs. New York: Council on Foreign Relations, Harper and Row, 1964.

Council for International Exchange of Scholars. "Annual Report to the Board of Foreign Scholarships, 1976-1977." Washington, D.C.: Council for International Exchange of Scholars, 1977.

East-West Center, Institute of Advanced Projects. The International Programs of American Universities: An Inventory and Analysis. East Lansing: Michigan State University, 1966.

East-West Resource Systems Institute. "The Interests and Activities of the East-West Resource Systems Institute, East-West Center." Honolulu: East-West Center, 1978.

Education and World Affairs. Internationalizing the U.S. Professional School. New York: Education and World Affairs, 1969.

_____. The University Looks Abroad: Approaches to World Affairs at Six American Universities. New York: Walker and Co., 1965.

Epstein, Harold. "Where Do Junior Colleges Fit In?" Junior Colleges Journal 37 (February 1967): 17-19.

Flack, Michael J. Sources of Information on International Educational Activities. Washington, D.C.: American Council on Education, 1958.

Furniss, W. Todd. "Implications of the International Education Act." International Educational and Cultural Exchange, Winter 1967, pp. 1-8.

General Accounting Office, Comptroller General. Strengthening and Using Universities as a Resource for Developing Countries. Washington, D.C.: U.S. Government Printing Office, 1976.

Gordon, J. King. "The IDRC and University Involvement in International Development." Presentation before meeting of Canadian University Presidents, Montreal, Quebec, February 22, 1978.

Harrington, Fred H. International Linkages in Higher Education: Feasibility Study. Draft Final Report to AID, February, 1978.

Hart, Henry C. Campus India: An Appraisal of American College Programs in India. East Lansing: Michigan State University, 1961.

Hartman, E. B., and Morgan, R. P. (eds.). Proceedings of Conference on University Education for Technology and Public Policy. St. Louis, Mo.: Washington University, 1976.

Hirschman, Albert O. Universities and Development Assistance Abroad. Washington, D.C.: American Council on Education, 1967.

Humphrey, Richard A. (ed.). Universities and Development Assistance Abroad. Washington, D.C.: American Council on Education, 1967.

IAESTE. International Association for the Exchange of Students for Technical Experience: Annual Report. Zurich: IAESTE, 1977.

The International Role of the University in the 1970's. Report of a conference sponsored by the University of Massachusetts at Amherst with the support of the International Council for Educational Development, May 17-19, 1973. Amherst: University of Massachusetts, 1973.

Jacobson, Robert L. "Community Colleges Seek a 'Global Perspective.'" The Chronicle of Higher Education (November 28, 1977): 5-6.

Julian, A., and Slattery, R. (eds.). Open Doors 1975/6 - 1976/7. New York: Institute of International Education, 1978.

Lavergne, Daly C. "A Role in International Education." Junior College Journal 38 (April 1968): 15-18.

Michigan State University. The International Programs of American Universities: An Inventory and Analysis. East Lansing: Michigan State University, 1958.

Moore, P. G. "International Education in the Seventies: Revolution or Turmoil on the Campus." International Educational and Cultural Exchange 6 (Summer 1970): 34-47.

National Research Council. Stewards for International Exchange: The Role of the National Research Council in the Senior Fulbright-Hays Program 1947-1975. Washington, D.C.: NAS, 1976.

O.A.S. Pan American Union. Inter-American University Cooperation: Survey of Cooperation in U.S. and Latin America. Washington, D.C.: OAS, 1968.

Organization for Economic Cooperation and Development. Aid to Education in Less Developed Countries: Note by the Secretariat. Paris: OECD, 1971.

Pearson, Lester B. Partners in Development: Report of the Commission on International Development. New York: Praeger, 1969.

Perkins, J. A. International Programs of U.S. Colleges and Universities: Priorities for the Seventies. New York: International Council for Educational Development, 1971.

Phillips, H. M. Educational Cooperation between Developed and Developing Countries. New York: Praeger, 1976.

_____. Planning Educational Assistance for the Second Development Decade. Paris: UNESCO/IIEP, 1973.

Richardson, John M., Jr. Partners in Development: An Analysis of AID: University Relations 1950-1966. East Lansing: Michigan State University Press, 1969.

Robinson, E. A. G., and Vaizey, J. E. (eds.). The Economics of Education. London: Macmillan, 1966.

Shields, J. J., Jr. Education in Community Development: Its Function in Technical Assistance. New York: Praeger, 1967.

Sinaver, Ernst M. The Role of Communication in International Training and Education: Overcoming Barriers to Understanding with the Developing Countries. New York: Praeger, 1967.

Singer, Derek S. "Developing Colleges for Developing Countries." Junior College Journal 39 (May 1969): 13-16.

Smith, Bruce L. Indonesian-American Cooperation in Higher Education. East Lansing: Michigan State University, 1960.

Spencer, Richard E., and Awe, Ruth. International Education Exchange: A Bibliography. New York: IIE, 1968.

Sutton, F. S.; Ward, P. C.; and Perkins, J. A. Internationalizing Higher Education: A United States Approach. New York: International Council for Educational Development, 1974.

Sutton, F. X. "Funding for International Education." New York: Ford Foundation, 1975.

Thompson, Kenneth W. "Higher Education for National Development. One Model for Technical Assistance," Occasional paper no. 5. New York: International Council for Educational Development, November 1972.

_____, and Fogel, Barbara R. Higher Education and Social Change: Promising Experiments in Developing Countries, vol. 1, Reports. New York: Praeger, 1976.

_____. Fogel, Barbara R.; and Donner, Helen E. Higher Education and Social Change: Promising Experiments in Developing Countries, vol. 1, Reports. New York: Praeger, 1976.

Tysse, Agnes. International Education: The American Experience. Metuchen, N.J.: Scarecrow Press, 1974.

U.S. Congress. House of Representatives. International Education: Hearings on H.R. 12451 and 12452. Washington, D.C.: U.S. Government Printing Office, 1966.

_____. House of Representatives. International Education: Past, Present, Problems and Prospects. Selected readings to Supplement H.R. 14643. Washington, D.C.: U.S. Government Printing Office, 1966.

_____. Senate. U.S. Foreign Service Scholarship Program. Washington, D.C.: U.S. Government Printing Office, 1971.

Ward, F. Champion (ed.). Education and Development Reconsidered: The Bellagio Conference Papers. New York: Praeger, 1974.

Weidner, Edward W. *The World Role of Universities.* New York: McGraw-Hill, 1962.

Witman, S. L. *Inter-Institutional Cooperation and International Education.* New York: Education and World Affairs, 1969.

Wood, Richard H. *U.S. Universities: Their Role in AID-Financed Technical Assistance Overseas.* New York: Education and World Affairs, 1968.

Woodcock, Lynda. "Influx of Foreign Students an Issue in Canada." *Chronicle of Higher Education,* March 13, 1978, p. 3.

World Congress of Comparative Education Societies. *Role and Rationale for Educational Aid to Developing Countries,* Proceedings. Ottawa, 1970.

ENGINEERING

Asimow, Morris, and McNown, John S. "Engineering Education and International Development." *Engineering Education,* November 1965, pp. 65-70.

Baranson, Jack. *Industrial Technologies for Developing Countries.* New York: Praeger, 1969.

Barrell, Lawrence L. "The Role of Engineering Technology Schools in International Development." Paper presented at Annual Meeting of American Society for Engineering Education, Vancouver, B.C., Canada, June 20, 1978.

Barry, Merton R. *AID-Wisconsin Engineering Project in India, Final Report.* Madison: University of Wisconsin, 1967.

_____. *Engineering Degree Development Program in Singapore, 1966-1967: Final Report to the Ford Foundation.* Madison, Wis.: University of Wisconsin, 1977.

Bernstein, Joel. "The U.S. Role in Poor Country Industrialization: Implications for American Engineering Schools." *TECHNOS* 1 (April - June 1972): 49-56.

Brembeck, Cole S. "Colleges of Education and Overseas Technical Assistance." *Comparative Education Review* XI(1): 87-97.

Brown, Gordon S., and Hoelscher, H. E. "Open Forum: Are We Mistraining Our Foreign Graduate Students?" Engineering Education 61 (December 1970): 272-275.

Caldwell, Lynton K. "Engineering and Development: International Initiatives." TECHNOS 6 (January-March, 1977): 4-15.

_____. "The Universities and International Technical Assistance: The Uses of Government Contracts." Journal of Higher Education 36 (May 1965): 266-273.

Colorado State University. Engineering Research Center. Water Management Research in Arid and Sub-Humid Lands of Less Developed Countries: Annual Reports 1969-1976. Fort Collins, Colo.: Colorado State University,

"Eastern Intercollegiate Conference on Industrialization of Underdeveloped Areas: Program, Conclusions, Recommendations." Sponsored by the Commission on Engineering Education at Tufts University, Medford, Massachusetts, May 1-2, 1964.

Eastman, Robert M. "An Experiment in International Engineering Education on a Regional Basis: The Middle East Technical University." TECHNOS 1 (October-December 1972): 13-20.

Education Development Center. "INELEC: The Institute." Newton, Mass.: Education Development Center, 1977.

_____. Kabul Afghan-American Program, 1963-1973. Newton, Mass.: EOS, 1973.

_____. Kanpur Indo-American Program, 1962-1972. Newton, Mass.: EOS, 1972.

Engelmann, Peter. "Engineering in Developing Countries." TECHNOS 1 (April-June 1972): 63-66.

English, J. M., and Collins, W. L. (eds.). Educating Engineers for World Development. Proceedings of a World Congress sponsored by the American Society for Engineering Education, International Division, June 10-12, 1975. Estes Park: ASEE, 1975.

Georgia Institute of Technology, Engineering Experiment Station. Employment Generation Through Stimulation of Small Industries: 211(d) Grant: Annual Reports, 1974-1978. Atlanta, Ga.: Georgia Institute of Technology.

Bibliography

_____. Analysis of Small Industry Development Methods and Techniques. Atlanta, Ga.: Georgia Institute of Technology, 1978.

_____. Fundacao Educacional Do Sul De Santa Catarina Activities (FESSC): Small-Scale Industry Grant. Atlanta, Ga.: Georgia Institute of Technology, 1977.

_____. Soong Jun University: Small-Scale Industry Grant. Atlanta, Ga.: Georgia Institute of Technology, 1977.

_____. Appropriate Technology Aspects of Solar Energy. Atlanta, Ga.: Georgia Institute of Technology, 1977.

_____. Industrialization Leading to the Master of Science Degree. Atlanta, Ga.: Georgia Institute of Technology, 1977.

_____. Stimulating the Growth of Small-Scale Industry: Final Report. Atlanta, Ga.: Georgia Institute of Technology, 1976.

_____. Financing Small-Scale Industry: The Tubarao Brazil Case. Atlanta, Ga.: Georgia Institute of Technology, 1976.

_____. Selected Aspects of Intermediate Technology. Atlanta, Ga.: Georgia Institute of Technology, 1976.

_____. Techniques and Methodologies for Stimulating Small-Scale Labor-Intensive Industries in Developing Countries: Proceedings. Atlanta, Ga.: Georgia Institute of Technology, 1975.

_____. Discussion Papers on the Problems of Science and Technology: Annual Symposium for AID. Atlanta, Ga.: Georgia Institute of Technology, 1975.

_____. An International Compilation of Small-Scale Industry Definitions. Atlanta, Ga.: Georgia Institute of Technology, 1975.

_____. Small-Scale Industry Development in Ecuador. Atlanta, Ga.: Georgia Institute of Technology, 1975.

_____. Finance and Small-Scale Industry Lending in the Philippines: 1967-1973. Atlanta, Ga.: Georgia Institute of Technology, 1975.

_____. Improving the Productivity of a Small Industry in Korea. Atlanta, Ga.: Georgia Institute of Technology, 1974.

_____. Engineering Experiment Station. <u>Small-Scale Industry Development in Paraguay</u>. Atlanta, Ga.: Georgia Institute of Technology, 197

_____. <u>Some Issues Related to the Impact of Micro-Development Projects</u>. Atlanta, Ga.: Georgia Institute of Technology, 1974.

Giral, J., and Morgan, R. P. <u>Appropriate Technology for Chemical Industries in Developing Economies</u>. Report of Foreign Area Fellowship Program Summer Research Training Project held at National Autonomous University of Mexico, July - August, 1972. St. Louis, Mo.: Washington University, 1972.

Glennan, T. Keith, and Sanders, Irwin T. <u>The Professional School and World Affairs: Report of the Task Force on Agriculture and Engineering</u>. New York: Education and World Affairs, 1967.

Goldschmidt, Victor W. "The Engineering Foreign Graduate Student: Some Statistics." <u>TECHNOS</u> 1 (April - June 1972): 39-42.

Goodman, Louis J., and Burian, Fredrich. "University-Community Interaction through Curriculum: A Case Study." <u>TECHNOS</u> 1 (October - December 1972): 5-12.

Grayson, L. P. <u>The Design of Engineering Curricula</u>. Paris: UNESCO, 1977.

Hazeltine, Barrett. "Strategies for Engineering Schools in Developing Countries." <u>TECHNOS</u> 6 (January-March, 1977): 16-26.

_____. "Advanced Engineering Curriculum Development in Zambia: Final Report to NSF for a SEED Grant." Providence, R.I.: Brown University, July 11, 1977.

_____. "Further Teaching Experiences in Zambia." <u>Engineering Education</u> 63 (April 1973): 537-538.

Henderson, Gregory. <u>Emigration of Highly-Skilled Manpower from the Developing Countries</u>. Research report no. 3. New York: UNITAR, 1970.

Hoelscher, H. E. "U.S. Engineering Education for Foreign Students." <u>TECHNOS</u> 1 (July-September 1972): 19-22.

International Council of Scientific Unions, Committee on Science and Technology in Developing Countries (COSTED). <u>Training of Engineers, Technologists and Technicians in Developing Countries</u>. Report prepared for UNESCO, Paris, 1975.

Lashmet, P.K. <u>Report on the Possible Use of VITA Requests as Educational Materials for Engineering School Curricula</u>. Schenectady: VITA, 1967.

Macagno, Enzo O. "Education of Exchange Engineering Students: Success, Impasse, Failure?" <u>TECHNOS</u> 1 (January-March 1972): 37-42.

Massachusetts Institute of Technology. <u>Technology Adaptation Program, Massachusetts Institute of Technology</u>. Cambridge, Mass.: MIT, 1976.

Mbele-Mbong, Samuel. "Discussion: U.S. Engineering Education for Foreign Graduate Students." <u>TECHNOS</u> 1 (July-September 1972): 26.

McNown, John S. "Discussion: U.S. Engineering Education for Foreign Students." <u>TECHNOS</u> 1 (July-September 1972): 23-25.

Miller, Charles L., and McGarry, Frederick J. "The M.I.T. Inter-American Program in Civil Engineering." Research report R64-36. Cambridge, Mass.: MIT, 1964.

Morgan, Robert P. "Technology and International Development: Is There A Role for U.S. Engineering Schools?" Paper presented at Annual AAAS Meeting, Washington, D.C., February 16, 1978.

_____. "Technology and International Development: New Directions Needed." <u>Chemical and Engineering News</u> 55 (November 1977): 31-39.

_____. "An International Development Technology Center." <u>Journal of Engineering Education</u> 60 (November 1969): 247-249.

_____. "International Directions for Engineering Education in the United States." Paper presented at the Conference on Engineering for International Development, Estes Park, Colorado, August 27 - September 1, 1967.

National Research Council. Board on Science and Technology for International Development. The Role of U.S. Engineering Schools in Development Assistance. Washington, D.C.: NAS, 1976.

National Science Foundation. Immigrant Scientists and Engineers in the United States: A Study of Characteristics and Attitudes. Washington, D.C.: U.S. Government Printing Office, 1973.

_____. "Scientists and Engineers from Abroad: Trends of the Past Decade, 1966-1975." Reviews of Data on Science Sources 28 (December, 1970).

OAS. Third Pan American Meeting on Graduate Engineering Education. December 9-13, 1969, St. Augustine, Trinidad and Tobago. Washington, D.C.: OAS, 1970.

Rao, K. N. "The Education and Training of Chemical Technicians." Paper presented at Commonwealth Conference on the Education and Training of Technicians, Huddersfield, England, 1966.

_____. "Technical Education in the Developing Countries." New York: Ford Foundation, 1965.

Rouse, Hunter. "Foreign Engineering Graduate Students: An Asset, A Responsibility, A Liability." TECHNOS 1 (January-March 1972): 43-45.

Seltzer, Norman, and Gannon, Joseph. Immigrant Scientists and Engineers in The United States. Final Report. Washington, D.C.: U.S. Government Printing Office, 1973.

Shen, Richard T. "Technical Assistance in Communities in Northeast Brazil by Woodson and Gilkeson (October-December 1972)." TECHNOS 3 (July-September 1974): 68-71.

Sherbourne, A. N. "Engineering Education and Development: Domestic and Overseas." TECHNOS 3 (January-March 1974): 5-24.

Sohns, Ernest R. "National Science Foundation Engineering Programs." TECHNOS 1 (April-June 1972): 57-61.

Soule, Theodore N. The Role of U.S. Engineering and Engineering Technology Colleges in International Development. M.S. thesis, Department of Technology and Human Affairs, Washington University, St. Louis, Missouri, 1979 (forthcoming).

_____, and Morgan. R. P. "Summary of ASEE International Activities Survey." Draft Report, August 1978.

Torda, T. Paul. "New Directions for Engineering Education: Suggestions for Developing Countries." Impact of Science on Society 27 (1977): 357-367.

UNESCO. International Conference on the Education and Training of Engineers and Higher Technicians, New Delhi, April 20-26, 1976. Paris: UNESCO, 1976.

University of Wisconsin. Engineering Experiment Station. Engineering Degree Development Program-Singapore: Final Report to the Ford Foundation, 1966-1976. Madison, Wis.: University of Wisconsin, 1977.

_____. AID-Wisconsin Engineering Education Project in India: Final Report. Madison, Wis.: University of Wisconsin, 1967.

Vaczek, L. "IIT's Approach to Engineering Education." Change, June 1973, pp. 27-30.

Wentworth Institute of Technology. "Summary Statement on Foreign Projects." (Boston: June 1978).

Woodson, Thomas T. and Gilkeson, Murray M. "Technical Assistance in Communities in Northeast Brazil, 1962-1968." TECHNOS 1 (October-December 1972): 27-37.

AGRICULTURE

Agency for International Development. Baseline Studies: A Conceptual Model for Analysis of Current Capacities and Development Needs of LDC Agricultural Research, Education and Extension Systems. Washington, D.C.: AID, 1978.

_____. Guideline for the Preparation and Submission of Proposals for Matching Formula Title XII University Strengthening Grants. Washington, D.C.: AID, 1978.

_____. Guidelines for the Role and Function of the Joint Committee on Agricultural Development. Washington, D.C.: AID, 1978.

_____. Report to the Congress on Title XII: Famine Prevention and Freedom From Hunger of the Foreign Assistance Act of 1961 as Amended. Washington, D.C.: AID, 1978.

_____, BIFAD. BIFAD: The First Year - A Progress Report. Washington, D.C.: AID, 1977.

Agency for International Development/India Office of Agricultural Development. Proceedings of the Fourteenth Annual Agricultural Conference. Bangalore, India, 1971.

Baker, C. B. "U.S. Perspectives on World Food Problems." Illinois Agricultural Economics 17 (July 1977): 1-6.

Baker, Marvel L. The Agricultural Universities in India. New Delhi: AID, 1964.

Berg, R. J.; Kitchell, R. E.; Schwab, P.; and Swan, A. Report of Review Team on Institutional Grants Program (211d). Washington, D.C.: AID, 1978.

Blase, Melvin G. Institutions in Agricultural Development. Ames, Iowa: Iowa State University, 1971.

Buddemeier, W. D. "Famine Prevention and Freedom from Hunger: Challenge and Responsibilities." Illinois Agricultural Economics 17 (July 1977): 13-16.

Castillo, Leopoldo S. "Graduate Training in the United State as Seen by a National from a Developing Country." Journal of Dairy Science 51.

Chandrasekhar, S. American Aid and India's Economic Development. New York: Praeger, 1965.

Collom, J.; Matteson, H.; and Zuidema, L. "The University Programming Role in AID Participant Training: Its Conduct and Support." Position paper of The Association of U.S. University Directors of International Agricultural Programs, June 1978.

Cromwell, Charles F., Jr.; Hagan, A. R.; Kroth, E. M.; Nolan, M. F. "An Assessment of the Agricultural Potential of Central Tunisia: Evaluations and Recommendations." Columbia, Mo.: University of Missouri, 1978.

Cummings, Ralph W., Jr. Food Crops in the Low-Income Countries: The State of Present and Expected Agricultural Research and Technology. New York: Rockefeller Foundation, 1976.

Development Associates, Inc. "A Seven Country Survey of the Roles of Women in Rural Development." Washington, D.C.: AID, 1974.

Dyck, Robert G., and Albritton, Robert B. (eds.). Summary Proceedings of the Conference on International Development: A Working Conference on University Action. Blacksburg, Va.: University International Programs, Virginia Polytechnic Institute and State University, 1976.

Ensminger, Douglas. Statement submitted to St. Louis Forum, 1979 U.N. Conference on Science and Technology for Development, U.S. Department of State and the National Research Council, January 23, 1978.

Esman, Milton J., and Blaise, Hans C. Institution Building Research: The Guiding Concepts. Pittsburgh, Pa.: University of Pittsburgh, 1966.

Fisher, John L. "Summary Report on the Conference of Women and Food." Unpublished. Conference sponsored by Consortium for International Development, Tucson, Arizona, January 9-11, 1978.

Freeh, LaVern A. "University of Minnesota, Suggested Policy and Guidelines Statement Relating to the University's World-Wide Mission and Responsibility." Draft Proposal, March 20, 1978.

Gautan, O. P.; Patel, J. S.; Sutton, T. S.; and Thompson, W. N. An Assessment of Progress to 1970: Punjab Agricultural University. New Delhi: Indian Council on Agricultural Research, 1970.

Glennan, T. Keith, and Sanders, Irwin T. The Professional School and World Affairs: Report of the Task Force on Agriculture and Engineering. New York: Education and World Affairs, 1967.

Guither, Harold D. "The Famine Prevention and Freedom from Hunger Amendment: Issues and Compromises in International Development Policymaking." Illinois Agricultural Economics 17 (July 1977): 7-12.

Hagan, Albert R. The Agricultural Development of Nepal. Special Report 189. Columbia, Mo.: University of Missouri, 1976.

Kriesberg, Martin. International Organizations and Agricultural Development. Washington, D.C.: U.S. Department of Agriculture, 1977.

Land Tenure Center. The Land Tenure Center Annual Report, 1976-1977. Madison, Wis.: University of Wisconsin, 1978.

_____. "A Brief Description of the Land Tenure Center Program, University of Wisconsin." Mimeographed, May 1977.

Lappe, Frances Moore, and Collins, Joseph. Food First: Beyond the Myth of Scarcity. Boston: Houghton Mifflin, 1977.

Larson, Vernon C. "Land-Grant Institutions as Change Agents." Paper presented at the Association of U.S. University Directors of International Agricultural Programs, Logan, Utah, June 20-23, 1978.

_____, and Medlin, Roger C. Sixteen Years in India: A Terminal Report. Manhattan, Kansas: International Agricultural Programs, Kansas State University, 1973.

Massey, H. F. "Suggestions of an Agronomist." Statement submitted to St. Louis Forum, 1979 U.N. Conference on Science and Technology for Development, U.S. Department of State and the National Research Council, January 23, 1978.

McDermott, J. K. "Propagation of the Land-Grant College, The Purdue-Vicosa Experience." Paper presented for the Midwest Council of Latin American Studies Association. East Lansing, Mich.: Michigan State University, 1966.

Michigan State University. "Highlights - International Activities." College of Agriculture and Natural Resources (June 1977-June 1978). Mimeographed, June 5, 1978.

Bibliography

Moseman, Albert H. Building Agricultural Research Systems in the Developing Nations. New York: Agricultural Development Council, 1970.

MUCIA. Higher Education in Agriculture in Nepal: The Report of a Preliminary Feasibility Study. Bloomington, Ind.: MUCIA, 1972.

National Research Council. Commission on International Relations. World Food and Nutrition Study. 5 vols. Washington, D.C.: NAS, 1977.

Nichols, Buford L. "Coordination of International Development with Relevant Agricultural and Nutrition Policies." Statement submitted to the St. Louis Forum, 1979 U.N. Conference on Science and Technology for Development, U.S. Department of State and the National Research Council, January 23, 1978.

Olver, E. F. Agricultural Engineering Education in Developing Countries. Urbana-Champaign, Ill.: University of Illinois, 1970.

Perez, Eduardo A. The Role of U.S. Universities in International Agricultural Development. Master's Thesis, Department of Technology and Human Affairs, Washington University, St. Louis, Missouri, January 1979.

Potter, Harry R. Criteria of Progress and Impacts of Technical Assistance Projects in Agriculture. Lafayette, Ind.: Purdue University, 1968.

Propp, Kathleen M.; Guither, H. D.; Regnien, E. H.; and Thompson, W. N. AID-University Rural Development Contracts 1951-1966. Urbana, Ill.: University of Illinois, 1968.

──────. The Establishment of Agricultural Universities in India: A Case Study of the Role of USAID-U.S. University Technical Assistance. Urbana, Ill.: University of Illinois College of Agriculture, 1968.

Purdue University, Committee on Institutional Cooperation. Building Institutions to Serve Agriculture. Lafayette, Ind.: Purdue University, 1968.

Read, Hadley. Partners with India: Building Agricultural Universities. Urbana, Ill.: University of Illinois, 1974.

Renne, Roland R. *Agricultural Universities in India*. Prepared for the Council of U.S. Universities for Rural Development in India. Urbana-Champaign, Ill.: University of Illinois, 1974.

Richardson, John M., Jr. *Partners in Development: An Analysis of AID-University Relations: 1950-1966*. East Lansing, Mich.: Michigan State University Press, 1967.

Rockefeller Foundation. *Strategy for the Conquest of Hunger*. Proceedings of a symposium. New York: Rockefeller Foundation, 1968.

Roskelley, R. W., and Rigney, J. A. *Measuring Institutional Maturity in the Development of Indigenous Agricultural Universities*. Utah State University/North Carolina State University, 1968.

Smuckler, Ralph H. "U.S. Cooperation with Emerging Centers for Science and Technology in Low and Middle Income Countries Including Regional Aspects." Paper presented at the Association of U.S. University Directors of International Agricultural Programs, Logan, Utah, June 20-23, 1978 (draft).

Streeter, Carroll P. *A Partnership to Improve Food Production in India*. New York: Rockefeller Foundation, 1969.

Thompson, William N.; Guither, H. D.; Regnien, E. H.; and Thompson, W. N. *AID-University Rural Development Contracts and U.S. Universities*. AID. Urbana, Ill.: University of Illinois, 1968.

Tinker, Irene, and Bramsen, Michele Bo (eds.). *Women in World Development*. Washington, D.C.: Overseas Development Council, 1976.

U.S. Congress, House. *Agricultural Research and Development: Background Papers*. Washington, D.C.: U.S. Government Printing Office, 1975.

University of Illinois, College of Agriculture. *Development of Improved Varieties of Soybeans Supporting Cultural and Marketing Practices for Production in the Tropics and Information Delivery Systems*, Annual Report. Urbana-Champaign, Ill.: University of Illinois, 1976.

_____. *Development of Improved Varieties of Soybeans*. Final Report. Urbana-Champaign, Ill.: University of Illinois, 1976.

———. The Sri Lanka Soybean Development Program. Urbana-Champaign, Ill.: University of Illinois, 1976.

University of Illinois/University of Puerto Rico. "International Soybean Program (INTSOY): Brief Outline Summary of Progress, 1973-1978." Urbana-Champaign, Ill.: University of Illinois, 1978.

Warnken, Philip F. Strategies for Technical Assistance. Columbia, Mo.: Department of Agricultural Economics, University of Missouri, 1968.

———. The Agricultural Development of Nicaragua. Special Report 168. Columbia, Mo.: University of Missouri, 1975.

Watts, Lowell H. "Linkage Between Science and Technology Bases and Agricultural Producers in LDC's." Paper presented at the Association of U.S. University Directors of International Agricultural Programs, Logan, Utah, June 20-23, 1978.

Wayt, William A. AID, Agriculture, and Africa: A Perspective on University Contract Projects. Columbus, Ohio: Ohio State University, 1968.

Whitaker, Morris D., and Boyd-Wennergren, E. "U.S. Universities and the World Food Problem." Science 194 (October 29, 1975).

White, T. Kelly. "Science and Technology: Institutional Development and the U.S. University." Statement submitted to St. Louis Forum, 1979 U.N. Conference on Science and Technology for Development, U.S. Department of State and the National Research Council, January 23, 1978.

Wilkening, Walter T. "Review of the University of Missouri India Programs: 1952-1972." Mimeographed. Columbia, Mo.: University of Missouri College of Agriculture, 1973.

Williams, T. T. "Strategies for Science and Technology Transfer for the Benefit of Disadvantaged Populations." Paper presented at the Meeting of the Association of U.S. University Directors of International Agricultural Programs, Logan, Utah, June 20-23, 1978.

Wisconsin University. Land Tenure Center. Annual Report, 1976/77. Madison, Wis.: University of Wisconsin, 1978.

Zuidema, L. W. "Key Elements in U.S. Training Programs for International Agriculturalists." Cornell International Agriculture. Mimeograph 51. Ithaca, N.Y.: Cornell University, 1975.

SCIENCE

Agency for International Development. Policies and Programs in Selected Areas of Science and Technology. Washington, D.C.: AID, 1973.

Atelsek, Frank J., and Gomberg, Irene L. International Scientific Activities at Selected Institutions, 1975-76 and 1976-77. Washington, D.C.: American Council on Education, 1978.

Baez, A. V. "Innovation in Science Education: Worldwide." Paris: UNESCO, 1976.

Blackett, P. M. A. "Science and Technology in an Unequal World." Development Digest 8 (July 1970): 73-77.

Boffey, Philip M. "Korean Science Institute: A Model for Developing Nations?" Development Digest 8 (July 1970): 73-77.

Callen, Earl, and Scadron, Michael. "The Physics Interviewing Project: A Tour of Interviews in Asia." Science 200 (June 2, 1978): 1018-1022.

Carleton, Robert H. The NSTA Story: A History of Ideas, Commitments and Actions 1944/1974. Washington, D.C.: National Science Teachers Association, 1976.

CIMMYT. CIMMYT. Mexico: International Maize and Wheat Improvement Center [n.d.].

_____. This is CIMMYT. Mexico. International Maize and Wheat Improvement Center, 1974.

Copeland, W. A. "Iran's Pahlavi University: A Decade of Cooperation with the University of Pennsylvania." International Educational and Cultural Exchange 7 (Summer 1971): 27-33.

Dedijer, Stevan. "Underdeveloped Science in Underdeveloped Countries." Minerva 2 (Autumn 1963): 61-81.

Bibliography

Denver Research Institute. New U.S. Initiatives in International Science and Technology. Report on the Workshop in Keystone, Colorado, April 13-16, 1977. Denver: Denver Research Institute, 1977.

Djerassi, Carl. "A Modest Proposal for Increased North-South Interaction Among Scientists." Bulletin of the Atomic Scientists 32 (Feburary 1976): 56-60.

_____. "A High Priority: Research Centers in Developing Nations." Bulletin of the Atomic Scientists 24 (January 1968): 22-27.

Drew, David E. Science Development: An Evaluation Study. Washington, D.C.: National Board on Graduate Education, 1975.

Dvorin, Eugene. "The Chile-California Experiment." Bulletin of the Atomic Scientists 21 (November 1965): 35-38.

Erk, L. Paper presented at Workshop, "A Seminar in Training for Development," sponsored by the AID/NAFSA Liaison Committee, July 14, 1977.

Greene, Michael. Physics in Latin America: Peru and Chile. College Park: University of Maryland, 1971.

Grobman, Arnold B. "Evaluation Abstracts: Factors Influencing International Curricular Diffusion." Studies in Educational Evaluation 2 (Winter 1976): 227-232.

Jones, Graham. The Role of Science and Technology in Developing Countries. Oxford: Oxford University Press, 1971.

Kidd, Charles V. "An Evaluation of the NSF-SEED Program: Scientists and Engineers in Economic Development." Report to National Science Foundation, November 1977.

King, Alexander. "UNESCO's First Ten Years." New Scientist 2 (1957): 15.

Kovda, Victor A. "Search for a U.N. Science Policy." Bulletin of the Atomic Scientists 24 (March 1968): 12-16.

Lopes, J. Leite. "Science for Development: A View from Latin America." Bulletin of the Atomic Scientists 22 (September 1966): 7.

Maybury, Robert H. Technical Assistance and Innovation in Science Education. New York: Wiley, 1975.

Meinwald, J.; Prestwich, G. D.; Nakanishi, K.; and Kubo, I. "Chemical Ecology: Studies from East Africa - ICIPE Laboratories in Nairobi." Science 199 (March 1978): 1167-1173.

Moravcsik, Michael J. "Science and the Developing Countries." A contribution to the U.S. Country Paper for the UNCSTED Conference, October 1977.

_____. Science Development: The Building of Science in Less Developed Countries. Bloomington, Ind.: PASITAM, 1975.

_____. "Foreign Students in the Natural Sciences: A Growing Challenge." International Education and Cultural Exchange 9 (Summer 1973): 45-56.

_____. "The Physics Interviewing Project." International Educational and Cultural Exchange 8 (Summer 1972): 16-22.

_____. "The Research Institute and Scientific Aid." International Development Review 13 (1971): 17-22.

_____. "Some Modest Proposals." Minerva 9 (1971): 54-65.

_____. "Some Practical Suggestions for the Improvement of Science in Developing Countries." Minerva IV (Spring 1966): 381-390.

_____. "Technical Assistance and Fundamental Scientific Research in Underdeveloped Countries." Minerva 2 (1964): 197-209.

Morrill, J. L., and Rao, K. N. "Science as Inquiry: Improvement of Secondary Education Science and Mathematics in Latin America." Discussion notes for Representatives Meeting, Latin America and Caribbean Program, Ford Foundation, November 1965.

Naraine, M. "Science for Progress: OAS Involvement in Science in Latin America." Nature 267 (May 1977): 298-299.

National Science Board. Science Indicators: 1976. Washington, D.C.: National Science Board, 1977.

National Science Foundation. "Links That Connect a Hemisphere." Mosaic 8 (November/December 1977): 44-50.

Bibliography

_____. "Small Projects; Large Impacts." Mosaic 8 (November/December 1977): 38-43.

_____. National Science Foundation Grants and Awards, Annuals, 1968-1976. Washington, D.C.: U.S. Government Printing Office.

Osborne, D. "The Use and Promotion of Science in Developing Countries." Minerva 9 (January 1971): 44-53.

Parthasarathi, Ashok. "Sociology of Science in Developing Countries: The Indian Experience." Development Digest 8 (July 1970): 78-82.

Pines, David. "On Building U.S. and World Science Through International Scientific Exchange." New Initiatives in International Science and Technology Workshop Reports, Denver Research Institute, 1977.

Rao, K. Nagaraja. "University Based Science and Technology for Development: New Patterns of International Aid." Impact of Science on Society 28 (1978/2): 117-125.

_____. "Collaboration in Science and Technology: An Inter-American Perspective." Issues in International Educational Report no. 4. New York: Institute of International Education, 1975.

_____. "Innovation, Adaptation and Diffusion of Reforms in Science Teaching in Latin America." Paper prepared for AAAS Meeting, Section on Education, Washington, D.C., December 27, 1970.

_____. "Financing Graduate Education and Research in Science and Engineering in Latin America." Paper presented at the First Pan-American Conference on Postgraduate Education in Engineering, Caracas, August 19-24, 1967.

Renetzky, Alvin, and Flynn, Barbara J. (eds.). NSF Factbook: Guide to National Science Foundation Programs and Activities. Orange, N.J.: Academic Media, 1971.

Sable, Martin H. Master Directory for Latin America. Los Angeles: University of California, Latin American Center, 1965.

Shils, Edward. Criteria for Scientific Development: Public Policy and National Goals. Cambridge, Mass.: MIT Press, 1968.

Skolnikoff, Eugene B. *Science, Technology, and American Foreign Policy.* Cambridge, Mass.: MIT Press, 1967.

Szmant, H. H. "Foreign Aid Support of Science and Economic Growth: Support of Chemistry in Latin America." *Science* 199 (March 1978): 1173-1182.

UNESCO. "Quantification of Scientific and Technological Activities Related to Development." Paris: UNESCO, 1976.

_____. *Program for Natural Sciences and Their Application to Development.* Paris: UNESCO, 1973.

_____. *World Summary of Statistics on Science and Technology.* Paris: UNESCO, 1970.

UNIDO. "The Need for Research Within Developing Countries." *Development Digest* 8 (July 1970): 64-68.

U.S. Congress. House. *Science, Technology and American Diplomacy.* Washington, D.C.: U.S. Government Printing Office, 1977.

_____. *International Activities of the Energy Research and Development Administration,* May 12, 1977. Washington, D.C.: U.S. Government Printing Office, 1977.

_____. *1977 National Science Foundation Authorization.* Washington, D.C.: U.S. Government Printing Office, 1976.

_____. *U.S. Scientists Abroad: An Examination of Major Programs for Non Governmental Scientific Exchange.* Washington, D.C.: U.S. Government Printing Office, 1974.

_____. *International Science Policy.* Compilation of papers presented for the 12th Meeting of the Panel on Science and Technology. Washington, D.C.: U.S. Government Printing Office, 1971.

_____. *International Science Policy: Proceedings of the 12th Meeting of the Panel on Science and Technology.* Washington, D.C.: U.S. Government Printing Office, 1971.

_____. Panel on Science and Technology. *International Science Policy,* January 26-28, 1971. Washington, D.C.: U.S. Government Printing Office, 1971.

_____. Policy Issues in Science and Technology. Washington, D. C.: U.S. Government Printing Office, 1968.

_____. Science, Technology and Public Policy During the 89th Congress. Washington, D.C.: U.S. Government Printing Office, 1967.

Weiss, Charles, Jr. "A Biophysicist Looks at Science in Africa." Development Digest 8 (July 1970): 55-63.

White, T. Kelley. "Science and Technology, International Development and the U.S. University." Paper prepared for the St. Louis Forum for the 1979 UNCSTED, January 23, 1978.

BRAIN DRAIN

Adams, Walter (ed.). The Brain Drain. New York: Macmillan, 1968.

Baldwin, George B. "Brain Drain or Overflow?" Foreign Affairs 48 (January 1970): 358-372.

Blume, Stewart. "Brain Drain: A Look at the Literature." Universities Quarterly 22 (June 1968): 281-290.

Chorofas, D. N. The Knowledge Revolution: An Analysis of the International Brain Market. New York: McGraw-Hill Book Co., 1970.

Committee on the International Migration of Talent. The International Migration of High-Level Manpower. New York: Praeger, 1970.

_____. Modernization and the Migration of Talent. New York: Education and World Affairs, 1970.

_____. Selected Publications and Research Related to the International Migration of Professional Manpower. New York: Education and World Affairs, 1968.

"Brain Drain in Developing Countries: Bibliography." Current Literature on Science of Science 6 (January 1977): 5-16.

Dedijer, Steven. "Migration of Scientists: A World-Wide Phenomenon and Problem." Nature 201 (1964): 964.

_____, and Svenningson, L. Brain Drain and Brain Gain. Lund, Sweden: Research Policy Program, University of Lund, 1967.

Glaser, William A. The Brain Drain: Emigration and Return. New York: Pergamon Press, 1978.

_____. The Brain Drain and Study Abroad. New York: UNITAR, 1974.

Godfrey, E. M. "The Brain Drain from Low-Income Countries." Journal of Development Studies 6 (April 1970): 235-247.

Henderson, Gregory. Emigration of Highly-Skilled Manpower from the Developing Countries. New York: UNITAR, 1970.

Johnson, Harry G. "The Economics of the 'Brain Drain': The Canadian Case." Minerva 3 (Spring 1965): 299-311.

Julian, A., and Slattery, R. (eds.). Open Doors 1975/6-1976/7. New York: Institute of International Education, 1978.

Kannappan, Subbiat. "The Brain Drain and Developing Countries." International Labour Review 98 (July 1968): 1-26.

Kidd, Charles. "Manpower Policies for the Use of Science and Technology for Development." Washington, D.C.: George Washington University, October 1978.

Library of Congress. Congressional Research Service. Brain Drain: A Study of the Persistent Issue of International Scientific Mobility. Washington, D.C.: U.S. Government Printing Office, 1974.

McKnight, Allen S. Scientists Abroad: A Study of the International Movement of Persons in Science and Technology. Paris: UNESCO, 1971.

Moravcsik, Michael J. "On the Brain Drain in the Philippines." Bulletin of the Atomic Scientists 27 (1971): 36.

Muriel, Amador. "Brain Drain in the Philippines." Bulletin of the Atomic Scientists 26 (1970): 38.

Bibliography

Myers, Robert G. Education and Emigration: Study Abroad and the Migration of Human Resources. New York: McKay, 1972.

_____. "Comments on the State of Research: 'Brain Drains' and 'Brain Gains.'" International Development Review 9 (December 1967): 4-9.

_____. "The 'Brain Drain' and Foreign Student Nonreturn: Fact and Fallacy in Definitions and Measurements." Exchange (Spring 1967): 63-73.

Psacharopoulos, George. "On Some Positive Aspects of the Economics of the Brain Drain." Minerva 9 (April 1971): 231-242.

Seers, Dudley. The Brain Drain from Poor Countries. Geneva: International Institute for Labour Studies, 1968.

UNITAR. The Brain Drain from Five Developing Countries: Cameroon, Colombia, Lebanon, The Philippines, Trinidad and Tobago. Research report no. 5. New York: UNITAR, 1971.

United Nations. General Assembly. Outflow of Trained Professionals and Technical Personnel at All Levels from the Developing to the Developed Countries, Its Causes, Its Consequences, and Practical Remedies for the Problems Resulting from It. New York: United Nations, 1968.

U.S. Congress. House of Representatives. Brain Drain: A Study of the Persistent Issue of International Scientific Mobility. Washington, D.C.: U.S. Government Printing Office, 1974.

_____. Scientific Brain Drain from the Developing Countries. Washington, D.C.: U.S. Government Printing Office, 1968.

_____. The Brain Drain of Scientists, Engineers and Physicians from the Developing Countries into the United States. Hearing before a Subcommittee on Government Operations. Washington, D.C.: U.S. Government Printing Office, 1968.

Watanabe, Susumu. "The Brain Drain from Developing to Developed Countries." International Labor Review 99 (1969): 401-433.

Index

ACTION, 99
Adams, Richard N., 102
Adelman, Irma, 225
Agency for International Development (AID), 1-4, 5, 9-12, 15, 16, 19-21, 22-24, 24-28, 31-33, 36-37, 45-46, 49-50, 51, 57-58, 60-61, 66, 69-70, 71-75, 80-83, 84, 89-90, 92, 94-98, 99-100, 105-107, 111-112, 113-118, 122-127, 130-131, 135-136, 140, 142-143, 153, 155, 184-185, 189-190, 193, 201-202, 204-206, 209, 213-214, 215-216, 219-223, 225, 228-229, 236-237, 238, 240-242, 243-246, 249, 251, 252, 257-258, 260-261, 263, 264, 266
   agricultural school involvement, 69-70, 71-75, 80-83, 84, 89-90, 92, 94-98, 99-100, 105-107, 111-112, 113-118, 122-127, 130-131, 135-136, 251, 252
   AID and the universities, 105-107
   AID-GAO exchange on 211(d) program, 9, 184, 237-238

Agency for International Development (cont'd)
   AID-graduate countries, 122-125, 228-229
   CIC-AID rural development research project, 112-117
   Development Support Bureau, 2-3
   engineering school involvement, 24-28, 31-33, 36-37, 45-46, 49-50, 57-58, 60-61, 66, 251
   Georgia Tech 211(d) program review, 36-37
   legislative mandate, 1-4, 5, 9-12, 15, 16, 19-21
   Office of Science and Technology (OST), 2, 11, 28
   policy issues and options, 219-223, 225, 228-229, 236-237, 238, 240-242, 243-246, 257-258
   science involvements, 140, 142-143, 153, 155
   spending on S&T for development, 219
   Technical Assistance Bureau, 2

Agency for International Development (cont'd)
  Title XII program (see International Development and Food Assistance Act)
  211(d) program (see Foreign Assistance Act)
  U. S. universities, future roles for, 184-185, 189-190, 193, 201-202, 204-206, 209, 213-214, 215-216
Agricultural Development Council, 94
Agricultural Trade and Development Assistance Act (PL 480), 6, 84,
Agriculture, international activities of U. S. universities in, 75-136, 251-252
  Concluding remarks, 136-137
  Current issues and questions
    AID-graduate countries, 122-123
    education in U. S. universities, relevance to foreign students, 128-130
    1890 (predomintly black) land-grant institutions, 135-136
    land-grant university approach, relevance of, 121-122
    "New Directions" and U. S. university involvement, 118-120
    program evaluation, 123-125
    science and technology, contribution of, 132-135
    states, support from, 126-128

Agriculture, international activities of U. S. universities in, (cont'd)
  Current issues and questions (cont'd)
    Title XII, implementation of, 125-126
    U. S. university involvement, rationale for, 130-132
    women, impact on, 120-121
  Early (pre-1958) involvement, 70-72
  Evaluations of U. S. university involvements
    analysis, 116-118
    CIC-AID Rural Development Research Project, 111-115
    Education and World Affairs, 105-111
    Gardner Report ("AID and the Universities"), 105-108
    Institute of Research on Overseas Programs, 101-104
    "The University look Abroad", 108-111
    World Role of Universities, (Weidner Report), 103-104
  Examples of programs
    early programs, 88-89
    International Soybean Program (INTSOY), 95-99, 251-259
    involvement in India, 89-92, 251
    Land Tenure Center, U. of Wisconsin, 92-94, 251
    Michigan State University, 99-100, 251
    University of Illinois, 95, 251, 259
  Historical background, 69-70
  Later (pre-1966) involvement, 71-73

Index

Agriculture, international activities of U.S. universities in, (cont'd)
  More recent involvements
    foreign students in U.S., 80-81
    research, relation to CGIAR, 78-80
    research and teaching activities, 77
    sources of funding, 84-88
    Title XII, 81-84
    211(d) grants, 81
    USDA-supported activities, 84
  Other types of involvements, 75-76
  Rural development contracts, 73-75
  Summary, 251-252
Agriculture, U.S. Department of, (USDA), 7, 12, 22, 70, 78, 84-87, 99-100, 114, 193, 219-220, 246, 249, 260-261, 265
  Tropical and Sub-Tropical Research and Training Program (TSRTP), 78, 85-86
  Office of Foreign Agricultural Relations (OFAR), 114
Allahabad Agricultural Institute, India, 95
American Association for the Advancement of Science (AAAS), 171, 203
American Council on Education, 77
American Society of Agricultural Engineers, 95
American Society of Agronomy, 125
American Society for Engineering Education (ASEE), 52, 56, 58, 65, 67, 203, 250

American Society for Engineering Education (ASEE), (cont'd)
  International Engineering Program Activities Survey, 56, 65
American Universities Consortium, 51-52
Anti-Nuclear Proliferation Act, 8, 22, 189, 192, 246
Appropriate technology, 226-229, 250, 260-261
Appropriate Technology International (ATI), 201-202, 215, 217, 255, 261
Arid Lands Agricultural Development Program (ALAD), 81
Arizona, University of, 28, 70, 137, 150
Arkansas, University of, 70
Asian Institute of Technology, (AIT), 24, 51, 137, 195, 261
Asian Vegetable Research and Development Center (AVRDC), Taiwan, 81, 98, 139
Asimow, Prof. Morris, 34, 255
Association for the Advancement of Appropriate Technology in Developing Countries (AAATDC), 171
Association of U.S. University Directors of International Agricultural Programs (AUSUDIAP), 118-119

Bandung Institute of Technology, Indonesia, 52
Barker, C., 182
Basic Needs ("New Directions"), 1, 2, 19-20, 93, 118, 214, 217, 225, 228, 249, 252
Bell, David, 114
Bengal Engineering College, India, 50

Berlinquet, Louis, 206
Bernstein, Joel, 204
Berry, Merton R., 60
Bhabba, H., 171
Billings, Bruce, 206
Birla Institute of Technology, India, 27
Blase, Melvin, 132, 195, 298
Board for International Food and Agricultural Development (BIFAD), 11-12, 84, 101, 126, 192, 243, 251
Board on International Scientific Exchange (BISE), 156
Board on Science and Technology for International Development (BOSTID), 147-148, 151, 155, 204
Borlaug, Norman, 75
Boyle, Neil, 34
"Brain drain", 18-19, 62
Brazilian National Research Council (CNPq), 148, 172, 193
Brookhaven National Lab, 202
Brookings Institution, 5, 7, 23, 96, 204, 215
Buddemeier, W. D., 130, 210

California Institute of Technology, 148
California State University at Los Angeles, 34-35
California, University of, 102, 141, 144, 149, 163
CARE, Inc., 86, 96
Carnegie Corporation of New York, 70, 102, 104
Carter, President Jimmy, 5, 205, 216, 229
Case, Senator Clifford, 5, 221
Ceará, University of, Brazil, 33

Center for Cultural and Technical Interchange (East-West Center), Hawaii, 18, 52, 71, 158, 176
Central Food Technological Research Institute (CFTRI), Mysore, India, 140
Centro Brasiliero de Pesquisas Fisicas, 148
Chile, Universidad de, 141, 150, 168
CIC-AID Rural Development Research Project, 111-119
Claremont University, 34
Collaborative Research Support Grants, 15
Colorado State University, 27, 51, 150
Colorado, University of, 144
Columbia University, 138, 150
Commerce, U. S. Department of, 8, 246
Committee for International Exchange of Scholars, 18
Committee on Institutional Cooperation (CIC), 111
Conclusions, 259-263
Conference on International Rural Development, 114
Consultative Group for International Agricultural Research (CGIAR), 80
Cooperative Research and Development, 27, 150-153, 250
Cornell University, 27, 51, 69, 70, 100, 109, 138
Costa Rica, Ministry of Agriculture, 80
Council of United States' Universities for Rural Development in India (CUSURDI), 95
Council on Science and Technology for Development, 195

Index 387

Criteria for successful U.S. university involvements, 207-208, 255-256
   in science, 172-173
Cumberland, Charles C., 102

Dahl, Prof. Norman, 48
Definitions, 10, 24, 52, 77, 137
   engineering, 52
   engineering technology, 52
   institution building, 24, 77
   institution strengthening, 11, 77
   resource base development, 10
   science, 137
Denver Research Institute (DRI), 202, 219
Department of Agriculture, U.S. (USDA), 7, 12, 22, 69, 78, 84-86, 99, 114, 193, 220, 245
   TSRTP, 78, 84
   OFAR, 114
Department of Commerce (DOC), U.S., 8, 246
Department of Energy (DOE), U.S., 8, 21, 193, 201, 220, 245, 249, 260-261, 265
Department of Health, Education, and Welfare (DHEW), U.S., 17, 246
   Office of Education, 8, 17
Department of the Interior (DOI), U.S., 8, 246
Department of Labor (DOL), U.S., 8, 246
Direccion Tecnica Interamericana Cooperative de Agricultura de Chile (DTICA), 102
Djerassi, Carl, 138, 148, 170-171, 181
Dunwoody Industrial Institute, 53

Eagleton, Senator Thomas F., 10
East-West Center (Center for Cultural and Technical Interchange), 18, 52, 71, 158, 176
Economic Cooperation Administration (ECA), 69-70, 114
Education and Training, 28-32, 80-81, 128-130, 156-163, 250
Education and World Affairs (EWA), 104-111
Education Development Center (EDC), 24, 31, 38, 42, 45, 46, 53, 179
Eilers, William, 91, 205
Energy, Department of, U.S., (DOE), 8, 22, 193, 201, 220, 245, 249, 260-261, 265
Engineering, international activities of U.S. universities in, 24-67, 250-251
   ASEE survey of U.S. engineering school international involvements, 54-55
   case studies of U.S. engineering school involvement, 33-53
      Georgia Tech Small Industries Program, 36-40, 250, 259
      AID review, 37
      evaluation, 40
      objectives, program elements, 36-37
      Kabul Afghan-American Program, 38, 41-45, 250
      analysis of, 43-45
      objectives and description, 41-42
      participant training, 42-45
      Kanpur Indo-American Program, 45-49, 250
      analysis of, 48-49
      background and objectives, 45-47

Kanpur Indo-American Program (cont'd)
  curriculum and program development, 47-48
  MIT Technology Adaptation Program, 51
  Other cases, 51-52
  Rural Industrialization Technical Assistance (RITA), 33-35, 250, 259
    description, 33
    World Bank evaluation, 34-35
  University of Wisconsin Programs
    Wisconsin in India, 50
    Wisconsin-Monterrey Tec, 50-51
    Wisconsin-Singapore, 50
  Concluding remarks, 65-67
  Current views of U.S. educators, others, 62-65
  Definitions, 52
  Engineering technology activity, 52-55
  Issues
    clout in Washington, lack of, 58
    continuity, lack of, 57
    funding, lack of, 55, 57
    immigration, brain drain, 105
    independent evaluations, lack of, 59
    LDC engineering education, nature of, 59-60
    political, 60-61
    red tape, 58-59
    U.S. education for LDC students, appropriateness of, 61
  Mechanisms for U.S. engineering school involvement, 32-33

Engineering, international activities of U.S. universities in (cont'd)
  Summary, 250-251
  Types of involvement
    cooperative R&D, 27, 250
    education and training, 28-32, 250
    foreign students in U.S., 28-32, 250
    institution building, 24, 27, 250
    U.S. faculty overseas, 32
    U.S. programs for foreign students, 30, 250
    U.S. resource base development, 27-28, 250
Engineering technology (defined), 52
Engineers Joint Council, 28
Engineers Council for Professional Development (ECPD), 52
Ensminger, Douglas, 119, 132-133
Environmental Protection Agency (EPA), U.S., 8
Escuela Superior de Agricultura "Antonio Narro," Mexico, 102
Evaluation of programs, 59, 101-118, 239-240, 263

Federal University of Paraiba, Brazil, 34
Federal University of Rio de Janeiro, Brazil, 148
Findley, Congressman Paul H., 12
Finn, Liam, 207
Florida, University of, 88, 153
Fogel, B.R., 171

Index

Food and Agricultural Act of 1977, 8, 84-85
Food and Nutrition Program (AID), 12
Food for Peace Act of 1966, 84-85
Ford Foundation, 46, 48, 50, 53, 70, 71, 78, 88, 104, 138, 141, 142-147, 168-169, 184, 205-206, 221, 222, 259
Foreign Assistance Act (FAA), 1-4, 5, 9-10, 12, 118, 243
  Section 211(d) of Title II, 1-3, 5, 9-10, 15, 21, 28, 32, 36-41, 51, 55, 57, 66-67, 81, 92, 94-95, 137, 153-154, 185, 192, 220, 232, 236, 244, 249-250, 258-260, 262, 265
Foreign Operations Administration (FAO), 114
Foundation for International Technological Collaboration (FITC), 7, 21, 23, 67, 117, 197, 204-206, 214, 215-217, 221, 222, 224, 229, 241-242, 245, 246, 249, 252, 255, 257, 258, 260-262, 263, 265
Fox, Thomas, 226-228
Fulbright-Hays Program, 157, 159, 189, 253, 265
Fulbright scholar exchange program, 8, 18, 22, 32, 142, 168
Fundacao Instituto Agronomico de Paraná (IAPAR), Brazil, 98
Funding for S&T for development, policy issues, 219-223, 256, 262
Funds for Overseas Research Grants in Education (FORGE), 172, 191, 254

Gardner, John W., 105-106, 108, 117, 205
Gardner Report, 105-108, 114

General Accounting Office (GAO), 9, 184, 238, 240
Georgia Institute of Technology (Georgia Tech), 11, 27-28, 30, 36-40, 61, 66, 181, 191
  Small Industries Program, 11, 27-28, 36-40, 209, 222, 236, 250, 259
Gilligan, John, 205
Glyde, Henry R., 32, 172, 211, 237, 244
Gomez, Mario, 61, 62, 198, 234
Gore, M.S., 183
Greene, Michael P., 153, 168-169
Grobman, Arnold, 145
Guayaquil, Universidad de, Ecuador, 142

Harrington, Fred H., 32, 171, 194
Harvard University, 155
Hathaway, Dale, 121, 129, 181
Hawaii, University of, 85 (see also East-West Center)
Hazeltine, Barrett, 59, 60, 61, 62
Health, Education, and Welfare, Department of, U.S. (see Department of Health, Education, and Welfare)
Heifer Project International, 99
Hollister, John B., 114
Houston, University of, 53, 54, 142
Howard University, 232
Humphrey, Senator Hubert H., 5, 12, 20

IBRD (World Bank), 5, 27, 32, 35-36, 80, 88, 195, 196, 261
Illinois, University of, 71, 79, 85, 95-98, 236, 251, 259

Immigration, visa, and passport regulations, 17-18
India League of America, 95
India, Ministry of Education, 48, 143
Indian Institute of Technology, Kanpur (IIT/K), 24, 31, 45-50, 60-61, 193, 250
Indian National Council for Science Education, 143
Indiana University, 109, 148
INELEC, Algeria, 25, 31, 53, 55, 65, 178, 188, 195, 199, 250
Information and Cultural Exchange Act, 17
Institute for International Education (IIE), 18, 33, 95, 191
Institute for Scientific and Technological Cooperation (ISTC), 23, 204, 263 (see also Foundation for Technological Cooperation)
Institute of Interamerican Affairs (IIAA), 69-70, 113
Institute of Research and Overseas Programs (IROP), 70, 101, 102
Institution building, 178-182, 250
  defined, 24, 77
Institution for Nutrition of Central America and Panama (INCAP), Guatemala City, 139
Institution strengthening (defined), 10, 77
  grants, 16
Inter-American Foundation, 94
Interfaith Center for Corporate Responsibility, 95
Interior, Department of, U.S., 8, 246
International Association for the Exchange of Students for Tecnical Experience (IAESTE), 159
International Center for Agricultural Research in Dry Areas (ICARDA), 81, 139
International Center for Insect Physiology and Ecology (ICIPE), Nairobi, Kenya, 138-139, 170, 253
International Center for Living Aquatic Resources Management (ICLARM), Philippines, 139
International Center for Tropical Agriculture (CIAT), Cali, Colombia, 81
International Communication Agency (ICA), 1, 8, 17-18, 22, 265
International Cooperation Administration (ICA), 5, 70, 88, 113
International Council of Scientific Unions (ICSU), 198
International Crop Research Institute for the Semi-Arid Tropics (ICRISAT), India, 81
International Development for Food Assistance Act, Title XII, 2, 5, 8, 10-16, 20, 56, 58, 66, 68, 77, 78, 81-83, 118, 120, 123-127, 133, 135, 137, 178, 181, 182, 185, 191-192, 197, 199, 202, 213, 220-221, 236, 242-244, 249-252, 254-255, 257-259
International Development and Humanitarian Assistance Act, 5
International Development Cooperation Act of 1978 (Proposed), 5
International Development Cooperation Administration (IDCA), 5, 204, 242, 263
International Development Foundation, 5, 22, 203

Index   391

International Development Research Centre (IDRC), Canada, 206, 222, 231, 247
International Education Act, 1, 8, 16, 22, 217, 243, 258, 265
International Education and Research Newsletter, Purdue University, 100
International Food and Agriculture Policy Institute, 122
International Foundation for Science (IFS), 191, 222
International Institute for Applied Systems Analysis (IIASA), 197
International Institute of Tropical Agriculture (IITA), 80, 98
International Labor Organization (ILO), 159
International Laboratory for Research on Animal Diseases (ILRAD), Nairobi, Kenya, 80
International Livestock Center for Africa (ILCA), Addis Ababa, Ethiopia, 80
International Maize and Wheat Improvement Center (CIMMYT), Mexico, 75, 80, 94, 132, 150
International Potato Center, Lima, Peru, 80, 100
International Programs of American Universities, 70, 71
International Rice Research Institute (IRRI), 80, 132, 163, 253
International Sorghum Research Network, 79
International Soybean Program (INTSOY), 79, 85, 96-99, 236
Iowa State University, 200

Kabul Afghan-American Program (KAAP), 38, 41-45

Kabul University (KU), Afghanistan, 24, 31, 38, 41-45, 60, 194
Kanpur Indo-American Program (KIAP), 45-49, 194
Kansas State University, 71, 131
Kelkar, Dr. P. K., 46
Kelleher, Alfred, 190
Kentucky, University of, 27, 52
Kidd, Charles V., 18-19, 60, 162, 187, 190, 235-236
Korean Institute of Science and Technology (KIST), 162

Labor, Department of, U. S., 8, 246
Land Tenure Center, University of Wisconsin, 92-94, 210
Legislative mandate,
   analysis, 19-22
   building an indigenous S&T base, 22
   general, 1-22
   international mandates for domestic agencies, 5-8
   pertaining to visa, passports, immigration and exchange, 17-18
   policies in foreign assistance legislation, 2-5
   specifically pertaining to education or universities, 8-17
   summary, 249
   table, 4
   U. S. agencies with international missions, 5
Legislative options
   domestic mission-oriented agencies, 246
   expanded Title XII authority, 243

Legislative options (cont'd)
  Foundation for International Technological Cooperation, 241-242
  funding authority for 211(d), 243-244
  international activity of the National Science Foundation, 243-245
  International Communication Agency, 247
  International Education Act of 1966, 242
  international organizations, 247
Lewis, Jean P., 9-10
Long, Congressman Clarence D., 2, 226
Long, Franklin A., 21, 171
Lucas, Barbara, 207, 240

Maryland, University of, 156, 168
Massachusetts Institute of Technology (MIT), 27-28, 30, 45, 51, 53, 62, 145, 153, 155, 196, 198
  Technology Adaptation Program (TAP), 30, 51
Massachusetts, University of, 68
Massey, H. F., 126
Mayaguez Institute for Tropical Agriculture, Puerto Rico, 84
Maybury, Robert H., 145
McNamara, Robert, 134
McNown, John, 34
Mechanisms, 31-33, 163-166, 190-197
Meinwald, J., 138
Metz, J. F., 100
Mexican National Institute of Livestock Research, 99

Michigan Partners of the Americas, The, 99
Michigan State University, 70, 71, 99-100, 110, 123, 149, 200, 251
Michigan, University of, 148, 162
Middle East Technical University, Ankara, Turkey, 60
Midwest Universities Consortium for International Activities (MUCIA), Inc., 75, 94, 143, 194
Miller, Hugh, 202
Minnesota, University of, 113, 128
Montana State University, 88
Monterrey Tec, Mexico, 30, 50, 147
Moravcsik, Michael Jr., 137, 163, 166-167, 175, 176, 186, 209, 219, 229, 238
Morgan, Robert P., 47, 225
Morrill Land Grant Acts, 134
Mutual Educational and Cultural Exchange Act, 16
Mutual Security Act, 16
Mutual Security Agency, 113

Nakanishi, Dr., 138
Naraine, M., 153
National Academy of Engineering (NAE), 24, 184
National Academy of Sciences (NAS), 24, 134, 143, 147, 148, 150, 155-156, 172, 183, 184, 193, 253, 259
National Advisory Committee of International Studies, 17
National Aeronautics and Space Administration (NASA), 8, 220, 246
National Association of State Universities and Land Grant Colleges (NASULGC), 59, 101, 111, 194, 202

National Autonomous University of Mexico (UNAM), 28, 147
National Defense Education Act (NDEA), 1, 8, 16
National Institute for Educational and Technical Cooperation (NIETC), 107, 118
National Institutes of Health (NIH), 149
National Research Council (NRC), 7, 76, 77, 82, 147
National Science Board, 193
National Science Foundation (NSF), 5, 6, 18, 26-28, 30, 31, 35, 99, 106, 143, 147-148, 149, 157, 159, 162, 183, 188, 190, 191, 193-194, 206, 209, 215, 220, 222, 238, 240-241, 245, 247, 249, 258, 260-261, 265
NATO, 197
"New Directions" (Basic Needs) 1973 Foreign Aid Bill, 2, 19, 93, 119, 213, 217, 225, 227, 249, 252
New International Economic Order (NIEO), 214, 231
Nichols, Buford L., 119
Njala University College, Sierra Leone, 95
North Carolina State University, 137, 154, 155
North Carolina, University of, 153
Northwestern University, 147, 153
Nutrition Center of the Philippines, Manila, 139

Objectives of study, 248
Office of Education, DHEW, 8, 17
Office of Naval Research, 105
Oklahoma State University, 52, 70
OPEC, Organization of Petroleum Exporting Countries, 19, 27, 29, 30-31, 58, 65, 178, 188, 217, 227, 229, 238, 247, 254, 257
Oregon, University of, 162
Organization of American States (OAS), 152-153, 157, 168, 253
  Regional Scientific and Technological Program (PRDCYT), 152-153, 176, 195

Pahlavi University, Iran, 138, 140
Peace Corps, 5, 71, 189, 200, 254-256, 261, 265
Peer review, 238
Pennsylvania State University, 71, 199
Pennsylvania, University of, 138, 140
Perez, Eduardo A., 16
Peterson Report, 5
Philippines, University of, Los Baños, 100
Physics Visiting Committee Project, 162
Pittsburgh, University of, 162
Policy Issues and Options, 218-247, 256-258
  bureaucratic vs professional approach
    small vs large projects, 237, 263
    peer review, 238
  evaluation
    defining objectives and methods, 239-240
    need for independent, 238-239
  funding
    control and administration, 222

Policy Issues and Options (cont'd)
  funding (cont'd)
    criteria for, 223-224
    distribution, 222-223
    level, 218-219
  legislative options, 241-247
  objectives
    AID graduate countries, 228-229
    appropriate technology, 226-228
    bilateral vs multilateral involvements, 230
    building an indigenous S&T base, 224-225
    mutual benefits, 229
    "New Directions" and basic needs, 224-225
    political considerations, 230
    self-reliance, 228
    U. S. minorities and international cooperation, 230
  summary, 257-258
  Why universities?
    Are universities relevant to development needs?, 235
    "brain drain", 235-236
    Does U.S. university resource base activity help development?, 236
    How do universities compare with other institutions?, 235-236
    What can universities do?, 231-232
Puerto Rico, University of, 79, 85, 97
Purdue University, 70, 79, 100, 127, 194
  Laboratory for the Applications of Remote Sensing (LARS), 100

RANN (Research Applied to National Needs), 6
Rao, K. N., 52, 65, 144, 145, 166-167, 171, 180, 196, 203, 239, 254
Read, H., 87-88, 97
Reader's Digest, 97
Recommendations, 263-266
  future studies and related activity, 263-265
  U. S. government initiatives in connection with UNCSTED, 264-265
Renne, Roland R., 90-91
Resource base development (defined), 9
Revelle, Roger, 190, 203
Richardson, John M., 113
Rockefeller Foundation, 70, 79, 88, 97, 155, 223
Roorkee, University of, India, 49-50
Rural Industrialization Technical Assistance (RITA) Program, 28, 33-35, 66, 181, 250, 259
Rutgers University, 190

Sabato, Jorge, 209
São Paulo, University of, Brazil, 148
Scenarios for future U.S. university involvement, 212-217, 256, 262
Schlie, T. W. et al. study, 219
Science, international activities of U. S. universities in, 137-177, 252-254
  analysis, conditions for success, 173-174
  cooperative research and development
    international research projects, 150

Index   395

Science, international activities of U. S. universities in (cont'd)
  cooperative research and development (cont'd)
    NAS activities, 150
    NSF cooperative science program in Latin America, 150
    OAS PRDCYT program, 151
    University of Wisconsin-Universidad de Chile program, 149
  current thinking on science involvements
    Djerassi, Carl, 170
    Greene, Michael P., 167-168
    Moravcsik, Michael, 163, 165
    Rao, K. N., 165
    Symant, H. Harry, 169-170
    other views, 170-171
  definitions
    science, 137
    basic vs applied science, 137
    scientific vs technological research, 137
  graduate and undergraduate education, exchanges, objectives of, 156-157
    foreign students and faculty to U. S., 157
    SEED program, 160
    U. S. students and faculty to LDCs, 157-160
  institution building,
    International Centers of Research Excellence, 138, 139
    University of California-Universidad de Chile, 141
    University of Houston-Universidad de Guayaquil, 141-142

Science, international activities of U. S. universities in (cont'd)
  institution building (cont'd)
    University of Pennsylvania-Pahlavi University of Iran project, 138, 139
  issues, 175-177
  mechanisms, 163-164
  resource base development
    MIT International Nutrition Policy and Planning Program, 154-155, 253
    North Carolina State 211(d) Tropical Soils Program, 154-155
    other involvements, 155-156
  science education improvement (undergraduate, secondary, graduate levels)
    Biological Sciences Curriculum study, 145-146, 253
    Ford Foundation graduate-level projects, 147
    MUCIA - Univ. Agraria del Peru, 143
    NAS-CNPq Chemistry Project in Brazil, 148, 253, 259
    NAS - Colciencias in Colombia, 146
    science programs in Central America, 142
    science programs in India, 142
    secondary-level projects supported by Ford Foundation, 143-144
  summary, 252-254
Scientists and Engineers in Economic Development (SEED) Program, 27, 30, 31, 160-161, 188, 190, 209, 244, 253, 255, 259, 265

Segal, Aaron, 19, 20, 21, 217, 242
Servicio Technico Interamericano de Cooperacion Agricola (STICA), Costa Rica, 87
Shearer-Izumi, Walter, 201
Shiraz Technical Institute, Iran, 52
Shiraz, University of, Iran, 140-141
Singapore, Institute of Industrial Research, 51
Singapore, University of, 51
Smuckler, Ralph, 123, 204
Soil and Water Development in Arid and Sub-Humid Areas, 82
Sonora, Universidad de, Mexico, 148
South-East Consortium for International Development (SECID), 135, 194
Sparkman, Senator John, 5, 220
Special Foreign Currency Program, 7, 84, 150
Stanford University, 52, 110, 144, 147
Stassen, Harold E., 113
Summary, 248-258
  objectives, 248
  legislative mandate, 249
  engineering, 249-251
  agriculture, 251-252
  science, 252-254
  future roles for U.S. universities, 254-256
  policy issues and options, 256-258
Sapporo Agricultural School, Japan, 68
Szmant, H. Harry, 169-170, 181, 191

Technical Assistance Bureau (AID), 2

Technical Cooperation Administration (TCA), 69-70, 113
Tennessee, University of, 72
Texas A&M University, 70, 79, 101
Thompson, K.W., 171
Title V of Foreign Relations Authorization Act (1978), 5
Tolman, Richard, 145-146
Truman, President Harry, 69-70, 251
Tulane University, 110
Tuskegee Institute, 135
Types of U.S. university involvments, 27-30, 182-190

U.N. Conference on Science and Technology for Development (UNCSTED), 23, 195, 202, 205, 218, 224-225, 228, 230, 236, 241, 243, 245, 246-247, 249, 256, 261, 265
  objectives, 23
U.N. Conference on Trade and Development (UNCTAD-V), 236
UNDP, 10, 26, 30, 50, 81, 96, 157, 197
UNESCO, 26, 30, 60, 145, 158, 197, 211
UN FAO, 10, 81, 85, 96, 99, 197
U.N. World Food and Nutrition Conference, 76
U.N. University, 155, 197, 230, 247, 253, 255, 259, 261, 265
U.N. Economic and Social Council, 158
UNICEF, 85, 96
Universidad Agraria del Peru, 144
Universidad de Chile, 141, 150, 168
Universidad de Sonora, Mexico, 149

Index 397

University of Arizona, 28, 70, 137, 150
University of Arkansas, 70
University of Ceará, Brazil, 33
University of California, 102, 141, 144, 149, 163
University of California at Los Angeles (UCLA), 33, 52, 163, 250
University of California at Riverside, 82
University of Colorado, 144
University of Florida, 88, 153
University of Hawaii, 85 (see also East-West Center)
University of Houston, 53, 54, 142
University of Illinois, 71, 79, 85, 95-98, 236
University of Kentucky, 27, 52
University of Maryland, 156, 168
University of Massachusetts, 68
University of Michigan, 148, 162
Unifersity of Minnesota, 113, 128
University of Nebraska, 81
University of North Carolina, 153
University of Oregon, 162
University of Pennsylvania, 138, 140
University of Petroleum and Minerals, Dhahran, Saudi Arabia, 24, 52, 55, 65, 195
University of the Philippines, Los Baños, 100
University of Pittsburgh, 162
University of Puerto Rico, 79, 85, 97
University of Roorkee, India, 49-50
University of São Paulo, Brazil, 148
University of Science and Technology, Kumasi, Ghana, 180
University of Shiraz, Iran, 140
University of Singapore, 51
University of Tennessee, 72
University of Wisconsin, 24, 30, 49-50, 54, 92-94, 110, 144, 149, 197, 210
   Land Tenure Center (LTC), 92-94, 210
University of Wyoming, 70
USDA (see Agriculture, Department of)
U. S. Engineering Team (USET), KAAP, 40-45
U. S. universities, future roles in S&T for development, 177-217, 254-257
   criteria for successful U. S. university involvements, 207-208
      conditions for success, 208-211
      limitations, 211-212
   mechanisms for future involvment
      bilateral programs for individuals, 190-192
      bilateral programs for institutions, 192-194
      consortia, councils, and networks, 194-195
      cooperative R&D programs, 197
      multilateral programs, 195-197
   new cooperation among U. S. institutions
      agriculture-engineering cooperation, 198
      Foundation for International Technological Cooperation, 202-206
      other linkages, 202

U. S. universities, future roles in S&T for development (cont'd)
  mechanisms for future involvement (cont'd)
    new cooperation among U. S. institutions (cont'd)
      professional society/organization linkages, 202
      university-ATI collaboration, 201
      university-community college/technical institute cooperation, 198
      university-industry collaboration, 201-202
      university-Peace Corps collaboration, 199-200
      university-PVO cooperation, 199
      university-research institute/national lab collaboration, 200
  scenarios
    all-out science and technology, 214-216
    modest increase, 216-217
    status-quo, 212-213
  summary, 254-257
  types of involvements
    cooperative R&D, 182-184
    education and training, 187-190
    institution building, 178-182
    LDC students to U. S., 187-190
    training programs, 189
    U. S. resource base development, 184-186
    U. S. students and faculty to LDCs, 188-189
Utah State University, 28
Uttar Pradesh (India) Agricultural University, 95

Volunteers in Technical Assistance (VITA), 199-200, 225, 255, 261, 265

Walbot, Virginia, 229
Waldman, George, 8
Walker, Robert, 171
Washington University in St. Louis, 100, 182
  Center for Development Technology (CDT), 197, 198-199, 210
  Department of Technology and Human Affairs, 197, 199
  plant biology program, 229
  School of Engineering, 185
  technology and international development activity, 30, 52, 62
Watts, Lowell H., 124
Weidner, Edward W., 102-103
Weir, William, 141
Wennergren, E. Boyd, 128
Wentworth Institute of Technology, Boston, 53, 54
Wharton, Clifford, 82
Whitaker, Morris D., 128
White, Kelly, 128
Wilburn, Adolph, 159, 188, 208, 232
Williams, T. T., 134-135
Wisconsin, University of, 24, 30, 49-50, 54, 92-94, 110, 144, 149, 197, 210
  Land Tenure Center (LTC), 92-94, 210
Witunski, M., 201, 205, 229
World Bank (IBRD), 5, 27, 32, 35-36, 80, 88, 195, 196
Wyoming, University of, 70

About the Authors

ROBERT P. MORGAN is Chairman of the Department of Technology and Human Affairs, and Director of the Center for Development Technology in the School of Engineering and Applied Science at Washington University in St. Louis, Missouri. He received the bachelors degree in chemical engineering from The Cooper Union, the masters and engineers degrees in nuclear engineering from M.I.T., and the Ph.D. in chemical engineering from Rensselaer Polytechnic Institute. He is a corporate member of Volunteers in Technical Assistance (VITA) and serves on the Committee on Science, Engineering and Public Policy of the American Association for the Advancement of Science. He is the author or co-author of some seventy papers, reports and articles spanning the fields of heat and mass transfer, nuclear reactor analysis, educational telecommunications, earth observation satellite applications, technology and international development, and the societal role and impact of technology. In 1978, Dr. Morgan received the Chester F. Carlson Award of the American Society for Engineering Education for his creative leadership and success in uniting elements of the social sciences, natural sciences and engineering to provide a new kind of education for technology and human affairs. Currently, his main professional interests are focused on science and technology for development. He has served as principal investigator of two studies as part of U.S. preparations for the 1979 UNCSTED.

ELLEN E. IRONS is an environmental engineer with Russell and Axon, Inc., St. Louis, Missouri, where she works in the fields of sanitary and water resource problems. She received the bachelors degree in civil engineering from Washington University, where she was active in the St. Louis Chapter of Science for the People. She is currently acting director of the Peacock Alley Art Center.

EDUARDO A. PEREZ is a project manager in the technical assistance department of Volunteers in Technical Assistance, Inc., a leading U.S. appropriate technology organization. He received his bachelors degree in civil engineering from Georgia Institute of Technology and his masters degree in technology and human affairs from Washington University. From 1975 through 1977 he served in the Peace Corps in Honduras in a variety of engineering assignments.

THEODORE N. SOULE is currently finishing work on the masters degree in technology and human affairs at Washington University. He received the bachelors degree in English literature at Washington University where he also did research in 1971-1972 at the Center for the Biology of Natural Systems in the program on ecology and international development. He co-edited the book International Development and the Human Environment: An Annotated Bibliography, published by Macmillan Information Inc.

AVA K. FRIED is Head of the Materials Access Division of the University of Cincinnati Medical Center Libraries. She received the bachelors degree in English from Russell Sage College, and masters degrees in education and in library science from the State University of New York at Albany and from the University of Rhode Island, respectively. After teaching language arts for two years in Massachusetts, she joined the Center for Development Technology, Washington University, as a staff associate from 1976 to 1978.

Pergamon Policy Studies

- No. 1 Laszlo—*The Objectives of the New International Economic Order*
- No. 2 Link/Feld—*The New Nationalism*
- No. 3 Ways—*The Future of Business*
- No. 4 Davis—*Managing and Organizing Multinational Corporations*
- No. 5 Volgyes—*The Peasantry of Eastern Europe, Volume One*
- No. 6 Volgyes—*The Peasantry of Eastern Europe, Volume Two*
- No. 7 Hahn/Pfaltzgraff—*The Atlantic Community in Crisis*
- No. 8 Renninger—*Multinational Cooperation for Development in West Africa*
- No. 9 Stepanek—*Bangledesh—Equitable Growth?*
- No. 10 Foreign Affairs—*America and the World 1978*
- No. 11 Goodman/Love—*Management of Development Projects*
- No. 12 Weinstein—*Bureacratic Opposition*
- No. 13 De Volpi—*Proliferation, Plutonium, and Policy*
- No. 14 Francisco/Laird/Laird—*The Political Economy of Collectivized Agriculture*
- No. 15 Godet—*The Crisis in Forecasting and the Emergence of the "Prospective" Approach*
- No. 16 Golany—*Arid Zone Settlement Planning*
- No. 17 Perry/Kraemer—*Technological Innovation in American Local Governments*
- No. 18 Carman—*Obstacles to Mineral Development*
- No. 19 Demir—*Arab Development Funds in the Middle East*
- No. 20 Kahan/Ruble—*Industrial Labor in the U.S.S.R.*
- No. 21 Meagher—*An International Redistribution of Wealth and Power*
- No. 22 Thomas/Wionczek—*Integration of Science and Technology With Development*
- No. 23 Mushkin/Dunlop—*Health: What Is It Worth?*
- No. 24 Abouchar—*Economic Evaluation of Soviet Socialism*
- No. 25 Amos—*Arab-Israeli Military/Political Relations*
- No. 26 Geismar/Geismar—*Families in an Urban Mold*
- No. 27 Leitenberg/Sheffer—*Great Power Intervention in the Middle East*
- No. 28 O'Brien/Marcus—*Crime and Justice in America*
- No. 29 Gartner—*Consumer Education in the Human Services*
- No. 30 Diwan/Livingston—*Alternative Development Strategies and Appropriate Technology*
- No. 31 Freedman—*World Politics and the Arab-Israeli Conflict*
- No. 32 Williams/Deese—*Nuclear Nonproliferatrion*
- No. 33 Close—*Europe Without Defense?*
- No. 34 Brown—*Disaster Preparedness*
- No. 35 Grieves—*Transnationalism in Politics and Business*

No. 36   Franko/Seiber—*Developing Country Debt*
No. 37   Dismukes—*Soviet Naval Diplomacy*
No. 38   Morgan—*The Role of U.S. Universities in Science and Technology for Development*
No. 39   Chou/Harmon—*Critical Food Issues of the Eighties*
No. 40   Hall—*Ethnic Autonomy—Comparative Dynamics*
No. 41   Savitch—*Urban Policy and the Exterior City*
No. 42   Morris—*Measuring the Condition of the World's Poor*
No. 43   Katsenelinboigen—*Soviet Economic Thought and Political Power in the USSR*
No. 44   McCagg/Silver—*Soviet Asian Ethnic Frontiers*
No. 45   Carter/Hill—*The Criminal's Image of the City*
No. 46   Fallenbuchl/McMillan—*Partners in East-West Economic Relations*
No. 47   Liebling—*U.S. Corporate Profitability*
No. 48   Volgyes/Lonsdale—*Process of Rural Transformation*
No. 49   Ra'anan—*Ethnic Resurgence in Modern Democratic States*